Lecture Notes in Civil Engineering

Volume 51

Lecture Notes in Civil Engineering (LNCE) publishes the latest developments in Civil Engineering—quickly, informally and in top quality. Though original research reported in proceedings and post-proceedings represents the core of LNCE, edited volumes of exceptionally high quality and interest may also be considered for publication. Volumes published in LNCE embrace all aspects and subfields of, as well as new challenges in, Civil Engineering. Topics in the series include:

- Construction and Structural Mechanics
- Building Materials
- Concrete, Steel and Timber Structures
- Geotechnical Engineering
- Earthquake Engineering
- Coastal Engineering
- Ocean and Offshore Engineering; Ships and Floating Structures
- Hydraulics, Hydrology and Water Resources Engineering
- Environmental Engineering and Sustainability
- Structural Health and Monitoring
- Surveying and Geographical Information Systems
- Indoor Environments
- Transportation and Traffic
- Risk Analysis
- Safety and Security

To submit a proposal or request further information, please contact the appropriate Springer Editor:

- Mr. Pierpaolo Riva at pierpaolo.riva@springer.com (Europe and Americas);
- Ms. Swati Meherishi at swati.meherishi@springer.com (Asia—except China—and Australia/NZ);
- Ms. Li Shen at li.shen@springer.com (China).

Indexed by Scopus

More information about this series at http://www.springer.com/series/15087

Kamal Jain · Kourosh Khoshelham ·
Xuan Zhu · Anuj Tiwari
Editors

Proceedings of UASG 2019

Unmanned Aerial System in Geomatics

Springer

Editors
Kamal Jain
Department of Civil Engineering
Indian Institute of Technology Roorkee
Roorkee, Uttarakhand, India

Kourosh Khoshelham
Department of Infrastructure Engineering
University of Melbourne
Melbourne, VIC, Australia

Xuan Zhu
School of Earth Atmosphere
and Environment
Monash University
Melbourne, VIC, Australia

Anuj Tiwari
Department of Civil Engineering
Indian Institute of Technology Roorkee
Roorkee, Uttarakhand, India

ISSN 2366-2557 ISSN 2366-2565 (electronic)
Lecture Notes in Civil Engineering
ISBN 978-3-030-37392-4 ISBN 978-3-030-37393-1 (eBook)
https://doi.org/10.1007/978-3-030-37393-1

This Springer imprint is published by the registered company Springer Nature Switzerland AG
The registered company address is: Gewerbestrasse 11, 6330 Cham, Switzerland

Contents

A Comparative Study of Drone and High Resolution Satellite Data for Detailed Land Use/Land Cover Mapping

Ajay Mathur, P. K. Litoria and B. Pateriya

Abstract Many a times, it is important for local authorities to have an inventory of land in a very detailed format for planning purposes and acquiring land for development. One such application was to prepare detailed land use/land cover map for developing a new township christened Aerotropolis near Mohali in Punjab for Greater Mohali Area Development Authority (GMADA). There are many choices of remote sensing data at hand which includes World View (WV)-II high resolution (HR) satellite data and the more recent drone data. In general, WV-II falls short to map some key features and needs extensive field work to support interpretation. Field work is both time consuming and costly and weather dependent. The present work focuses on the comparison of two data sets to identify drones potential for use in detailed mapping. The study included five villages covering 20 km^2 area and located just outside the 5 km limit of Chandigarh International Airport as is required under the Govt. Policy to fly drones. Necessary permissions were also sought from District Magistrate and Superintendent of Police SAS Nagar with information to Airport Authority. A Quadcopter drone with a spatial resolution of 5 cm was flown around 24th April 2018 and data compared with WV-II data of 50 cm spatial resolution of vintage of 5 May 2018. Apparently, WV-II data was marked with shadow effect unlike drone data and therefore, drone could clearly demarcate vacant plots, alleys, buildings, village boundary, drain, trees, electrical line and poles and provided a means for detailed mapping and is, therefore, a fit candidate as a choice for planning and acquiring land for developing new townships.

Keywords Drones · GMADA · High resolution satellite data

1 Introduction

The acquisition of land is inevitable for developmental purposes. One of the concerns in development is acquiring land compulsorily and owners with majority of them farmers are to be adequately compensated. The farmers are deprived of their land

A. Mathur (✉) · P. K. Litoria · B. Pateriya
Punjab Remote Sensing Centre, PAU Campus, Ludhiana 141004, India
e-mail: mathurajay12@gmail.com

K. Jain et al. (eds.), *Proceedings of UASG 2019*, Lecture Notes in Civil Engineering 51,
https://doi.org/10.1007/978-3-030-37393-1_1

and delay in compensation makes them feel cheated. Farmers for instance, whose land was acquired for construction of Yamuna Expressway in Uttar Pradesh, India have agitated for better compensation. This is the story across India as regards land acquisition. Therefore, there is a need to strike a balance between the two competing goals i.e., acquiring land and paying adequate compensation. This is very important in Punjab as it is an agrarian state and majority of people are into farming. It is, therefore, imperative that neither the acquisition of land should be at the cost of interest of farmers nor the development should be hampered in the process.

The compensation to be awarded is based on market value of land. However, due to delay in land acquisition, there is great variation in the land of prices as on the date of notification of acquiring land and date of paying compensation and this leads farmers to feel aggrieved and leads to lot of litigation and thereby affecting the project both in monetary and time frame.

The way out is to have a negotiated settlement with the farmers keeping in view the element of market rate as well. To achieve this, a very detailed land use/land cover map is required with a cutoff date. The detailed land use map would help not only for planning purposes but also help to arrive at market value based on existing land use and land cover and a judicious way of negotiated settlement with the farmers. For example, a patch of land located nearby to each other with a wasteland cover or an orchard cannot have the same market rate. The map can also be used to freeze the land category as on the date of acquisition of the remotely sensed data to ward off any changes in land use by the farmers to attract higher compensation and thus avoid unnecessary increases in the cost of acquisition. The detailed land use/cover map if overlaid with cadastral information helps a lot in paying compensation to the intended owners of the land and avoids a lot of time and litigation.

Many a projects has witnessed problems of land acquisition in India and to ward off such issues GMADA needed a detailed land use/land cover map overlaid with Cadastral information to consult for arriving at compensation to be paid to farmers for acquiring land for developing a new township christened Aerotropolis near Mohali in Punjab. Many a project of the Urban bodies in Punjab such as Master plans have been prepared using High Resolution satellite data WV-II in particular. Many sensors have been designed with spatial resolutions from about 1 km to sub-meter for identifying Land cover types [1]. The present study focuses on comparing WV-II data with Drone data for comparing detailed mapping of the land. Drones are now being increasingly used in civil applications such as mapping [2], Surveying and cadastral mapping [3–6] and disaster management [7, 8]. Normalized Difference Vegetation Index (NDVI) calculation [9], cultural heritage [10, 11]. In fact, Geospatial data play an important role in an estimated 80% of our daily decisions [12].

2 Objectives

1. Mapping of detailed land use/cover to assist in finalising market value of land.
2. Overlaying cadastral data on the detailed land use/cover map to understand ownership to assist in paying compensation to the legitimate owner of the land.

3 Study Area

Five villages namely Kurari, Patron, Siaun, Bari, Matran, Kishanpura, Azizpur and Sekhan Majra falling in SAS Nagar (Old name Mohali) district (Fig. 1) of Punjab were the focus of the study. The study area was very close to Chandigarh International Airport and is therefore, witnessing a lot of development. The Government of Punjab through its urban Body of the area GMADA was interested in acquiring land in the budding area for developing a new town township.

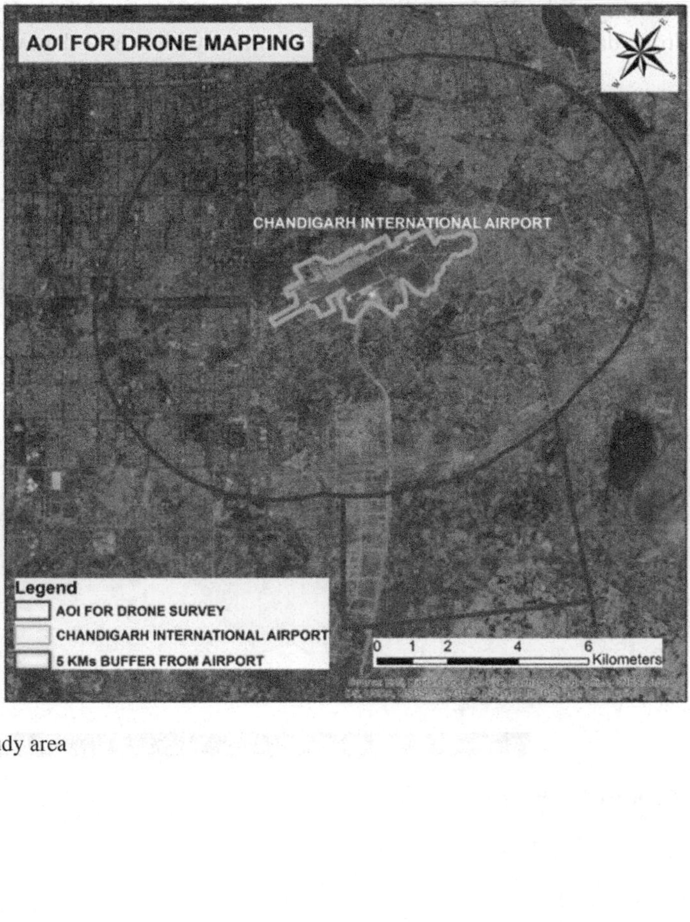

Fig. 1 Study area

4 Methodology

The study area was very close to Chandigarh International Airport so a buffer of 5 km was drawn around the airport to avoid any area closer than 5 km from the airport as per the policy of the Government of India. The area in question was also not in direct line of flight (ascent or descent). To acquire very high resolution georeferenced drone data also referred to as remotely piloted Aircraft [2], necessary permissions to fly drone were also taken from District Magistrate and Superintendent of Police SAS Nagar with information to Airport Authority. Weather conditions conducive to fly drones especially wind was taken into account to finalise the dates of flight. The path of the flights were predefined for the study area and drones flown from six locations to cover the study area by RSI Softech India Pvt. Ltd., Hyderabad, the company entrusted with the job. A Quadcopter drone with a spatial resolution of 5 cm was flown around 24 April 2018 and data compared with WV-II data of 50 cm spatial resolution of vintage of 5 May 2018. The area in question was mainly under agriculture but was chosen to develop the township as it was located in the built-up category of the master plan of the area i.e., the land could be utilized for built-up (Fig. 2).

The processing of the 3D drone data was carried by RSI Softech Ltd., Hyderabad. Only 2D drone data was used in the study to compare with WV-II data which is 2D only.

For detailed mapping visual interpretation was undertaken and for relating this ground category with ownership details, cadastral map of the villages were overlaid on the two datasets. The process (Fig. 3) involves digitization of the hardcopy musavis of the villages in AUTOCAD as per the given dimension on the musavis.

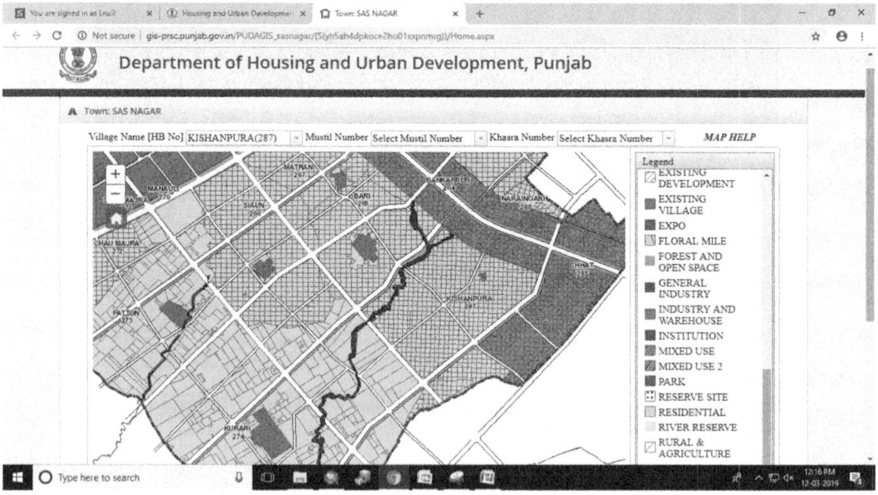

Fig. 2 Master plan of study area

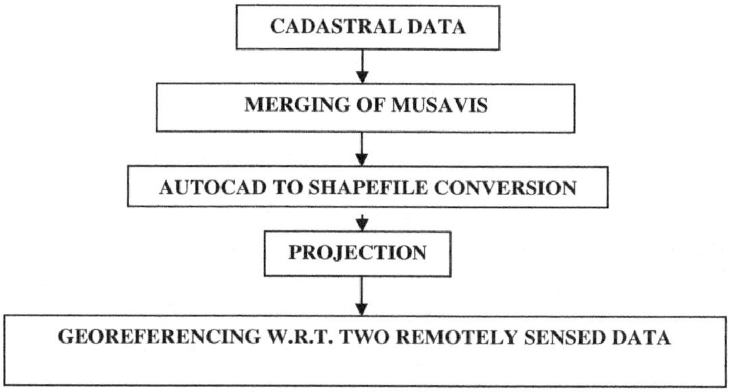

Fig. 3 Process of overlaying cadastral data over satellite data

The AutoCAD version was then converted to shape file and overlaid on the datasets (drone and satellite) by translating (x, y) the vectors, rotating it and scaling it to fit on the two datasets.

5 Results and Discussion

The whole process of acquiring datasets especially drone, a lot of administrative work was undertaken before the drone could be flown and the same are enumerated in Table 1.

Processed 2D drone data along with WV-II data was visually interpreted and key results are placed in Table 2.

Table 1 Administrative comparison

S. no.	Issue	WV-II satellite data	Drone data
1.	Acquisition	Government agency can acquire data from NRSC Hyderabad	Government agency can acquire subject to certain conditions
2.	Permissions	User license (single) has to be signed	Permissions from district administration required to fly drone
3.	Limitation of coverage	No	Limitations to fly near borders, airports and defence installation(s)
4.	Risk	No risk involved	Risk of accident is always there
5.	Cost	Less costly	Very costly

Table 2 Comparison of interpretation

S. no.	Mapping	Comparison in interpretation between WV-II satellite data and drone data
1.	Built-up and roads	Boundary of each building can be clearly demarcated with the drone data unlike satellite data. Roads can be clearly demarcated with drone data as well as the type is very easy to distinguish with drone data (Fig. 4)
2.	Wastelands	Even details within wasteland such as partly covered with swampy water could be easily distinguished in the drone data (Fig. 5)
3.	Crops	With drone data even crop type can be inferred (Fig. 6)
4.	Plantations	With drone data even trees can be counted unlike satellite data where the crown of the trees merges (Fig. 7)
5.	Drains	With drone data, drains could be interpreted in its entire stretch. However, with satellite data at locations it appeared to be only shadow and the course could not be interpreted where there was crowding of trees (Fig. 8)
6.	Electric lines and poles	With drone data both electric lines and poles are very visible but not possible with satellite data (Fig. 9)

Fig. 4 Demarcation of buildings and roads through drone and WV-II satellite data

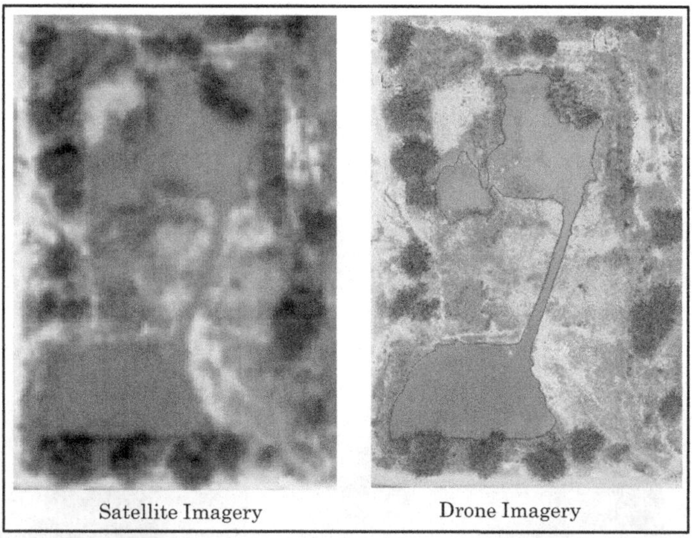

Fig. 5 Demarcation of wastelands through drone and WV-II satellite data

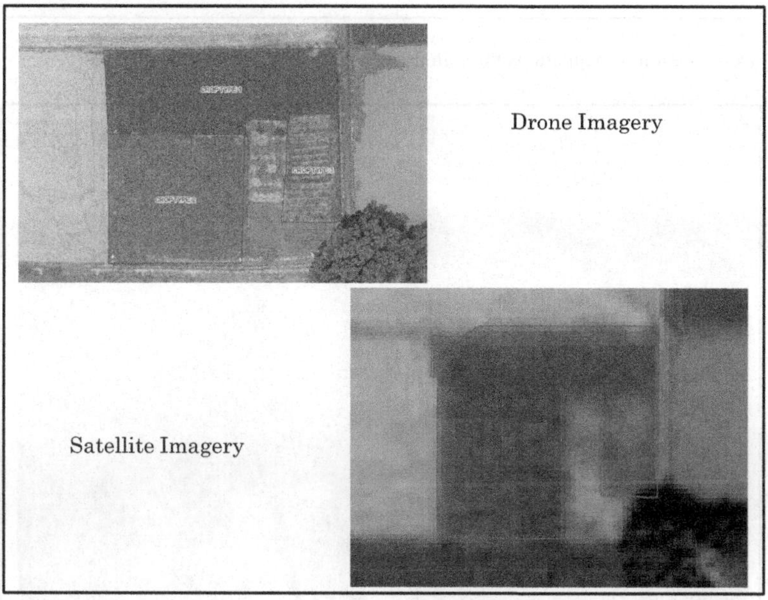

Fig. 6 Demarcation of crops through drone and WV-II satellite data

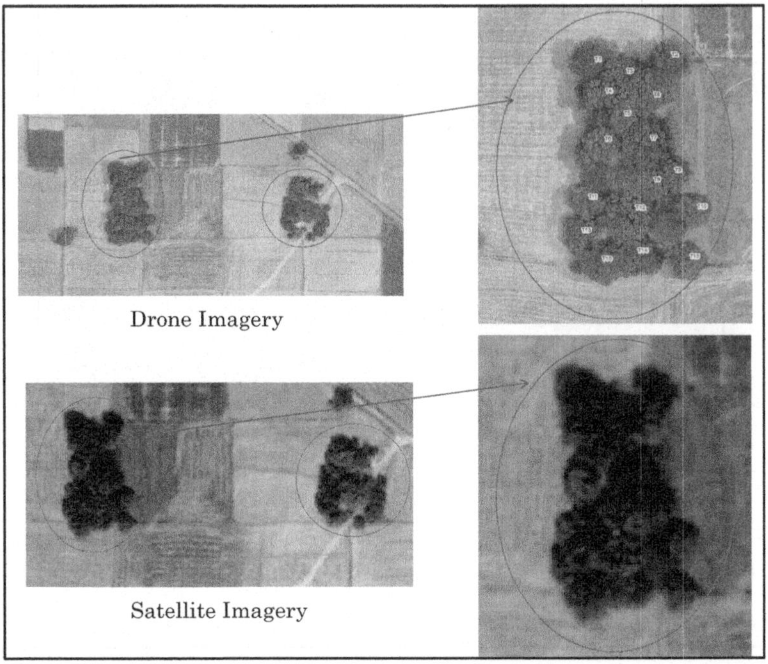

Fig. 7 Demarcation of plantations through drone and WV-II satellite data

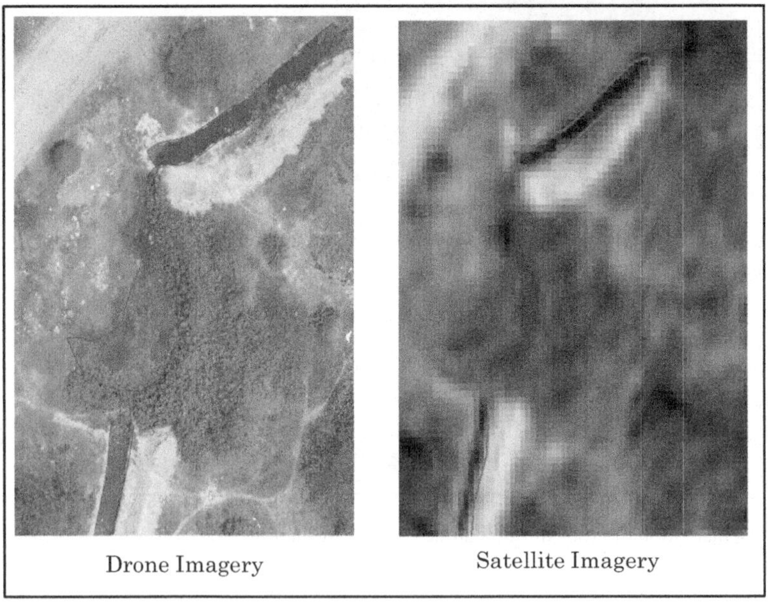

Fig. 8 Demarcation of drains through drone and WV-II satellite data

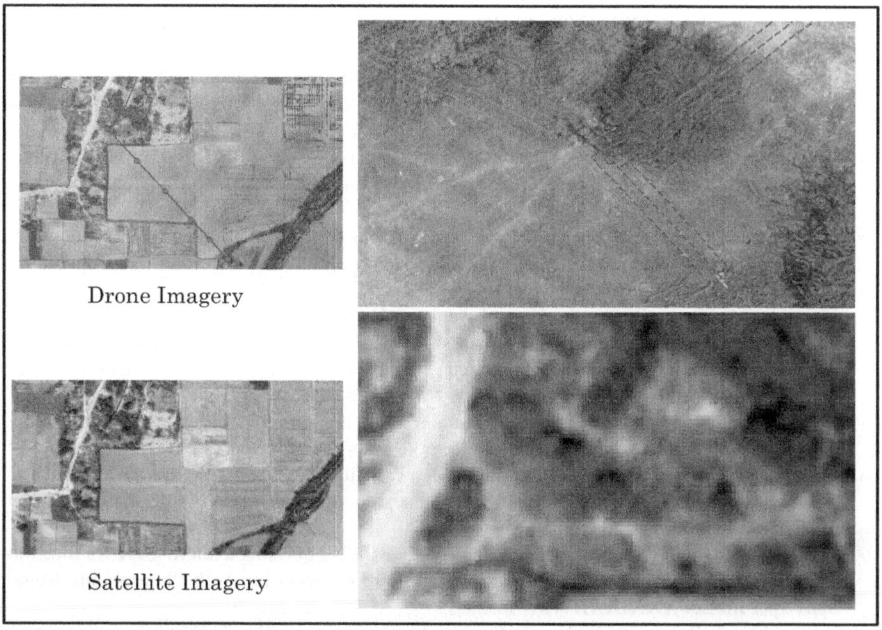

Fig. 9 Demarcation of electric lines and poles through drone and WV-II satellite data

6 Conclusions

Visual Interpretation of the WV-II satellite and drone data reveals that drone data was able to clearly aid in demarcating vacant plots and alleys in built-up areas unlike WV-II satellite data where shadow effects obscured the features. Likewise boundaries of plots can be clearly demarcated in agricultural lands where shadow of trees makes it difficult to interpret with satellite data. Trees and plantations in the fields were also very clear in drone image. Drains could be completely demarcated with drone data unlike satellite data where at many locations the drain was confused for shadows and at places could not be interpreted because of surrounding trees. Electrical lines and poles which could not be demarcated with satellite data were markedly very clear to demarcate with drone data. Given the clarity of features that can be easily demarcated by drone data without much a field work as compared to High Resolution satellite image of WV-II, drone data is a fit candidate as a choice for preparing detailed Land use/Land cover map for planning and acquiring land for developing new townships.

References

1. Sun H, Wu J, Xu X (2016) A method to determine appropriate spatial resolution for hard image classification. J Indian Soc Remote Sens 44(1):11–19

2. Nex F, Remondino F (2014) UAV for 3D mapping applications: a review. Appl Geomat 6(1):1–15
3. Barnes G, Volkmann W, Sherko R, Kelm K (2014) Drones for peace: part 1 of 2 design and testing of a UAV-based cadastral surveying and mapping methodology in Albania. In: World Bank conference on land and poverty, Washington DC, USA, 24–27 Mar 2014
4. Cramer M, Bovet S, Gultlinger M, Honkavaara E, McGill A, Rijsdijk M, Tabor M (2013) On the use of RPAS in national mapping—the EUROSDR point of view. ISPRS Int Arch Photogramm Remote Sens Spat Inf Sci XL-1/W2:93–99
5. Cunningham K, Walker G, Stahlke E, Wilson R (2011) Cadastral audit and assessments using unmanned aerial systems. In: UAV-g: conference on unmanned aerial vehicle in geomatics, Zurich, Switzerland, 14–16 Sept 2011
6. Manyoky M, Theiler P, Steudler D and Eisenbeiss H (2011) Unmanned aerial vehicle in cadastral applications. ISPRS Zurich 2011 Workshop, 14–16 September 2011, Zurich, Switzerland. International archives of the photogrammetry, remote sensing and spatial information sciences, vol XXXVIII-1/C22. https://doi.org/10.5194/isprsarchives-XXXVIII-1-C22-57-2011
7. Choi K, Lee I (2011) A UAV based close-range rapid aerial monitoring system for emergency responses. ISPRS Int Arch Photogramm Remote Sens Spat Inf Sci XXXVIII-1/C22:247–252
8. Molina P, Pares M, Colomina I, Vitoria T, Silva P, Skaloud J, Kornus W, Prades R, Aguliera C (2012) Drones to the rescue! Unmanned aerial search missions based on thermal imaging and reliable navigation. Inside GNSS 7:36–47
9. Lucieer A, Robinson S, Turner D, Harwin S, Kelcey J (2012) Using a micro-UAV for ultra-high resolution multi-sensor observations of Antarctic moss beds. Int Arch Photogramm Remote Sens Spat Inf Sci XXXIX-B1:429–433
10. Remondino F, Barazzetti L, Nex F, Scaioni M, Sarazzi D (2011) UAV photogrammetry for mapping and 3D modeling—current status and future perspectives. Int Arch Photogramm Remote Sens Spat Inf Sci XXXVIII1/C22:25–31
11. Rinaudo F, Chiabrando F, Lingua AM, Spanò AT (2012) Archaeological site monitoring: UAV photogrammetry can be an answer. Int Arch Photogramm Remote Sens Spat Inf Sci XXXIX(B5):583–588
12. Heipke C, Woodsford PA, Gerke M (2008) Updating geospatial databases from images. In: Advances in photogrammetry, remote sensing and spatial information sciences: 2008 ISPRS congress book. CRC Press, Boca Raton, pp 355–362

Assessment of Low-Cost Unmanned Aerial Systems for Engineering Surveys

Amit Rana, Inshu Chauhan and Gaurav Bhatt

Abstract Unmanned Aerial Systems are increasingly becoming popular to carry out purposes which were previously done manually or through sophisticated machines. From the extent of non-technical works like video recording to photogrammetric analysis, they have been deployed for variety of applications. Yet, the usage of UAS remains largely unexplored for engineering surveys, the gap which this research tries to fulfill. Engineering surveys have specific requirements with regard to precision, accuracy and efficiency. Assessment of a low-cost UAV i.e. DJI Mavic Air is being done on 3 parameters vis-a-vis (i) accuracy of distance and area measurements in comparison to traditional surveying instruments, (ii) time required to carry out such surveys and (iii) associated costs. Also, comparison of free trials of software such as Pix4D Mapper and Agisoft Metashape has been done to inform readers who wish to use low-cost UAS for engineering surveys. Lastly the applications of the approach are discussed to the extent to which low-cost UAS are reliable enough for conducting surveys in engineering projects.

Keywords UAV · Engineering survey · Low cost · Theodolite

1 Introduction

Unmanned Aerial Systems are increasingly becoming popular to carry out purposes which were previously done manually or through sophisticated machines. From the extent of non-technical works like video recording to photogrammetric analysis, they have been deployed for variety of applications. Yet, the usage of UAS remains largely unexplored for engineering surveys, the gap which this research tries to fulfill. Engineering surveys have specific requirements with regard to precision, accuracy and efficiency which can be improved by using low-cost UAV's.

A. Rana (✉) · I. Chauhan · G. Bhatt
G.B. Pant Institute of Engineering and Technology, Pauri Garhwal, India
e-mail: ranaamit96@gmail.com

K. Jain et al. (eds.), *Proceedings of UASG 2019*, Lecture Notes in Civil Engineering 51,
https://doi.org/10.1007/978-3-030-37393-1_2

1.1 Unmanned Aerial Systems

Unmanned Aerial Systems or UAS are known under various different names and acronyms, such as "Unmanned Aerial Vehicle" (UAV), "aerial robot" or simply "drone," with "UAV" and "drone" being the most popular terms [3]. Unmanned inspection, surveillance, reconnaissance and mapping of inimical areas were the primary military aims. In the field of geomatics, the first study was made out nearly thirty years ago. Now, the recently launched UAVs have been widely popular in the field of Geomatics and thus became a major platform for data acquisition. UAV photogrammetry [2, 4] has recently opened various new applications in close-range aerial surveys, introducing a cheaper alternative to the older methods of photogrammetry and topographical mapping or detailed 3D recording of ground information and being a valid complementary solution to terrestrial acquisitions [6]. After the arrival of better gyroscope and GPS technology the overall performance, especially the endurance, payload, and flexibility for various applications of UAV systems, have been improved [7]. However, applications in civil engineering are beginning to be explored even though UAV's have the potential to provide more cost- and task-efficient ways to conventional approaches. Low-cost UAVs with consumer grade sensors/camera are increasingly becoming popular for rapid data acquisition and effortless surveying for engineering purposes. Low-cost UAV's can be adjudged as the ones which cost in the range of 800–1000 USD.

1.2 Engineering Surveys

Engineering surveys mainly refers to the measurement of various parameters needed before and during an engineering project which could be construction of a building, road, dam, bridges or some other structural feature. In the early stage of construction, we need to analyze the terrain scientifically for any cutting, excavation or filling work or for producing alignment. After analyzing the related data, we make the construction plan to ensure that the construction can proceed smoothly. Inaccurate engineering survey data not only affects the quality of the construction, but also the construction cycle, which in turn will even seriously affect the safety of people's lives and exaggerates cost. Surveying applications are mostly relying on GPS, laser scanning, Robotic Total Station (RTS), and tachymetry. Also, there are aerial technologies available, but their use depends on the survey area and size of the terrain which is to be surveyed. They, have potentially high measurement errors, are limited in range, very labor intensive and costly and are time consuming to perform [7]. The use of new technology like UAV's in engineering surveys can effectively improve the quality and efficiency of surveys. Although the measurement of data by traditional surveying instruments like theodolite and total station can be accurate too if

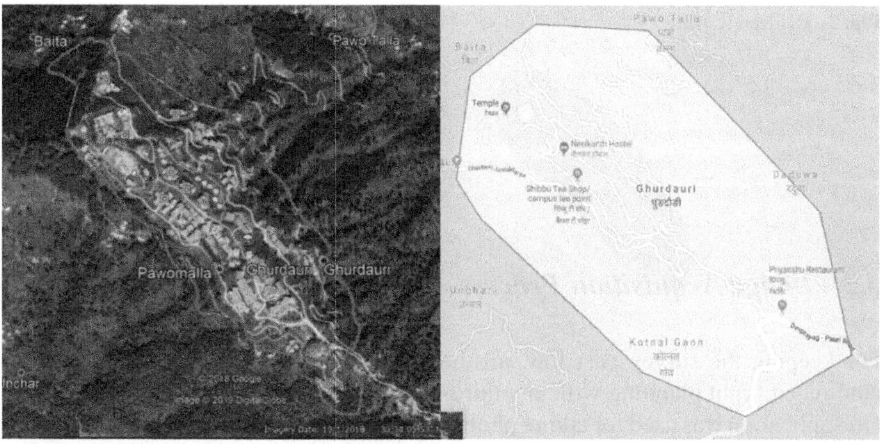

Fig. 1 Google image and Google map of the study area

done properly, yet the actual operation is more complex. This makes the measurement time longer and the working efficiency low which can be overcome by using low-cost UAV's.

2 Study Area

Due to regulatory issues, data could be acquired within the boundaries of our institution. Hence, the campus of G.B. Pant Institute of Engineering and Technology (Fig. 1), itself was taken the study area. The campus of G.B. Pant Institute of Engineering and Technology, Pauri Garhwal, Uttarakhand is spread in an area of 69 acres. The place was built in 1992. Ultimately, the site chosen for study was the administrative block of the institute for analysis and measurements as the whole campus could not be done within a stipulated time.

3 Data Acquisition and Analysis

3.1 Instrument Used i.e. UAS

The UAV used was DJI Mavic Air (Fig. 2), which is a micro drone weighing 430 g with a cost of less than $1000 USD. The camera used in the UAV is 12 MP 1/2.3″ CMOS sensor with an 85° field of view. The UAV is a quadcopter type with a 2375 mAh battery that lasts for around 20 min on a single flight.

Fig. 2 DJI Mavic Air

3.2 Image Acquisition Process

For keeping the survey cost low, all the flights were done manually without any automated flight planning with any third-party application. However, the native DJI Go application was used for taking photographs and flying the drone. Simultaneously, with aerial photography, x, y, coordinates, and elevation of ground control point (GCP) were stored in the metadata of each photograph taken. The UAV was flown at a height of 32 m above from the ground. For keeping the photographs well-lit and thoroughly exposed, the ISO and shutter speed were further adjusted and RAW images in the .dng format were captured. The results of this data acquisition are 93 aerial photographs in the overall study area (administrative building) in which each photograph was having 70–80% overlap with the consecutive photograph.

3.3 Analysis

3.3.1 Methodology

The captured photographs from the low-cost DJI Mavic Air were processed with the help of Pix4D Mapper and Agisoft Metashape softwares and DEM was generated. The basic idea was to keep the survey cost as low as possible so, we used trial versions of Pix4D Mapper and Agisoft Metashape Professional softwares.

Firstly using the Agisoft Metashape software, the images were aligned with a key point limit of 40,000 and a tie point limit of 4000. This was done by using the location and elevation data stored in the images. Further a dense cloud was prepared in a moderate depth filtering and a mesh was generated with the sparse cloud data. After this, textures were generated on this mesh. Initially, before creating DEM, a tiled model was generated with a pixel size of 0.011 m, along with a tile size of 256. A DEM was than built with geographic projection i.e. WGS 84 (EPSG::4326), along with interpolation enabled and a resolution of 0.022 m. The total size in pixels was 8058 × 7963. The DEM created was then used for calculating the area and the elevation of a part of the study area. The ruler tool of Metashape was used to measure the elevation of different points and the area of region (a) in Fig. 3.

Fig. 3 Processed DEM of Agisoft Metashape software

Next, the DEM was created using Pix4D Mapper. Pix4D Mapper doesn't support direct import of .dng raw files, therefore we have to firstly convert those aerial photographs to a separate .jpg format. Initial processing was carried out after importing the images. Then, key point extraction and image matching, sparse point cloud was generated. Through During this process, the parameters of camera calibration were also computed. No, ground control points were imported for proper georeferencing. The bundle block adjustment method was used to adjust the whole block of photos and errors on GCP and CP were examined. Further, dense point cloud with mesh were generated as shown in Fig. 4.

Comparison of the process of obtaining DEM from Agisoft Metashape with Pix4D Mapper. As the main purpose of our study was to keep the survey cost as low as possible, we tried with the trial version of the softwares, however, for Pix4D Mapper the trial was not fully activated due to a technical glitch, we could only use a part of it. Due to these limitations, we could not compute the DSM, Orthomosaic and the index measurements of the area effectively in Pix4D Mapper. So overall it can be said that AgiSoft MetaShape was more effective, accurate and cost friendly with regard to creation of DEM and further measurements.

Comparison of measurements from DEM created from UAV with traditional surveying instruments. The values obtained from the UAV and theodolite data were compared with the recorded measurements of the points of the study area as shown in Table 1. Accuracy of elevation was determined by calculating the percentage of differences in the elevation obtained from theodolite and the elevation obtained from the DEM with respect to the recorded measurements of the points.

Area measurements were also done with the help traditional surveying instruments (240 m^2), this area along with the area calculated with DEM (232 m^2) was compared

Fig. 4 Processed DEM of Pix4D Mapper

Table 1 Comparison of elevations through different techniques

Point	Recorded data (m)	Elevation obtained from theodolite (m)	Elevation obtained from UAV data (m)	% of accuracy obtained by theodolite	% of accuracy obtained by UAV data
Point 1	3	2.74	2.842	91.33	94.66
Point 2	4.5	4.28	4.373	95.11	97.17
Point 3	8	8.30	8.251	96.25	96.86
Point 4	11	10.46	10.788	95.09	98.07

with the area of same region obtained with the help of Google Earth (236 m^2). However, both the comparison of the elevation and area obtained from the UAV data shows that UAV's are good in terms of accuracy and there is not much difference as compared to traditional surveying instruments.

4 Conclusion and Applications

4.1 Conclusion

The DJI Mavic Air being a micro-drone has the advantages of having low cost, operation, and maintenance costs, with a flexibility to take off and land in very small projects. It has another benefit that it can be easily controlled in both autonomous and pilot mode, even in the harsh environmental conditions, for example, strong winds. By the above field investigation, we can see that accuracy of elevation and area measurements done by DEM created by UAV's image data are better or at par with the traditional surveying instruments. But the time taken to carry out the measurements by traditional surveying instruments came out be five times more than the time taken by UAV image acquisition plus the processing. Also, the manpower required was significantly more than UAS in which one person was able to handle and take the data as well as process the UAV data. This field measurement proves that UAV's in the future could well compliment and in long term replace traditional surveying instruments.

Yet we cannot be so overoptimistic as limitations of the micro-drones i.e. limited range (typically a few hundred meters) and flight time (up to 20–30 min) pose difficulty in the applications requiring survey of a large area or cadastral mapping. Also, the small or medium format cameras which are generally employed on low-cost and small payload systems, enforce the acquisition of a higher number of images (in our case 93 images of the administrative building alone were taken and in total 496 images for the whole campus) in order to achieve the same image coverage at a comparable resolution [6]. This may increase the processing time of data to a considerable extend, again limiting the use of low-cost UAV's to smaller areas. In these conditions, development of one's own methodology or automated and reliable orientation software will have to be used to reduce the processing time.

4.2 Applications in Civil Engineering

Most of the papers in literature related to Unmanned Aerial Systems offer a comprehensive review of applications that are possible with UAV's. Our work here provides a corroboration with the help of field work done. Based on that, we suggest that UAS can be used for following applications in Civil Engineering [5]:

Structures. The low-cost UAV's can be used for creating building inventory data. Also, they can be used for seismic risk assessment, observing structural damage and establishment of reliable seismic fragility databases for buildings and infrastructure.
Transportation. The applications in transportation engineering are the following but not limited to monitoring and maintenance of bridges, pavement condition assessment, 3D evaluation of rural road distresses and traffic management.
Disaster Response. UAV's can be employed for the initial assessment that includes assessing induced damage in the disaster zone, as well as the condition of the transportation network in a locality [1, 5]. But low-cost UAV's due to their less flight time cannot be used for more-detailed assessments regarding civilians needing assistance, disturbed critical transportation links, and critical power and communication infrastructure. These applications will require at least mini-drones.
Construction Management. An emerging field in construction management is Building Information Modelling or BIM integrated with GIS. The low-cost UAV's can be employed for the same due to their efficiency and speed.

Surveying and Mapping. A UAV provides a flexible and economical mobile platform for acquiring spatial information on objects of interest. Compared with those obtained by traditional aerial or satellite platforms, UAV-acquired datasets have better resolution in both temporal and spatial aspects. Also, as seen by our study, accuracy of elevation and area measurements done by DEM created by UAV's image data are better or at par with the traditional surveying instruments. So low-cost UAV's can be easily used for surveying smaller areas, delineating alignments of linear structures etc.

Acknowledgements The above work was carried out in the partial fulfillment of final year B. Tech. Project. We express our sincere thanks towards Prof. MPS Chauhan, Head, Civil Engineering Department and our other 4 team members i.e. Abhishek Singh, Jyoti Dangi, Shubham Singh Bisht and Akashdeep Singh Manral without their support this work would have not been achieved.

References

1. Adams SM, Friedland CJ (2011) A survey of unmanned aerial vehicle (UAV) usage for imagery collection in disaster research and management. In: 9th international workshop on remote sensing for disaster response, vol 8
2. Colomina I, Aigner E, Agea A, Pereira M, Vitoria T, Jarauta R, Pascual J, Ventura J, Sastre J, Brechbühler de Pinho G, Derani A (2007) The uVISION project for helicopter-UAV photogrammetry and remote-sensing. In: Proceedings of the 7th geomatic week, vol 2023, Barcelona, Spain, Feb 2007
3. Colomina I, Molina P (2014) Unmanned aerial systems for photogrammetry and remote sensing: a review. ISPRS J Photogramm Remote Sens 1(92):79–97
4. Eisenbeiss H (2004) A mini unmanned aerial vehicle (UAV): system overview and image acquisition. Int Arch Photogramm Remote Sens Spat Inf Sci 36(5/W1):1–7
5. Liu P, Chen AY, Huang YN, Han JY, Lai JS, Kang SC, Wu TH, Wen MC, Tsai MH (2014) A review of rotorcraft unmanned aerial vehicle (UAV) developments and applications in civil engineering. Smart Struct Syst 13(6):1065–1094
6. Nex F, Remondino F (2014) UAV for 3D mapping applications: a review. Appl Geomat 6(1):1–5
7. Siebert S, Teizer J (2014) Mobile 3D mapping for surveying earthwork projects using an unmanned aerial vehicle (UAV) system. Autom Constr 1(41):1–4

Comparing Sensors for Feature Extraction

Vickyson Naorem, Kamal Jain, Mahua Mukherjee and Kumar Abhishek

Abstract The extraction of features from coarse resolution satellite imagery is reliable for regional scale of the area, primarily, and applicable to a relatively smaller area to some extent. Very high-resolution (VHR) satellite imagery from space borne data is useful for small-scale regions. However, the price is very costly and depends on the cloud cover. The data required during the rainy season for flood studies cannot be used at all. On the other hand, aerial borne data is emerging to replace the space borne data due to easier accessibility, availability and also for a better resolution. The conventional methods to extract features itself have challenges and limitations due to pixel-based classification. The high spectral variability within classes reduces the accuracy of VHR image in pixel-based classification. Therefore, we used the object-based image analysis (OBIA) techniques in the study for overcoming such conventional difficulties. In order to compare the classification with a different source of data throughout the study, we have demonstrated the comparison of image classification using different sources of data for feature extraction. In this study, we used Unarmed Aerial Vehicle (UAV) data of Chingrajpara slum area in the state of Chhattisgarh, India having various morphological features. Firstly, we segmented the images of the study area to enhance the classification accuracy to compare with the results of space borne data of Mumbai slum area in the state of Maharashtra, India. We applied in different source of data to extract the features of formal buildings, vegetation, roads and informal settlements subjected to the availability of features in the subset. The result of the classification using UAV data is comparatively better and getting more than 90% accuracy as compared to the accuracy results of space borne

V. Naorem (✉) · K. Jain · M. Mukherjee
Centre of Excellence in Disaster Mitigation and Management (CoEDMM), Indian Institute of Technology Roorkee, Roorkee, India
e-mail: vnaorem@dm.iitr.ac.in

K. Jain
e-mail: kjainfce@iitr.ac.in

M. Mukherjee
e-mail: mahuafap@iitr.ac.in

K. Abhishek
Department of Architecture and Planning, Indian Institute of Technology Roorkee, Roorkee, India
e-mail: kabhishek@ar.iitr.ac.in

K. Jain et al. (eds.), *Proceedings of UASG 2019*, Lecture Notes in Civil Engineering 51,
https://doi.org/10.1007/978-3-030-37393-1_3

data. Since the accuracy has been depending upon location-specific, the sensors having better classification accuracy can be suggested for further classification having same specified features in other locations. Furthermore, the results can be used for monitoring and rapid digitization purposes in digital repository and for disaster risk reduction, especially.

Keywords Classification accuracy · Sensors · Disaster risk reduction · Feature extraction · OBIA and UAV data

1 Introduction

The extraction of features from coarse resolution satellite imagery is limited only for global and regional level. There is no benefit for research having a local level from coarse resolution imagery. The Sendai Framework for Disaster Risk Reduction 2015–2030 (SFDRR) considers and guides to achieve the targets using science and technology to local level to manage disaster risk for resilience [1–3]. With increasing very high resolution (VHR), satellite imagery from space borne data is useful for small-scale regions; one can make use of such imagery for local level study. However, the prices are very costly, not easy to access and it depends on the cloud cover as well. Furthermore, the data required during rainy season for flood studies cannot be used at all. So, the users can leverage to some extent. On the other hand, aerial borne data is emerging to replace the space borne data due to easier accessibility, availability and better resolution. However, conventional methods to extract features itself are the main challenging task due to pixel-based classification. In this context, we practised with different sensors in the object-based image analysis (OBIA) classification using UAV data in the paper to highlight the importance of comparing source of data to choose conducive purposes for different domains. Additionally, from such UAV data we can generate the data of elevation [4–6]. Eventually, the processes helped a lot in feature extraction of vegetation and building extraction. For example, change detection and rapid mapping processes for monitoring disaster risk could set a better preparedness and resilience for disasters. In this context, the processes of using OBIA with different sensors has a lot to contribute to extract data into information rather than using conventional methods of pixel-based classification. This article highlights the needs to leverage UAV data rather than space borne data like VHR imagery because the objects required to be classified during pixel-based classification are smaller than pixel of the imagery. The results contributed to false positives which in turn led to lesser accuracy and mixing of different feature classes [7–9]. This study compares and identifies conducive sensors to play a greater role in feature extraction in the context of remote sensing. This paper also highlights the importance of UAV data by comparing the classification accuracy performed with the space borne data conducted in another paper [10].

2 Materials and Methods

The dataset includes the images of a slum area with sub-urban features which are taken for Chingrajpara slum area in the state of Chhattisgarh, India. The source of data was given by Jain [11]. The dataset had 85 images with average GSD of 2.19 cm and area coverage of 0.130 km^2. This dataset was used for OBIA classification to compare sensors for feature extraction. And the dataset for VHR satellite imagery used for comparison was from DigitalGlobe i.e. WorldView imagery data, where we were only using for the accuracy results for comparison. We used few sensors because the focus was to highlight the importance of UAV data as compared to space borne data in this paper. And we have taken the limited subset because of the limitations in data and computer configuration to run the processes.

In this study, we have skipped filtering the remote sensing data acquired from the space borne source in the pre-processing stage because the dataset was of third party from Jain [11]. So, we have not included the flow of radiometric correction and pan-sharpening processes in the initial stage of data processing.

The methodology of the paper has been shown in Fig. 1. To initiate the above classification, we have conducted the first step of OBIA classification i.e. segmentation, and then, the image has converted into segmented objects to enhance the classification accuracy where the segmented objects has been chosen as pixel in case of OBIA classification which is the advantage of OBIA. Then, we applied in different

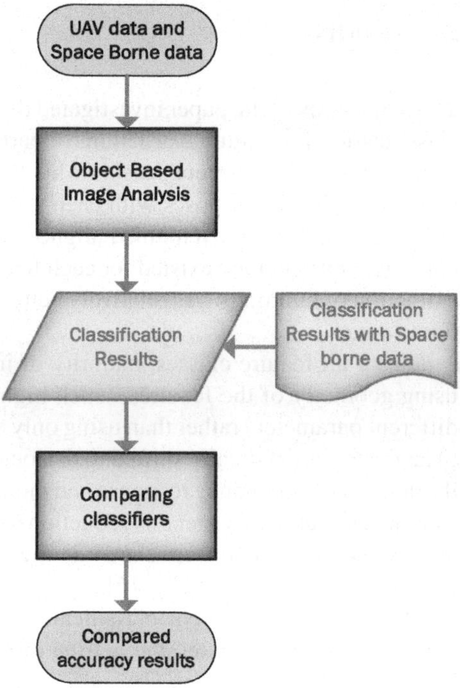

Fig. 1 Flowchart of the methodology

source of data to extract the features of formal buildings, vegetation, open spaces and informal settlements subjected to availability of features in UAV data. In each feature class, we used threshold parameters in the threshold classification where existence of vegetation was used first. Then classified regions of vegetation were merged. While classifying vegetation, we used greenness of green band due to lack of near infrared red band in UAV data. Then, in the second criteria, we used the heights of feature having more than 2.85 m to assign the class of formal buildings. In third, we used unclassified segmented objects having less than 2.85 m in heights and less than 9 m^2 in area was assigned as feature class of informal settlements. Then, we again merged the classified informal buildings. The rest of unclassified image were assigned and merged as open spaces. This is how we have finished the whole OBIA classification after using and applying criteria of rule sets in order to compare the classification accuracies with different sensors for the study area. Then we have demonstrated the comparison of image classification using different sensors for feature extraction with the classification accuracy results of the paper from de Naorem et al. [10]. In this study, we used Unarmed Aerial Vehicle (UAV) data with different morphological features configuration to check the robustness of the rule sets. Eventually, the results of the study highlighted the conducive purposes followed by the conclusion. The results could very much help in risk communication for disaster risk reduction and translate the risk perception into the spatial results of the classification for risk communication for different stakeholders [12–14] (Fig. 1).

3 Results

The main study of the paper investigated the comparison of different sensors in OBIA classification for feature extraction. In particular, the study compared how different source of data i.e. different sensors while OBIA classification has currently depicted its own conducive purposes with each sensor. The authors found more than 90% of classification accuracy remained aligned with each other. However, the variation of classification accuracy existed for each feature class of different sensors. The feature extraction of buildings was relatively better in UAV data as compared to that in space borne data.

Out of all feature classes, majority of image-based parameters were engaged in using geometry of the features which highlighted the importance of exploration of different parameters rather than using only spectral and related parameters. The scale of analysis has drastically shifted to the local level which would help tremendously in the domain of emergency response and monitoring by rapid digitisation processes, for example, targets in disaster risk reduction focusing data in the local level using science and technology which were shown in the list of targets from Sendai Framework for Disaster Risk Reduction [15, 16].

The results of different classifications using UAV data in Fig. 2 and Table 1 were comparatively better and getting more than 90% accuracy as compared to the

Fig. 2 Images showing OBIA classification using UAV data with more accuracy and crispier boundary

Table 1 Table showing OBIA classification with more than 90% accuracy using UAV data

		Ground truth				
		Formal	Informal	Open spaces	Vegetation	
Classified image	Formal	*50*				50
	Informal	2	*46*	2		50
	Open spaces		3	*45*	2	50
	Vegetation			1	*49*	50
		52	49	48	51	*200*
Overall accuracy		95%				
Kappa		0.93				

The italics values are true positives which have to be always achieved higher in values in order to get higher accuracy in classification and also to reduce the false positives

accuracy of space borne data (Fig. 3) provided in a previously published research paper by the primary author [10].

Fig. 3 Images showing OBIA classification using WorldView imagery (space borne data). *Source* de Naorem et al. [10]

4 Discussion

Overall, for the entire study area, the authors found using more image-based parameters of geometry rather than spectral related parameters which enhance the classification accuracy and also for the particular classes of buildings—formal and informal. In contrast, UAV data lacks consistency as it is obtained through lesser accurate cameras and sensors but are more affordable which is limited to the exploration of more image-based parameters. An important finding from this paper is that the classified image which was recently obtained using UAV data shows *better user-friendliness to apply the image-based parameters* and generates *crispier boundary* of feature classes as compared to previous classified image from space borne data, highlighting the complexity of this issue. And rule sets having better classification accuracy can be adopted for location having similar characteristics features of the area because of the segmentation process is always depending upon the location-specific i.e. segmented objects change place to place. Overall results help in identifying the conducive purposes for each sensors and robustness of the rule sets to be able to apply in the image having similar characteristics features of the area.

5 Conclusions

The main study investigates comparison of different sensors of space borne data and UAV data in OBIA classification for feature extraction. The variation of classification accuracy happened for each feature class of different sensors. The feature extraction of buildings was relatively better in UAV data as compared to that in space borne data due to the existence of geometric parameters in the data itself which allows the image-based parameters to perform more than the previous classification. In contrast, study compares and shows significance of community participation in building urban flood resilience that mobile app-based data shows through more informative and in detailed manner than literature-based data. An important finding from this paper has claimed that classified image that recently performed using UAV data shows better user-friendliness and crispier boundary of feature classes. Furthermore, the results suggest that the presence of unique sets of new image based parameters can be used to achieve targets of SFDRR by feature extraction of data for disaster risk reduction.

Acknowledgements We would like to especially thank research scholars from the Geomatics Department for their support and valuable inputs. We also acknowledge Dr. Kamal Jain for providing UAV data in this paper. Last but not the least, we would also like to thank Computer Lab, CoEDMM, IIT Roorkee for enabling us to perform the analysis for this paper.

References

1. UNISDR (2015) Sendai framework for disaster risk reduction. Sendai, Japan
2. NIDM (2013) National workshop on Uttarakhand disaster 2013: lessons learnt
3. NIDM (2014) Toolkit for urban planning in disaster management. New Delhi
4. Fernandez Galarreta J (2014) Urban structural damage assessment using object-oriented analysis and semantic reasoning. ITC—Faculty of Geo-Information Science and Earth Observation, University of Twente
5. Xue J, Su B (2017) Significant remote sensing vegetation indices: a review of developments and applications. J Sens 2017
6. Feng Q, Liu J, Gong J (2015) UAV remote sensing for urban vegetation mapping using random forest and texture analysis. Remote Sens 7(1):1074–1094
7. Martha TR, Kerle N, van Westen CJ, Jetten V, Kumar KV (2011) Segment optimization and data-driven thresholding for knowledge-based landslide detection by object-based image analysis. IEEE Trans Geosci Remote Sens 49(12):4928–4943
8. Ardila JP, Bijker W, Tolpekin VA, Stein A (2012) Context-sensitive extraction of tree crown objects in urban areas using VHR satellite images. Int J Appl Earth Obs Geoinf 15:57–69
9. Kohli D, Warwadekar P, Kerle N, Sliuzas R, Stein A (2013) Transferability of object-oriented image analysis methods for slum identification. Remote Sens 5(9):4209–4228
10. de Naorem V, Kuffer M, Verplanke J, Kohli D (2016) Robustness of rule sets using VHR imagery to detect informal settlements—a case of Mumbai, India. In: GEOBIA 2016: solutions and synergies
11. Jain K (2019) Urban slum dataset for Chingrajpara slum area. Chhattisgarh
12. Schelfaut K, Pannemans B, van der Craats I, Krywkow J, Mysiak J, Cools J (2011) Bringing flood resilience into practice: the FREEMAN project. Environ Sci Policy 14(7):825–833
13. Seidelsohn K, Voss M, Krüger D (2018) Urban disaster resilience and security. Springer International Publishing, Cham
14. Ajibade I, McBean G, Bezner-Kerr R (2013) Urban flooding in Lagos, Nigeria: patterns of vulnerability and resilience among women. Glob Environ Change 23(6):1714–1725
15. Etinay N, Egbu C, Murray V (2018) Building urban resilience for disaster risk management and disaster risk reduction. Procedia Eng 212(2017):575–582
16. Witt E, Lill I (2018) Methodologies of contemporary disaster resilience research. Procedia Eng 212:970–977

Survey in Closed Environments Through UAS Technology. Methodological Approaches to the Study and Image Processing of Religious Furnishings

R. Valenti and E. Paternò

Abstract The present study deals with survey in closed environments with UAS technology and with the solutions of problems related to the adoption of innovative approaches for the investigation and knowledge of historical Cultural Heritage. In particular, the altar dedicated to St. Ignatius of Loyola in the Church of Collegio dei Gesuiti in Siracusa has been surveyed. Designed around 1756 by the sculptors Domenico Battaglia and Giovan Battista Marino, it is placed in the transept of the church and with its imposing presence it is the emblem of the great Baroque season in Sicily. The survey campaign developed two levels of analysis: on the one hand, the UAS experimentation for the image processing and on the other, TLS implementation for the collection of metric data. The integration of the two approaches leads to the creation of a detailed 3D model. More specifically, the low cost UAS platform used in the present experimentation lacking a GPS system can fly at a maximum altitude of 5 m. Therefore, given the great height of the architectural work, the main problem for the achievement of objectives was to find a solution to overcome such limits. This is fundamentally important in order to provide a large amount of information able to increase the collected data with the traditional systems of terrestrial surveying. Thus, the point cloud obtained from such an integration can be converted into a continuous high-resolution model, a higher density point cloud which can be a useful support for any further determination.

Keywords 3D survey · Photogrammetry · 3D modelling · Baroque architecture · Cultural Heritage

1 Introduction

The present study is the expression of the evolution of technological advancement. In particular, the technological equipment available in the field of Geomatics is of great

---

R. Valenti · E. Paternò (✉)
Special Educational Structure of Architecture, DICAR, University of Catania, Syracuse, Italy
e-mail: patermanu@hotmail.it

R. Valenti
e-mail: ritam.valenti@gmail.com

K. Jain et al. (eds.), *Proceedings of UASG 2019*, Lecture Notes in Civil Engineering 51,
https://doi.org/10.1007/978-3-030-37393-1_4

help to the survey of Cultural Heritage during the data acquisition phase. Methodological strategies are therefore directed towards the solution of problems connected with instrument adjustment to all practical situations. In order to give a precise answer to the specific needs of survey, it is important to provide appropriate solutions able to favour knowledge organization and culture dissemination. Being aware that a systematic approach to concepts remains constant over the years while technology innovation grows rapidly in its ability to process more and more complex algorithms. The present study is the fruit of a synthesis between applied research in architectural surveying and historical analysis of the investigated Cultural Heritage and takes an innovative approach aimed at the understanding and visualization of the specific critical path. The relevant aspect of the research deals with the implementation of UAS technology for integrated architectural survey indoors (see Fig. 1). Remotely piloted aircraft systems have usually been implemented in recent survey campaigns especially for the environmental monitoring of specific areas, for the archaeological survey aimed at the 3D rendering of cultural sites and for the other outdoor uses.

The need to catalogue, analyse and manage data of indoor small-scale Cultural Heritage has led towards the implementation of non-invasive integrated survey systems (TLS and UAS). Such technologies provide the collection of a big amount of data which, appropriately managed, in the post-processing phase can make it possible to create detailed outputs of every complex structures through the construction of 3D multipurpose models. Finally, the experimentation has been conducted on the occasion of the study of the imposing altar dedicated to St. Ignatius of Loyola in the Church of Collegio dei Gesuiti in Siracusa started in 1744. Its final virtual visualisation has been coherently processed together with the subsequent geometric analysis to find out any possible contact or contamination with the Roman models of the treatises.

Fig. 1 DJI Spark and an example of UAV digital image

2 Related Works

Recently, the scientific community has shown a great interest in light UAV (Unmanned Aerial Vehicles) platforms implementing them in different fields such as engineering, architecture and archaeology [1–3]. In particular, the acquisition of digital images through UAS systems for the survey of historically important architecture and archaeological sites can generate, quickly and at a low cost, the necessary data for the production of HD models for documentation and dissemination of Cultural Heritage. Such models represent the basis for the creation of multimedia products able to involve wider audiences. Several studies show the importance of integrating 3D survey methodologies (TLS and UAV photogrammetry) for the understanding and documentation of architecture. An integrated investigation may reduce problems connected with the survey of complex or oversized structures obtaining better results and more detailed representations [4–6]. Nowadays, a lot of researchers have positive reactions about the combination of nadir and oblique UAV imagery [7, 8]. Shifting the axis of the lens it will be possible to survey and measure the higher parts of those artefacts not easily feasible with the common survey techniques such as terrestrial laser scanning and terrestrial photogrammetry. Thus, it will be possible to obtain complete models especially in the case of articulated and oversized architectural structures. Although all the advantages offered by new technologies, the growing use of drones has determined the introduction of some regulations which define their use, strongly restricting the flight areas. In order to overcome such limits, very light and small platforms (micro-UAV) have been developed. The present study has employed a drone weighing less than 300 g for the architectural investigation of a baroque altar in a closed environment.

2.1 Case Study

The Laboratory of Representation has launched a survey campaign of the Ecclesiastical Heritage of south-eastern Sicily long ago focusing on baroque religious furnishings (chapels, altars, altar frontals) which are small scale pieces of architecture. In particular, the Cultural Heritage the project has concentrated on presents complex morphological characteristics which are expression of the conformative dynamism of that historical period (see Fig. 2).

Instrumental survey integrated with TLS and UAS technologies is at the moment the most appropriate modus operandi for these works, producing high-quality 3D visualization. The simple management of the models is important for the generation of 2D and 3D images which will provide information about dimensions and forms necessary for furthers historical or conceptual determinations. The present case of study conducted an experimentation with the altar dedicated to St. Ignatius of Loyola, placed at the rear of the left wing of the cross in the Church of Collegio dei Gesuiti in Siracusa. It was started in 1744 by sculptors Giovanni Battista Marino, Ignazio

Fig. 2 Examples of religious furnishings

Marabitti, Domenico Battaglia and derives from the Roman models designed by Friar Andrea Pozzo for the altar of the Blessed Luigi in the Church of St. Ignatius in Rome [9]. The wall altar is imposing for the monumental characteristics and the richness of all its elements. The four spiral columns, the broken tympanum with two symbolic statues and two groups of cherubs and the rich decorations are all expressions of the Baroque art of the period. Archive documents [9] (the contract dated 7th July 1744 between the church rector and G. B. Marino) report Marino's obligations to follow the "project of the venerable Chapel of St. Luigi, made by friar Andrea d'Apuzzo". Materials, the green marble of the Solomonic columns in particular, faithfully reflect Andrea Pozzo's instructions written under the drawing on his Treatise (1693–98). Compared with Pozzo's drawing, the studied altar presents two clear differences: the absence of decoration which winds the columns with golden bronze vine branches and the presence, in the centre, of a niche with the statue of St. Ignatius instead of panel with a high relief. The first difference probably derives from a certain impoverishment due to the abandonment and improper use of the church for a long period of time. Actually, archive documents, reported by Agnello [10] mention payments for the carving of 20 leaves, certainly for the column decoration. The second difference comes from one of the project ideas that Marabitti, who made the statue, certainly knew and was inspired to Pozzo's drawing of the altar of St. Ignatius in the Church del Gesù in Rome. A comparison between Pozzo's drawings and the altar in Siracusa required precise survey procedures in order to produce digital images appropriate to the research. The grandeur of the altar, more than 13 m high, and its formal complexity suggested a methodology for the data acquisition based on UAV especially for the highest parts.

3 TLS and UAV Technology: Problems and Resolutions

The present research has applied innovative 3D survey technology to examine the architectural structure previously described in order to create a HD digital model. Digital data have been acquired during a 3-day survey campaign (see Fig. 3).

In the course of the first day instrumental survey procedures have been conducted with Faro CAM2 Focus. Three station points were determined so to make three high-resolution scans to obtain a denser and smoother point cloud. The second day of work was dedicated to the artefact digitization with the images from the drone. The two low cost drones used in the research were two DJI Spark exemplars whose weight was appropriately reduced and they were also modified to satisfy Art. 12, paragraph 5, of ENAC (Italian Civil Aviation Authority) regulations on RPAS. They come with a mounted camera equipped with a stabilized z-axis gimbal with 1/2.3″ CMOS sensor able to shoot 12 MP images and record 1080p 30fps videos. Although RPAS operating indoors are not subject to ENAC regulations except for what is stated at Art. 10, paragraph 7, it was anyway decided to use harmless drones due to their lighter dimensions which seemed appropriate to the position and complexity of the case study. Anyway, a series of problems connected with the implemented

Fig. 3 Survey activity with UAS and TLS technologies

technology soon arose. Since the operations were performed indoors, GPS couldn't track satellites so the two mini RPAs flew in ATTI Mode. This way, the drones automatically keep their altitude through the optical sensors placed under them but the position is not stabilized making the flight more difficult and the pilot has to readjust it. Surveying operations were carried out under particular conditions. Low light and homogeneous flooring have further increased difficulties keeping the drone steady. Moreover, missing a GPS connection the drone maintains its altitude of 5 m from its launch spot according to DJI factory standards. Then, it was necessary to take alternative procedures to overcome the obstacle in order to survey the highest parts of the altar avoiding an incomplete final model. Taking inspiration from some videos on the web showing micro-drones flying indoors, in sports arenas or cinemas, and where the drone, taking off from the highest altitude, maintains it drifting low, a launch platform was built to reduce the limit of 5 m allowing it to reach the maximum altitude of 13 m, which is the height of the surveyed structure (see Fig. 4). The platform was made of double corrugated synthetic material, light and extremely stable. Moreover, side flaps were added in order to stop the drone from slipping. This structure was anchored to a metric rod which could be stretched as far as 5 m, fixed to a tripod. Raising the launch platform of about 6–7 m from the ground has, on the one hand, allowed the drone to take off from there. The vehicle was able to detect the starting point at zero altitude on the platform and from there it could climb 5 m reaching a height of 12–13 m. However, on the other hand, it was difficult to control the drone because it, trying to read the floor through its sensors and to keep its level, couldn't be radio-controlled.

The survey campaign ended on the third day. The operations required the simultaneous presence of a UAV pilot and an assistant giving the pilot indications about the flight and the presence of possible obstacles. Two missions have been accomplished on this occasion in manual mode, and about 160 shots were taken, with an overlaying of 70%. Photos were taken in two modes. In the first case, the drone drifted from left to right along an ideal line parallel to the architectural structure at a distance of about 4 m and with the focal plane parallel to the front. Thus, through UAV an unusual type of photogrammetry was used. A photogrammetry where photos are taken keeping the focal plane parallel to the studied item that, in the present case, develops vertically

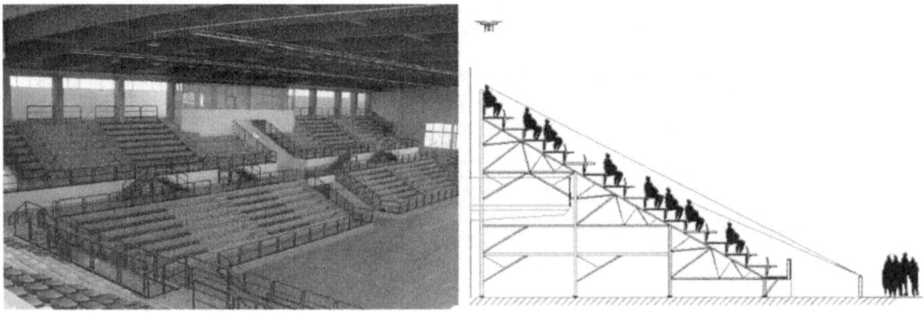

Fig. 4 UAS technology indoors

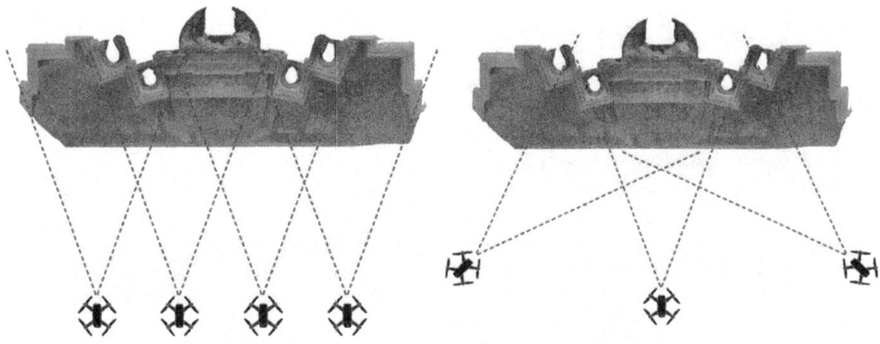

Fig. 5 UAV positions during photogrammetric survey

along the z axis, instead of horizontally as its usually happens in photogrammetry. The second flight was performed at a smaller distance and making the drone rotate around the structure in order to survey the lateral sides hidden by jutting out portions (see Fig. 5).

However, because of the uncomfortable position of the altar, standing between two pillars, it was difficult and sometimes impossible to keep the RPA close to the wall to capture the most internal parts. As for the jutting out top parts, changing the inclination of the lenses and the focus, it was possible to survey in detail the structure integrating the acquired data with those of the laser scanning device.

3.1 Data Processing

During the post-processing phase two point clouds were generated. The first one was produced manually aligning laser scanning data, the second one was produced through structure from motion technique (SfM). SfM is a technique which reconstructs a 3D scene from a set of digital photos recovering automatically the key points present in several photos and defining their spatial coordinates. A "sparse" point cloud is produced and subsequently made denser. 3D Zephyr Aerial software, used in the Laboratory. It automatically integrates the point cloud produced with laser scanning technology and previously processed through Cyclone. After aligning the two dense point clouds it is possible to convert them to triangle meshes and, finally, the seamless model obtained will be transformed into a 3D textured model (see Fig. 6).

Fig. 6 Data acquisition and some views of the final model produced by photogrammetry and TLS

4 3D Modelling and Formal-Geometric Analysis

The digital model obtained through integrated survey methodology allowed the visualization of predefined views of the architectural structure in order to start the second phase of the research aiming at the understanding and interpretation of architecture. In particular, the DSM (Digital Surface Model) was of fundamental importance for the graphical analysis of the altar. Following a UAV photogrammetric workflow it provided higher data density which generated High resolution ortophotos and perspective views of the highest parts, too. These last ones were then scaled in order to compare them with Andrea Pozzo's drawings and to assess their building correspondence according to what is stated on the published documents (see Fig. 7). The results of the research confirm the almost totally superimposition of the altar of St. Ignatius of Loyola with Pozzo's drawing of the altar of the Blessed Luigi in the Church of St. Ignazio del Collegio Romano, as far as the architectural structure is concerned. In particular, in the drawing the four spiral columns are slightly more slender than those in Siracusa. It is also clear the correspondence of the central niche holding the statue of St. Ignatius with the one drawn by Pozzo for the altar of St. Ignatius in the Church del Gesù in Rome (see Fig. 7). It can be said that the altar in Siracusa, compared with the Roman examples, is a contamination, rather mysterious, of the two altars drawn by Pozzo. Even if the documents dating back to 1744 make explicit reference to the altar of the Blessed Luigi, its construction is different being influenced by the altar of St. Ignatius in the Church del Gesù in Rome. It seems that the workers left a trace of their thoughts joining together the two highest examples of Jesuit art. The study emphasizes how the applied methodology combines graphical analysis, historical research and survey integrated activities with the aim of promoting sector-wide scientific approaches. The combination of technological and instrumental innovation led to theoretical knowledge supported by digital image processing technology.

5 Conclusions

The main objective of the research was to find innovative solutions in order to test the potentialities offered by the new survey and 3D rendering systems and to overcome the limits of the adopted instruments. The launch platform of variable height adjustable to the dimensions of the architectural structure was extremely important to collect precise data. It is worth saying that the adopted instruments are not always designed for the purpose dealing with survey activities so their use necessarily requires some adaptations to the specific tasks. Therefore, the specialist is a

◄**Fig. 7** Drawing and 3D model: 3D comparison of the altar of St. Ignazio di Loyola in the Collegio dei Gesuiti in Siracusa and Andrea Pozzo's drawings of the altars of St. Ignatius in the Church del Gesù in Roma and Blessed Luigi in the Church of St. Ignatius del Collegio Romano

key figure to take the appropriate methodological approach. In conclusion, the use of mini UAVs to survey close range objects indoors has given positive results in relation to expectations and the subsequent use of data for 3D processing. Therefore, if flight restrictions and safety rules limit the use of UAVs exclusively for top photo and video shooting, in the specific case of architectural survey such measures represent a real obstacle to the planned activities. From the conducted research also emerges that the mini UAVs are developing interesting scenarios for the digital rendering of Cultural Heritage indoors. The instruments used in the survey campaigns, notably UAS, being designed for purposes not specifically connected with the present research, would require a deeper exchange of information between designers and commissioners.

Acknowledgements The authors wish to thank Engineer Pietro Pistone, managing director of the Regional Civil Protection Department in Siracusa who granted permission to conduct survey on the Church of Collegio in Siracusa; Dr. Tommaso Palermo for the bibliographical research at the Regional University Library in Catania; Mr. Domenico Triolo Puleio for the full support in the implementation of the UAV take-off platform.

Translation by Dr. Melania Grancagnolo.

This work has partially been financed by the University of Catania within the project "Piano della Ricerca Dipartimentale 2016–2018" of the Department of Civil Engineering and Architecture. The conducted study within the research program "Representation of Baroque religious architecture between geometry and symbology. From large to small scale, from façades to altars" (project PON-NEPTIS) is the result of the collaboration of all the authors. In particular, Rita Valenti wrote paragraphs 1, 2.1, 4, 5; Emanuela Paternò wrote paragraphs 2, 3, 3.1.

References

1. Remondino F, Barazzetti L, Nex F, Scaioni M, Sarazzi D (2011) UAV photogrammetry for mapping and 3d modeling—current status and future perspectives. ISPRS Int Arch Photogramm Remote Sens Spat Inf Sci XXXVIII-1/C22:25–31. https://doi.org/10.5194/isprsarchives-XXXVIII-1-C22-25-2011
2. Bolognesi M, Furini A, Russo V, Pellegrinelli A, Russo P (2014) Accuracy of cultural heritage 3D models by RPAS and terrestrial photogrammetry. ISPRS Int Arch Photogramm Remote Sens Spat Inf Sci XL-5:217–224. https://doi.org/10.5194/isprsarchives-XL-5-113-2014
3. Carnevali L, Ippoliti E, Lanfranchi F, Menconero S, Russo M, Russo V (2018) Close-range mini-UAVs photogrammetry for architecture survey. ISPRS Int Arch Photogramm Remote Sens Spat Inf Sci XLII-2:113–119. https://doi.org/10.5194/isprs-archives-XLII-2-217-2018
4. Fiorillo F, Remondino F, Barba S, Santoriello A, De Vita CB, Casellato A (2013) 3D digitization and mapping of heritage monuments and comparison with historical drawings. ISPRS Int Arch Photogramm Remote Sens Spat Inf Sci II-5/W1:133–138. https://doi.org/10.5194/isprsarchives-XL-5-W4-431-2015
5. Balletti C, Guerra F, Scocca V, Gottardi C (2015) 3D integrated methodologies for the documentation and the virtual reconstruction of an archaeological site. ISPRS Int Arch Photogramm

Remote Sens Spat Inf Sci XL-5/W4:215–222. https://doi.org/10.5194/isprsarchives-XL-5-W4-215-2015
6. Majid Z, Ariff MFM, Idris KM, Yusoff AR, Idris KM, Aspuri A, Abbas MA, Zainuddin K, Ghani ARA, Ardi (2017) Three-dimensional mapping of an ancient cave paintings using close-range photogrammetry and terrestrial laser scanning technologies. ISPRS Int Arch Photogramm Remote Sens Spat Inf Sci XLII-2/W3:215–222. https://doi.org/10.5194/isprs-archives-XLII-2-W3-453-2017
7. Chiabrando F, Lingua A, Maschio P, Teppati Losèa L (2017) The influence of flight planning and camera orientation in UAVs photogrammetry. A test in the area of rocca San Silvestro (LI), Tuscany. ISPRS Int Arch Photogramm Remote Sens Spat Inf Sci XLII-2/W3:163–170. https://doi.org/10.5194/isprs-archives-XLII-2-W3-163-2017
8. Vacca G, Furfaro G, Dessì A (2018) The use of the UAV images for the building 3D model generation. ISPRS Int Arch Photogramm Remote Sens Spat Inf Sci XLII-4/W8:217–223. https://doi.org/10.5194/isprs-archives-XLII-4-W8-217-2018
9. Agnello G (1930) Arte Gesuitica. In: Per l'arte sacra, Roma, VII, Lug–Set 1930, pp 78–83
10. Agnello G (1928) Architettura gesuita. La chiesa del Collegio di Siracusa. In: Per l'arte sacra, Roma V, Gen–Feb 1928, pp 7–16

Integration of Lidar Data in Topographical Feature Extraction from Very High-Resolution Aerial Imagery

Kuldeep Chaurasia

Abstract Geospatial technology has been demonstrated as a reliable and efficient tool for monitoring of the land cover pattern for vast geographical areas. Although, the demand for the various thematic layers including landcover maps at finer scale has got increased for various applications such as urban studies, forestry and disaster management. In this paper, the utilization of LiDAR data for urban land cover classification of aerial imagery has been discussed. The study area has been classified into seven land-use/cover classes based on the textural, and spectral features using object-oriented classification approach. The applicability of various texture measures based on the gray level co-occurrence matrix along with the effect of varying pixel window has also been discussed. The classification results indicate that homogeneity texture image generated using 3 * 3 window size is best suitable for extraction of various topographical objects. The suitability of the various textural features has also been investigated. The LiDAR data has been found best suitable for identification of small objects such as buildings, trees and vehicles over aerial imagery. The overall accuracy of the classification has been obtained as 87.21% with the kappa coefficient of 0.84. The outcome of the study can be effectively utilized for disaster management applications such as evacuation planning, damage assessment, and post-flood recovery effort.

Keywords Remote sensing · Feature extraction · LiDAR · Image classification · Textural feature

1 Introduction

Nowadays, Aerial and ground-based LiDAR data is widely used in a variety of applications including terrain modeling [1], virtual city creation [2], image classification [3]. The invention of LiDAR technology has provided an approach through which a detailed description of 3D structures of the object surface can be obtained. 'Point cloud' which is the range scan output of a LIDAR system comprised of a large number

K. Chaurasia (✉)
Bennett University, Greater Noida, Uttar Pradesh, India
e-mail: kuldeep@bennett.edu.in

K. Jain et al. (eds.), *Proceedings of UASG 2019*, Lecture Notes in Civil Engineering 51,
https://doi.org/10.1007/978-3-030-37393-1_5

Table 1 Details of the datasets used for analysis

S. No.	Data	Spatial resolution (m)	Date of acquisition
1	Aerial image	0.05	13/03/2011
2	LiDAR	0.10	13/03/2011

of points with x, y, z coordinate information. With the advancement of technology, LiDAR system offers progressively denser points coverage of the surfaces [4]. Physical composition and characteristics of land elements on the Earth surface is regarded as Land cover [5]. Distribution of land cover plays a vital role and has a significant impact in climate and environment, which necessitates the mapping of land cover patterns at global, regional and local scales to help scientists and other organizations for better monitoring of the changing world. Yan et al. [6] have presented a review of the urban land cover classification in which LiDAR data were used to improve the accuracy of classification. In this study, high spatial resolution color imagery along with the NDVI image and LiDAR data have been used to extract different types of land cover classes towards improvement in the classification accuracy.

2 Study Area and Data Resources

The LiDAR dataset used in the current research belongs to a part of the urban region in Zeebruges, Belgium. The upper left scene coordinates 51° 20′ 00″ N, 03° 12′ 37.55″ E and lower right scene coordinate 51° 19′ 53.05″ N, 03° 12′ 50.42″ E. The LiDAR data was collected at a pulse rate of 125 Hz. The point density for the LiDAR sensor was approximately 65 points/m^2. The Raw LiDAR dataset consists of about 1 million points with an average point spacing of 10 cm. The aerial imagery and LiDAR data acquisition were carried out by Royal Military Academy, Belgium. The aerial imagery was ortho rectified and processed into the coordinate system (Universal Transverse Mercator) similar to LiDAR data set. The details of the areal imagery and LiDAR data have been summarized in Table 1.

A subset of a scene covering an urban region of size 4189 × 4056 has been selected to test the classification algorithm. The region is a residential area with smaller size houses, trees, grassland and roads. The color image of the selected urban region along with the digital surface model of the same area has been shown in Fig. 1.

3 Methodology

Before performing any operation on the LiDAR data point cloud, it is required to carry out preprocessing of point cloud for better results [7, 8]. In the first step, outlier points due to noise were eliminated by creating the grids of points into 10 × 10 cm

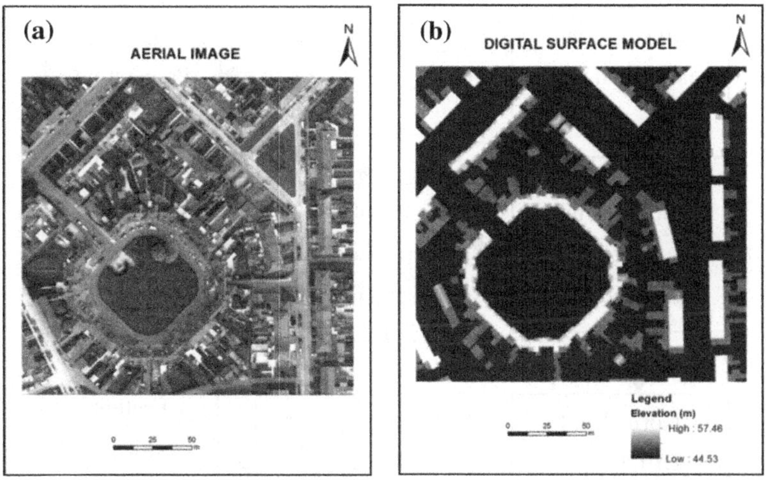

Fig. 1 **a** Aerial color image and **b** LiDAR point cloud

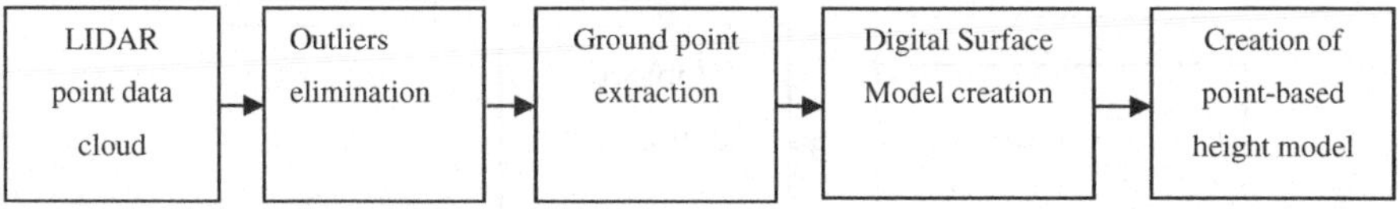

Fig. 2 LIDAR point cloud preprocessing

tile size followed by ground point extraction to obtain the digital surface model as shown in Fig. 2.

The NDVI image has been generated from aerial data to monitor the vegetation cover and tree canopy. The typical vegetation class displays a higher NDVI value as compared to other land cover classes. The formula for calculating NDVI image is shown in Eq. (1).

$$NDVI = \frac{(NIR - RED)}{(NIR + RED)} \tag{1}$$

where, NIR and RED represents the spectral reflectance measured at near- infrared and visible red bands respectively. After preprocessing of LiDAR points, the next step was data preparation in which RGB color image was stacked with the DSM of the same region. In this study, aerial imagery has been segmented using multi-resolution segmentation method. The suitable scale of segmentation for the input data has been identified as 20. As a result of segmentation, the smaller segments are needed to be merged to obtain the complete object as one segment. For this purpose, region merging approach has been adopted with merging parameter as 10. The steps followed in the implementation methodology are shown in Fig. 3.

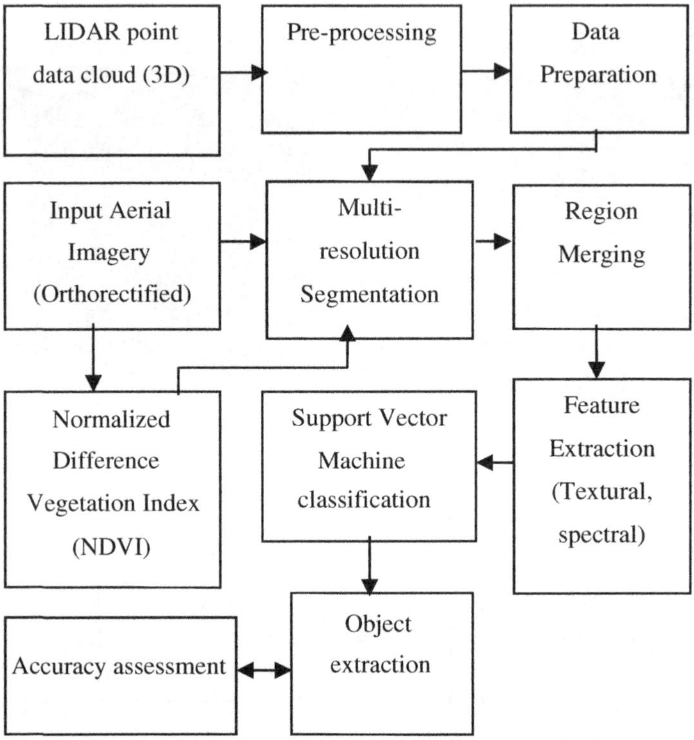

Fig. 3 Methodology

4 Results and Discussions

In this study, aerial imagery has been classified into eight classes viz. road, red rooftop, gray rooftop, concrete rooftop, grassland, tree, fallow land and vehicles using support vector machine algorithm. The input aerial imagery has been classified based on the homogeneity feature, NDVI and LiDAR point cloud derived Digital surface model. The homogeneity feature has better identified the roads and concrete rooftops which are having the same reflectance but different texture. NDVI feature has better capture the vegetation information. LiDAR point derived DSM has been used to capture the objects based on their elevation information. Accuracy assessment of classified images has been conducted with reference to Google Earth Image. Overall accuracy and Kappa coefficient of classified images as shown in Fig. 4, are computed with the help of the confusion matrix. It has been observed that textural, spectral and NDVI features improves the classification accuracy and enable the classifier to extract the smallest feature such as vehicles and trees from the aerial imagery. The overall accuracy of the classified image is 87.21% with a kappa coefficient 0.84.

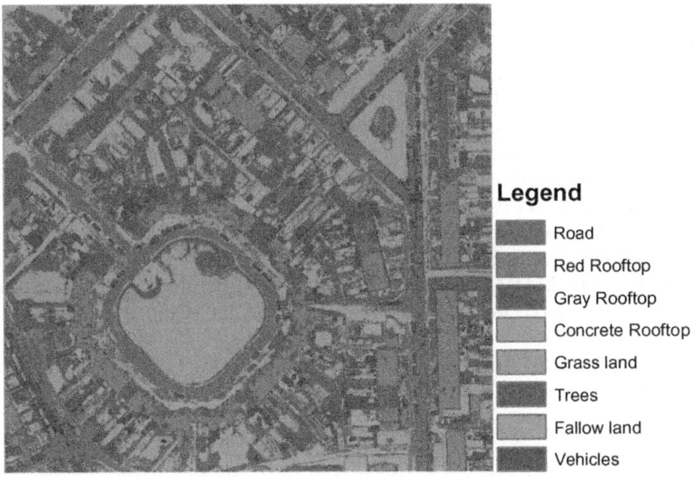

Fig. 4 Classified imagery

5 Conclusions

The fusion of LiDAR data with aerial imagery brings out important outcomes to be useful for numerous land resources and water management applications. In this paper, the utilization of LiDAR data for urban land cover classification of aerial imagery has been discussed. The study area has been classified into 7 land-use/cover classes based on the textural, and spectral features using object-oriented classification approach. The LiDAR data has been found best suitable for identification of small objects such as buildings, trees, and vehicles over aerial imagery. The overall accuracy of the classification has been obtained as 87.2149% with the kappa coefficient of 0.8429. The features from multiple textural methods belonging to different class can be incorporated to further improve the classification accuracy.

Acknowledgements The author would like to thank the Department of Communication, Information, Systems & Sensors (CISS) of the Belgian Royal Military Academy (RMA) and IEEE GRSS Image Analysis and Data Fusion Technical Committee for providing the data used in this study.

References

1. Vosselman G (2000) Slope based filtering of laser altimetry data. Int Arch Photogrammetry Remote Sens 33:935–942
2. Frueh C, Zakhor A (2003) Constructing 3d city models by merging ground-based and airborne views. In: 2003 IEEE computer society conference on computer vision and pattern recognition, 2003. Proceedings, pp II-562–9 vol 2

3. Blair JB, Rabine DL, Hofton MA (1999) The laser vegetation imaging sensor: a medium-altitude, digitisation-only, airborne laser altimeter for mapping vegetation and topography. ISPRS J Photogrammetry Remote Sens 54:115–122
4. Bandyopadhyay M, van Aardt JAN et al (2013) Classification and extraction of trees and buildings from urban scenes using discrete return LiDAR and aerial color imagery. SPIE defense, security, and sensing. International Society for Optics and Photonics, vol 8731
5. Cihlar J (2000) Land cover mapping of large areas from satellites: status and research priorities. Int J Remote Sens 21(6, 7):1093–1114
6. Yan WY, Shaker A, El-Ashmawy N (2015) Urban land cover classification using airborne LiDAR data: a review. Remote Sens Environ 158:295–310
7. Lodha SK, Fitzpatrick DM, Helmbold DP (2007) Aerial lidar data classification using expectation-maximization. In: Electronic imaging 2007. International Society for Optics and Photonics
8. Uzar M, Yastikli N (2013) Automatic building extraction using LiDAR and aerial photographs. Bol de Cienc Geodesicas 19(2):153–171

Automatic Extraction of Roads from UAV Using Thresholding and Morphometric Parameters

R. Sree Ram, Srinivasa Raju Kolanuvada and M. Shanmugam

Abstract Segmentation of surface elevation points derived from high-resolution UAV imagery is used to identify terrain and non-terrain points. Grey level thresholding of radiance values associated with terrain points enables extraction of impervious surfaces like Roads and Platforms. A morphometric based algorithm along with threshold length to width ratio is used to extract more precisely the roads out of other ground features. The experimental results show that this method can detect better linear features. Some comparatively small mask and boundaries are detected with changes in the local statistics by the automatic methods. Further, the discontinuity is detected and eliminated with morphometric parameters to minimize the effects of interfering objects and shadow effects. The extracted road network is evaluated for both spatial and semantic accuracy by comparing with the existing road network. This method demonstrated better extraction of linear features like roads than conventional image classification methods.

Keywords UAV · Thresholding · Grey level radiance · Morphometric parameters

1 Introduction

Extraction of information from the aerial images is a difficult process, but has numerous applications for understanding the natural or manmade events like the disaster monitoring (earthquakes, floods, vegetation fires, etc.), crop monitoring in precision agriculture, border surveillance, traffic monitoring, and so on. Road extraction through segmentation plays an important role in urban planning and challenging task from high-resolution aerial imagery like UAV imagery. Different image processing

R. Sree Ram (✉) · S. R. Kolanuvada · M. Shanmugam
Institute of Remote Sensing, Anna University, Chennai, India
e-mail: sreeram720@gmail.com

S. R. Kolanuvada
e-mail: ksrajuirs@gmail.com

M. Shanmugam
e-mail: shan_mc50@yahoo.com

K. Jain et al. (eds.), *Proceedings of UASG 2019*, Lecture Notes in Civil Engineering 51,
https://doi.org/10.1007/978-3-030-37393-1_6

techniques have been adopted by researchers to extract accurate road information. Texture analysis is one of the technique used to detect and segment the area of interest especially roads from aerial or UAV images but the choice of region or interest depends on the specific application of the extracted features. The supervised learning approach is to detect road textures and features using a neural network found promising in feature extraction by many researchers [1]. To extract the urban and suburban area roads, many algorithm based image processing techniques have been introduced in recent years. Increasing use of UAV images for geology, land resource management and military applications has impending need for efficient automatic and semi-automatic feature extraction techniques. A human interpreter can detect boundaries and regions to some extent with mean intensity but in the automatic method boundaries are detected using the local/focal statistics [9]. Before extraction of features, image preprocessing techniques like binarization, thresholding, resizing, normalization etc. are applied on to imagery for improving radiometric and geometric characteristics of imagery. Subsequently, feature extraction techniques are adopted to extract features of interest to help in classifying and recognition of objects/features from the imagery [10]. Feature extraction techniques are used in various image processing applications like character recognition, object recognition. Research in the past indicate in the field of automatic road extraction indicate improvements in the automation, accuracy and efficiency. These algorithms include multisource fusion technique, supervised classification with the spectral and geometrical attribute, a priori knowledge and expert system [2]. In the field of road extraction notable outcome with higher accuracy is reported but does not satisfy the requirements of practical ground applications [6]. The standard of automation, accuracy, efficiency also improved for the application purposes. In this paper, we propose a technique for road extraction from UAV images with a high degree of automation and compatible precision. The road network has broken line segments due to tunnel or shadow of the tree hence continuity is used as characteristic in the algorithm in this research [3]. We give an abbreviated introduction for image stitching and road (line) detection procedure first. This paper provides description of experiment to extract roads from aerial or UAV image using new algorithm and discusses the experimental results and concludes the paper.

2 Study Area

The study area is urban Slum dataset for Chingrajpara slum area in Bilaspur (Chattisgarh) (Fig. 1). It lies in latitude, longitude 22.0899° N, 2.1703° E.

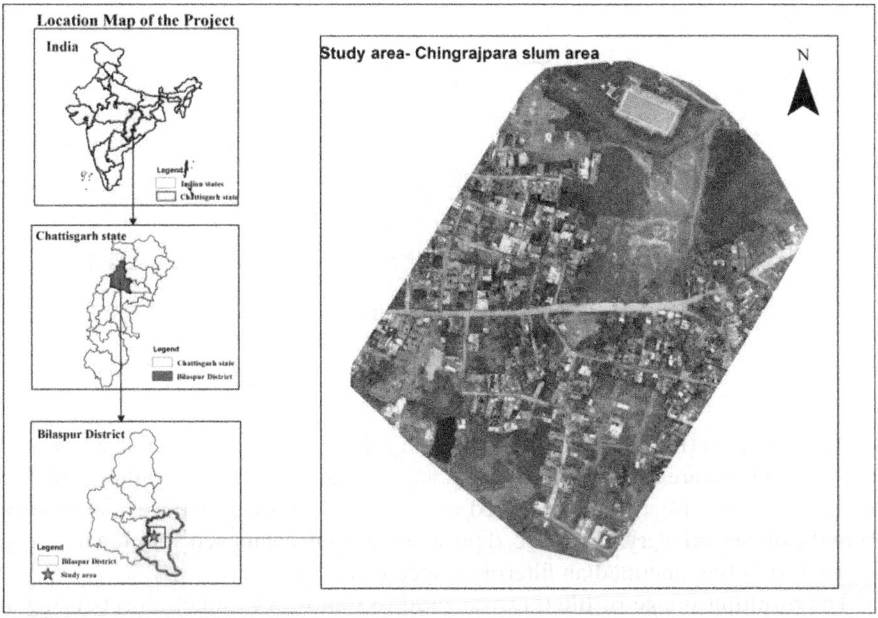

Fig. 1 Study area and its location

3 Methodology

In the UAV image, the threshold values of the road lie within a range. The steps adopted in the algorithm includes Creation of grayscale image, adjustment of threshold values in the image [4], Conversion into a binary image using Otsu's method and finally removal of unwanted noise and shadows using a median filter and morphological parameters. Median filtering was used to remove random noise [5]. This filtering method is observed to be more efficient than another filtering method as it is a nonlinear operation. The method carried out in this study is explained in Fig. 2.

The range of threshold values is defined as 0.5 to 0.85(0.9) because most of the roads in the region have Grayscale values within region as derived from image statistics. The image is then converted into a grayscale image using Binary Masking. The binary image thus obtained, is used for generating a filtered image for enhancing the road features and removal of random noise from the image. The image is further enhanced using size thresholds to eliminate small irregular features.

4 Results

The results derived in various stages of the algorithm is presented in Fig. 3a–d.

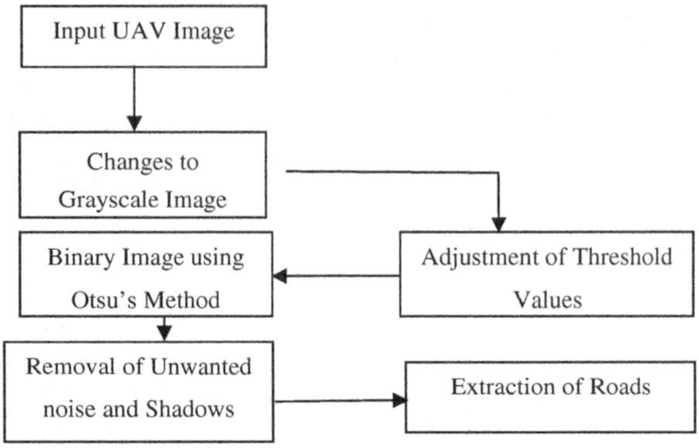

Fig. 2 Methodology

After filtering (Fig. 3d) and binary masking (Fig. 3c) the random noise and other non-relevant features are removed from the image dataset. Of the available filtering methods, median filtering has resulted in higher efficiency to extract road network from the binary imagery. Figure 3c, d present the results achieved after binary image after thresholding and median filtering respectively.

The resulting image of filtering was used to remove the small objects using an algorithm to remove dispersed pixels with continuity less than 4 pixels. The results are shown in Fig. 3e. The thresholding has eliminated the small objects which are normally noise. Similarly the thresholding also removed small stretch of road segment in left centre part of image where in a discontinuity of road due to obstructions like trees in exaggerated due to removal small isolated segment of road. Hence thresholding places a major role on accuracy of the resulting image.

Morphological operation on the image dataset is used detects other unwanted pixels and objects considering the linearity of road features. This operation is based on the foreground and background of the given image and to remove unwanted pixels and it can be done in MATLAB using bwmorph() and remove function [7]. The resulting image after the morphological operation is presented Fig. 3f. The parking area in the top left portion of imager is simplified during the operation, which can help in discriminating the linear roads from rectangular parking areas.

The refined dataset indicating the roads after applying morphological operation is used to identify of edges of roads by using SOBEL operation. The SOBEL has demonstrated the efficiency in linear edge detection as represented in Fig. 3g. The outcome of SOBEL operation resulted in identification of road edges and further line thinning will result in vectorization of road features from the high resolution UAV imagery.

The accuracy of road extraction is estimated by overlaying the extracted road network on to grayscale image to find the accuracy of the road extraction. The final image is shown in Fig. 3h. The visual evaluation of extracted road network versus of the image indicates good correlation and accuracy of extraction. However, it is to be noted that the discontinuity of road in several locations due to existence of avenue trees, shadow of building is implicit in the output. Varying the size of morphological

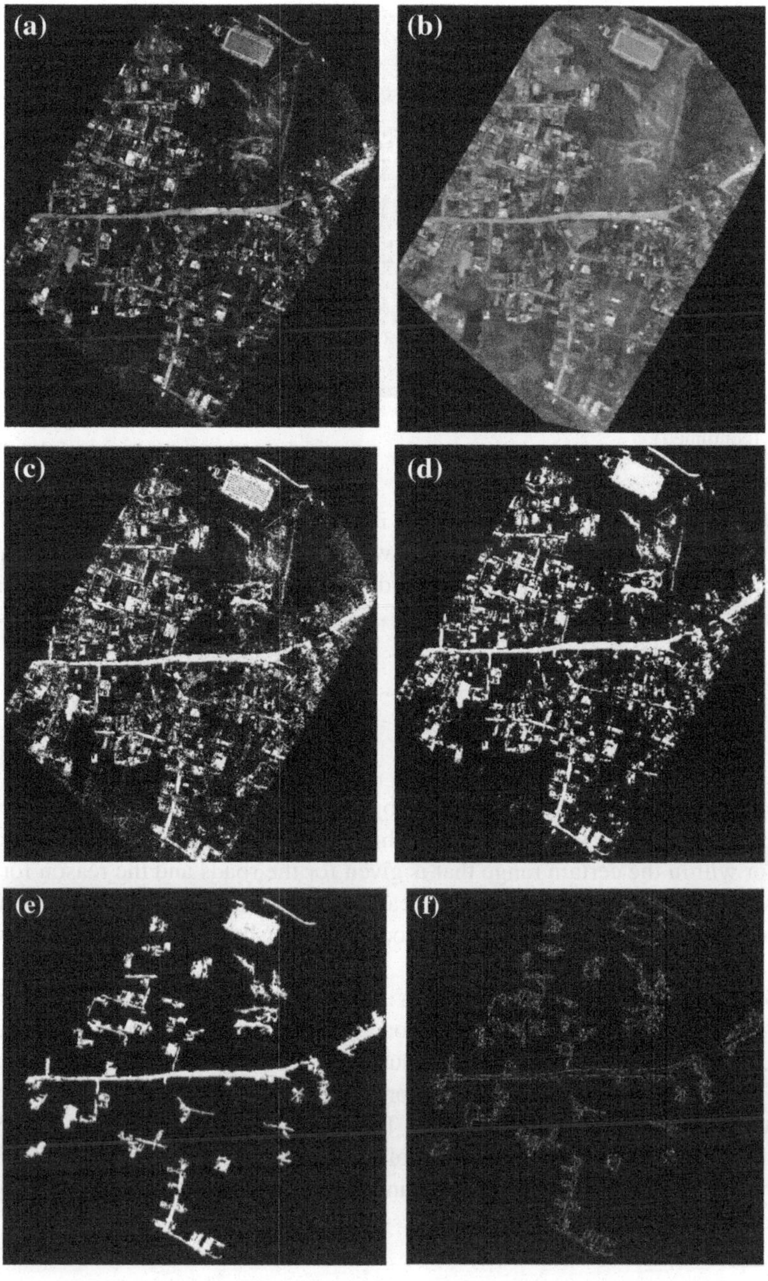

Fig. 3 **a** Binary image, **b** grayscale image, **c** binary image after thresholding, **d** median filtered image, **e** removing the unwanted components (pixel less than 3), **f** morphological operation process, **g** edge detection (sobel), **h** overlay on grayscale image

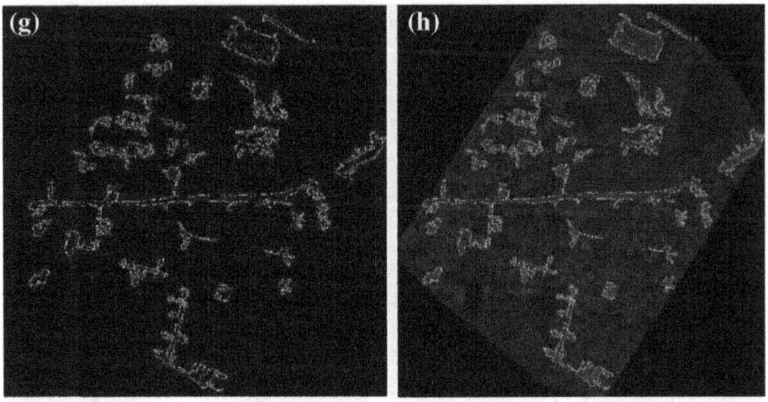

Fig. 3 (continued)

operation window may sometimes results in filling up of the discontinuities and at the same time may cause creating road links which are not existing on ground. Hence the selection of morphological operation window is critical in accuracy of the extraction process.

5 Conclusions

The final image clearly shows that some objects other than roads as well as some roads are not detected. This is because those objects are also having the same reflectance or color within the certain range that is given for the roads and the reason for some undetected roads because the study area is slum and the drainage and others just overlap the road and the color of the road has vanished. The objects detected are small houses, barren lands etc.

Road extraction from high-resolution images is a challenging area due to its complexity. Road extraction identifies the road pixels from high-resolution images and can be used in a variety of applications such as map implementation, traffic management, vehicle navigation, crop estimation etc. [8] this paper uses threshold method for the current study of road extraction. The detection of objects other than roads is due to the same pixel intensity values of the roads. Usage of artificial intelligence and digital elevation models (DEM) can also include removing unwanted objects accurately.

References

1. Ichim L (2018) Road detection and segmentation from aerial images using a CNN based system. In: 2018 41st international conference on telecommunications and signal processing (TSP), pp 1–5
2. Lu D, Weng Q (2007) A survey of image classification methods and techniques for improving classification performance. Int J Remote Sens 28(5)
3. Cao Y, Yan L (2010) Automatic road network extraction from UAV image in mountain area. In: 2012 5th international congress on image and signal processing, Chongqing, pp 1024–1028
4. Feng W, Yundong W, Qiang Z (2009) UAV borne real-time road mapping system. In: 2009 joint of urban remote sensing event, Shanghai, pp 1–7
5. Shi WZ, Zhu CQ, Wang Y (2001) Road feature extraction from remotely sensed image: review and prospects. Acta Geodaet Cartogr Sin 30(3):237–242
6. Dal Poz P, Zanin RB, do Val GM (2006) Automated extraction of road network from medium- and high-resolution images. Pattern Recogn Image Anal 16(2):239–248 (Pubitemid 44072322)
7. Chaudhuri D, Kushwaha NK, Samal A (2012) Semi automated road detection from high resolution satellite images by directional morphological enhancement and segmentation techniques. IEEE J Sel Top Appl Earth Obs Remote Sens 5(5)
8. Guo B, Shao Y, Hu X, Di L (2010) Application of a fast linear fixture detector to road extraction from remotely sensed imagery. IEEE J Sel Top Appl Earth Obs Remote Sens 99:1–6
9. Kong H, Audibert JY, Ponce J (2010) General road detection from a single image. IEEE Trans Image Process 19:2211–2220
10. Zhou H, Kong H, Wei L, Creighton D, Nahavandi S (2001) On detecting road regions in a single UAV image. IEEE Trans Intell Transp Syst 18(7):1713–1722

Detection of Water Body Using Very High-Resolution UAV SAR and Sentinel-2 Images

Ojasvi Saini, Ashutosh Bhardwaj and R. S. Chatterjee

Abstract The extent of water body has far-reaching effects on agriculture, flood control, and ecological studies. Synthetic Aperture Radar (SAR) imaging technique can be operated in all weather, day and night circumstances. Due to the numerous advantages of SAR imaging technique over other conventional image acquisition practices, it has been used for the detection of the waterbody. Subsets of quad-pol, georeferenced (L-band) SAR imagery of UAV platform (provided by JPL, NASA) of Mondah, Gabon region and optical imagery by Sentinel-2 of the same region is used for the extraction of the water body. After preprocessing of UAV SAR image, Yamaguchi Decomposition was carried out and volume scattering image array (T33) has been used for the extraction of the waterbody. T33 array element of the coherency matrix represents volume back-scattering responses from the area of acquisition. Since the surface of the water body (either smooth or rough water surface) shows negligible volume back-scattering, water bodies can be easily delineated using thresholding and then applying the SVM classification method. The area covered by water reflects most of the radiations falling in the Green color frequency range and strongly absorbs Near-Infrared part of the electromagnetic spectrum. Taking advantage of this unique behavior of water surface while interacting with the electromagnetic spectrum, Normalized Difference Water Index (NDWI) is used for the extraction of waterbody from Sentinel-2 optical image. Finally, the SVM classified outcomes for extracted water area from both the images were compared. The harmonizing information from the, (UAV SAR and Sentinel-2 multi-spectral) images have been used for the quick and precise recognition of waterbody.

Keywords Waterbody · UAV SAR · VHR · Sentinel-2 · NDWI · SVM

O. Saini (✉) · A. Bhardwaj · R. S. Chatterjee
Indian Institute of Remote Sensing (IIRS), Dehradun, India
e-mail: ojasvi21nov@gmail.com

A. Bhardwaj
e-mail: ashutosh@iirs.gov.in

R. S. Chatterjee
e-mail: rschatterjee@iirs.gov.in

K. Jain et al. (eds.), *Proceedings of UASG 2019*, Lecture Notes in Civil Engineering 51,
https://doi.org/10.1007/978-3-030-37393-1_7

1 Introduction

Water is one of the most important resources on the earth, making this planet habitable. Chemically, water is the condensed form of water molecules, present in a liquid state. In the context of this paper, waterbody on the earth surface is a well-defined collection of water molecules on the macroscopic scale with a certain geographic and topographic boundary. According to Oxford's definition, waterbody is the body of water forming a physiographical feature such as ocean, lakes, reservoirs, rivers, and ponds [1]. Any geographical, biological or chemical change in water bodies directly affects the ecology of the earth. Therefore, study and extraction of the geography of water bodies have become a significant need for the growth of the human race and safety of this planet. Transportation, agriculture, aquaculture and several other vital necessities of growing human are directly associated and influenced with the knowledge of the geographical extent of water bodies. Variation in the rate of climate change and natural disaster monitoring like flood, tsunami, and cloudburst, etc. can be easily supervised with the prior and post knowledge of waterbody extent [2].

Synthetic aperture radar (SAR) remote sensing is a growing and developing field for the extraction of features on the earth surface from hundreds of kilometers apart. SAR imaging techniques are capable of acquiring data in cloudy weather, as well as, being an active sensor it can work in both day and night circumstances [3–5]. As far as the extraction of waterbody features is concerned, polarimetric SAR plays an important role because it is less time consuming, do not need expensive conventional ground surveys and easily extractable low backscattering signature of water bodies are present in SAR images [6]. Besides these advantages, some constant challenges such as speckle, layover, shadow, and foreshortening are always present in SAR images [7], due to which extraction of a waterbody is not a straightforward task. To deal with this problem, along with the SAR image, Sentinel-2 multispectral image is also used for confirmation of results and for addressing misclassification of the water body.

Polarimetric decompositions using fully polarimetric SAR data may be used to get scattering based characteristics of the targets from the area of acquisition. Different type of scattering responses can be obtained from the SAR return signal, such as volume scattering response from dense vegetation, double-bounce scattering from buildings (man-made structures) and single-bounce scattering from smooth surfaces (roads, land with dry soil, water surfaces). Any particular scattering matrix component from the decomposed coherency matrix of fully polarimetric SAR data alone may also be used for the extraction of information from the SAR data.

T33 coherency matrix component (volume scattering component) has been used for the extraction of a water body features because all type of water surfaces (calm or rough) show very low volume back-scattering response in SAR return signal. Therefore, the T33 coherency matrix component is suitable for the extraction of the water body. In volume scattering component of the decomposed SAR data, pixels less than a certain backscattering value can be interpreted as water bodies. After

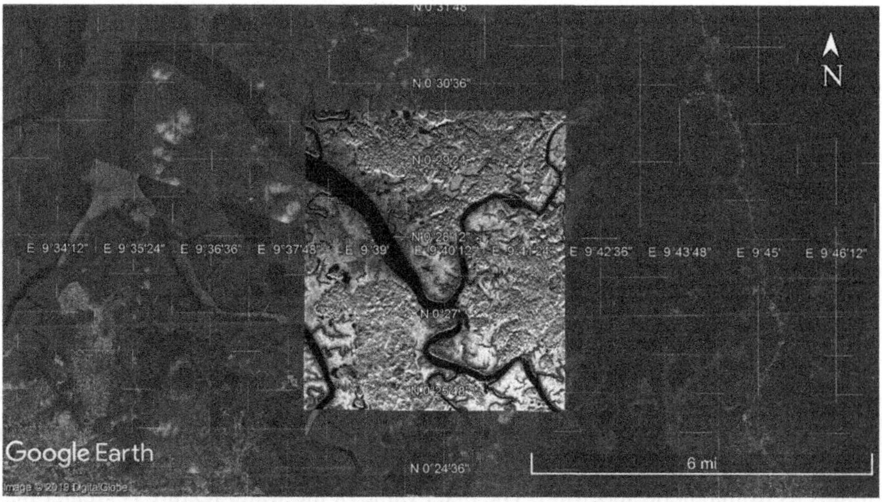

Fig. 1 The study area of Mondah forest, Gabon

applying the thresholding, SVM classifier has been used to classify the threshold image in two classes (Water Body and Background).

2 Study Area and Data Sources

The study area is shown in Fig. 1 as a part of Mondah forest which is legally protected the coastal forest in Komo-Mondah Department, northwest Gabon. The forest is somewhat flooded and comprises of a rich ecosystem. The forest is under huge environmental pressure due to the proximity of Libreville city.

Quad-pol, georeferenced (L-band) SAR imagery of UAV platform (provided by JPL, NASA) of Mondah, Gabon region and optical imagery by Sentinel-2 of the same region were used for the extraction of the water body. The UAV SAR image was acquired on March 06, 2016 and Sentinel-2 multispectral image was acquired on February 05, 2016.

3 Methodology

The T33 coherency matrix component (volume scattering component) of Yamaguchi Decomposition contains information on volume scattering responses from the target [8]. Water bodies with smooth as well as rough surfaces offer negligible volume back-scattering to the incoming waves from the polarimetric SAR sensors [9]. T33 component of coherency matrix image array derived from fully polarimetry SAR data, represent the targets offering multiple scatterings with bright pixels. Since the calm water body surface behaves as specular reflectors, they do not behave as a target offering multiple scatterings and can be easily extracted because of dark appearance in the imagery [10, 11]. Even the rough water surface does not offer the multiple scattering to the incoming SAR waves. This special backscattering property of the water body in decomposed T33 image makes it suitable for the extraction of smooth and rough water surfaces from the imagery. In T33 Coherency matrix component, smooth surface other than water surfaces like dry sand, roads and some man-made features may also show low back-scattering responses and can be misclassified as water body [12]. To deal with this problem, Sentinel-2 multispectral image is also used for confirmation of the extent of water body obtained from SAR data.

NDWI is an effective method for the extraction of water bodies, applicable for multispectral optical images. The NDWI can enhance the water information effectively in most cases [9, 13, 14]. The area covered by water reflects most of the radiation falling in the Green color frequency range and strongly absorbs Near-Infrared part of the electromagnetic spectrum. The pixel value of the water feature is greater than 0.5 in the NDWI image [15].

Taking advantage of this unique behavior of water surface, while interacting with the electromagnetic spectrum, Normalized Difference Water Index (NDWI) is used for the extraction of waterbody from Sentinel-2 optical image. Finally, results obtained from both the (SAR and Sentinel-2 optical) images were compared and Analyzed to validate the extraction of the area covered by water (Fig. 2).

a. Preprocessing of UAV SAR data

The step in the preprocessing of georeferenced fully polarimetric SAR images after sub-setting the region of interest is to remove speckle as shown in Fig. 2. Speckle is inherent in SAR images and arises due to random fluctuations in the return signals from the targets having sizes smaller than a single image processing element. Enhanced Lee speckle filter is used due to its ability to preserve edges in the imagery, making it easy to identify even the banks of the narrow water passages [16].

All the four polarimetric speckle filtered images (HH, HV, VH, and VV) were then used for the generation of Yamaguchi decomposed coherency matrix element images. Among all the coherency matrix element images T33 (Volume scattering component) was used for the extraction of water bodies due to its ability to depict both calm as well as turbulent water surface by dark pixels in the imagery.

b. Extraction of water area

Extraction of water area from UAV SAR data

Water area is easily extractable from SAR images because it appears as a low backscattering feature. Training data samples for two different type of classes (pixels with extremely low backscattering values depicting water bodies and pixels with comparatively high backscattering values depicting background) have been chosen

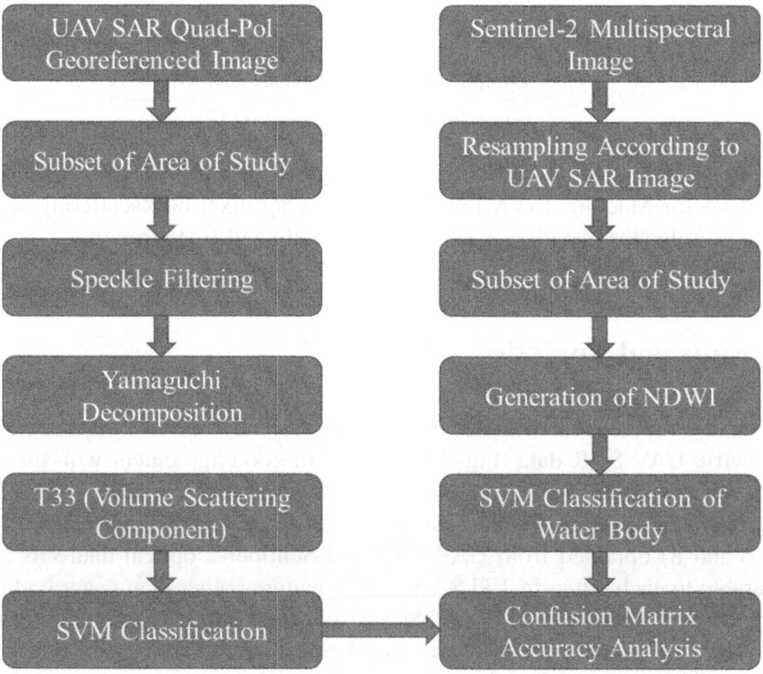

Fig. 2 Flow chart of the methodology followed

for SVM supervised classification algorithm with kernel type "Linear". Linear kernel SVM algorithm has been used because it gives the best performance among radial basis function kernel, polynomial Kernel and sigmoid kernel [17].

Using the dot product between the input (x) and each support vector (x_i), the equation for the prediction of a new input for **linear kernel** is calculated as follows:

$$f(x) = B(0) + \sum (a_i * (x, x_i)) \tag{1}$$

This equation involves calculating the inner products of a new input vector (x) with all support vectors in training data. The coefficients B(0) and a_i (for each input) must be estimated from the training data by the learning algorithm.

Water area and background were classified from the image and the accuracy assessment was done using the confusion matrix method by giving ground truth ROIs.

Extraction of water area from Sentinel-2 multispectral image

The NDWI can enhance the water information effectively. The area covered by water reflects most of the radiation falling in the Green color frequency range and strongly absorbs Near-Infrared part of the electromagnetic spectrum. NDWI has been calculated by using Green (band-03) and Near-Infrared (band-08) of the Sentinel-2 image as shown in Eq. (2).

$$NDWI = \frac{Green(Band - 03) - NIR(Band - 08)}{Green(Band - 03) + NIR(Band - 08)} \tag{2}$$

Taking NDWI image as the input image, water bodies have been extracted, using Support Vector Machine (SVM) classifier by giving pixel backscattering values as training samples for two classes namely Waterbody and Background.

4 Results and Analysis

The area of extracted water bodies using T33 coherency matrix component of fully polarimetric UAV SAR data (Figs. 5 and 6) is in good agreement with the area of extracted water bodies using Sentinel-2 multispectral (optical) data, exploiting the concept of NDWI. Water bodies are recognized in both the final resulted images (Figs. 6 and 8) obtained from UAV SAR and Sentinel-2 optical data sets and are comparable to each other. In T33 SAR image, features other than water bodies may also appear with dark or completely black pixels due to their low volume scattering. In the dark background of classified water bodies, some features are misclassified (white pixels) as water bodies. By comparing the results for the extent of water bodies from both the images, a rough validation of the water body extracted from UAV SAR data was done.

a. **Analysis before removing misclassified water body features from UAV SAR image**

$$\text{Area of extracted water bodies using UAV SAR image} = 7{,}047{,}743\ \text{m}^2 \tag{3}$$

$$\text{Area of extracted water bodies using Sentinel-2 optical image} = 6{,}101{,}714\ \text{m}^2 \tag{4}$$

$$\text{Absolute error in the extracted area} = (7{,}047{,}743 - 6{,}101{,}714)\ \text{m}^2 = 946{,}029\ \text{m}^2 \tag{5}$$

$$\text{Relative absolute percentage error} = 13.42\% \tag{6}$$

b. **Analysis after removing misclassified water body features from UAV SAR image**

$$\text{Area of extracted water bodies using UAV SAR image} = 6{,}333{,}521\ \text{m}^2 \tag{7}$$

Confusion Matrix: E:\F drive data\NISAR_project\uav sar\Yamaguchi3_Vol12_class

Overall Accuracy = (2985/3049) 97.9010%
Kappa Coefficient = 0.9541

Class	Ground Truth (Pixels) Water_Body	Background	Total
Unclassified	0	0	0
Water_Body	1049	52	1101
Background	12	1936	1948
Total	1061	1988	3049

Class	Ground Truth (Percent) Water_Body	Background	Total
Unclassified	0.00	0.00	0.00
Water_Body	98.87	2.62	36.11
Background	1.13	97.38	63.89
Total	100.00	100.00	100.00

Class	Commission (Percent)	Omission (Percent)	Commission (Pixels)	Omission (Pixels)
Water_Body	4.72	1.13	52/1101	12/1061
Background	0.62	2.62	12/1948	52/1988

Class	Prod. Acc. (Percent)	User Acc. (Percent)	Prod. Acc. (Pixels)	User Acc. (Pixels)
Water_Body	98.87	95.28	1049/1061	1049/1101
Background	97.38	99.38	1936/1988	1936/1948

Fig. 3 The classification accuracy (UAV SAR image) using the confusion matrix method

$$\text{Area of extracted water bodies using Sentinel-2 optical image} = 6{,}101{,}714\,\text{m}^2 \quad (8)$$

$$\text{Absolute error in the extracted area} = (6{,}333{,}521 - 6{,}101{,}714)\,\text{m}^2 = 231{,}807\,\text{m}^2 \quad (9)$$

$$\text{Relative absolute percentage error} = 3.79\% \quad (10)$$

After masking out misclassified waterbody pixels in UAV SAR the relative absolute percentage error reduced to 3.79 from 13.42%.

The water body classification accuracy in case of UAV SAR imagery was achieved 97.9% and the kappa coefficient was achieved 0.95 from the confusion matrix method. On the other hand waterbody classification accuracy in case of Sentinel-2 imagery, after performing NDWI was achieved 99.36% and the kappa coefficient was achieved 0.98 (Figs. 3, 4, 5, 6, 7 and 8).

5 Conclusion

As shown in confusion matrix accuracy analysis, NDWI is an effective method for the extraction of the water surface from multispectral (optical) satellite data but due to the limitation of electromagnetic waves lying in the optical region of the electromagnetic spectrum that they are not able to penetrate through clouds and

Confusion Matrix: E:\F drive data\NISAR_project\uav sar\New_Folder\NDWI_class

Overall Accuracy = (3248/3269) 99.3576%
Kappa Coefficient = 0.9855

	Ground Truth (Pixels)		
Class	Water_Body	Background	Total
Unclassified	0	0	0
Water_Body	2175	0	2175
Background	21	1073	1094
Total	2196	1073	3269

	Ground Truth (Percent)		
Class	Water_Body	Background	Total
Unclassified	0.00	0.00	0.00
Water_Body	99.04	0.00	66.53
Background	0.96	100.00	33.47
Total	100.00	100.00	100.00

Class	Commission (Percent)	Omission (Percent)	Commission (Pixels)	Omission (Pixels)
Water_Body	0.00	0.96	0/2175	21/2196
Background	1.92	0.00	21/1094	0/1073

Class	Prod. Acc. (Percent)	User Acc. (Percent)	Prod. Acc. (Pixels)	User Acc. (Pixels)
Water_Body	99.04	100.00	2175/2196	2175/2175
Background	100.00	98.08	1073/1073	1073/1094

Fig. 4 The classification accuracy (Sentinel-2 image) using the confusion matrix method

haze, optical satellites are not suitable to work in all weather conditions. A small change in weather can affect the feasibility of meaningful data acquisition from any optical sensor of remote sensing satellite. In any disaster or emergency situation, the optical remote sensing technique cannot be trusted for giving quick information about the target. Keeping this problem in prime focus, the area of water bodies was extracted from SAR satellite sensor (here UAV SAR) and results obtained from SAR image were compared with the extracted results obtained from Sentinel-2 optical satellite data of the same region.

SAR data can be exploited to extract water body due to its ability to be able to work in all weather and day-night conditions. Results of water body extraction were in good agreement with the resulted water body extent from optical Sentinel-2 data.

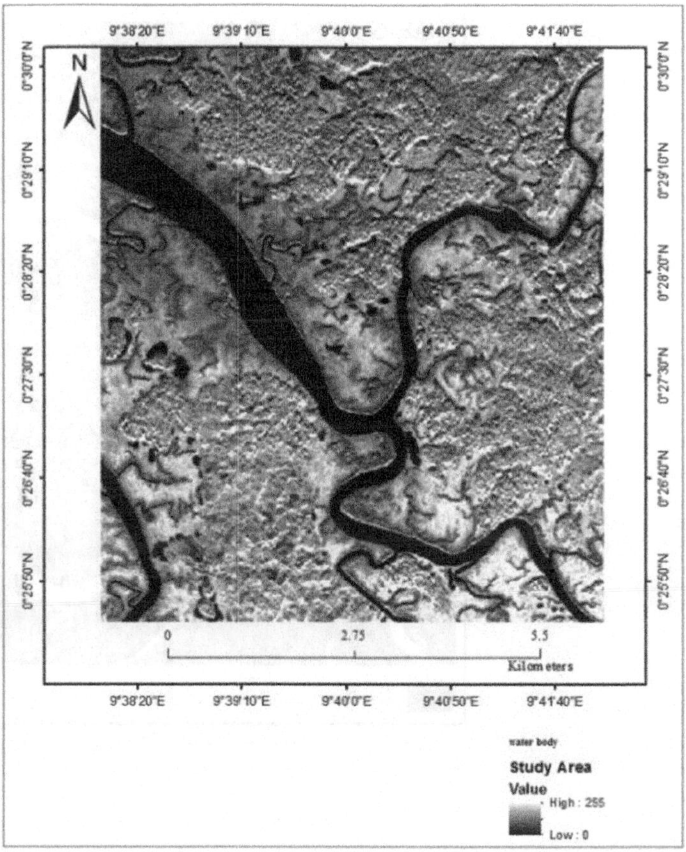

Fig. 5 T33 coherency matrix image of Yamaguchi decomposition of the study area

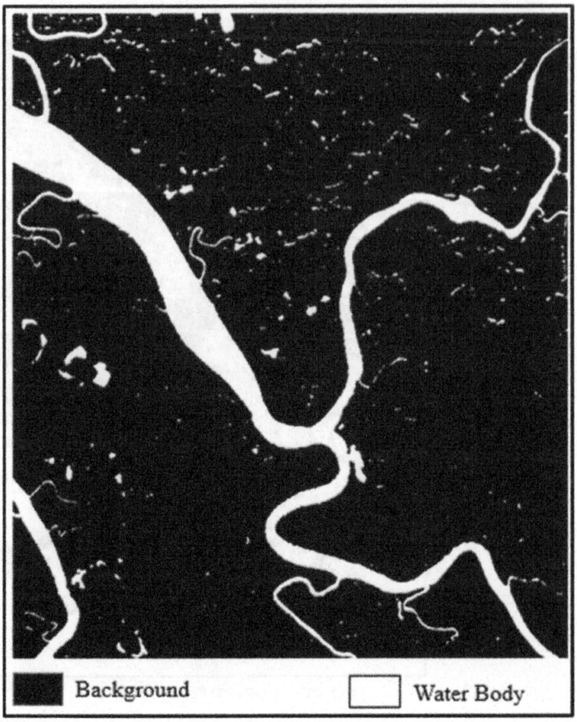

Fig. 6 Classification result of water body from UAV SAR imagery

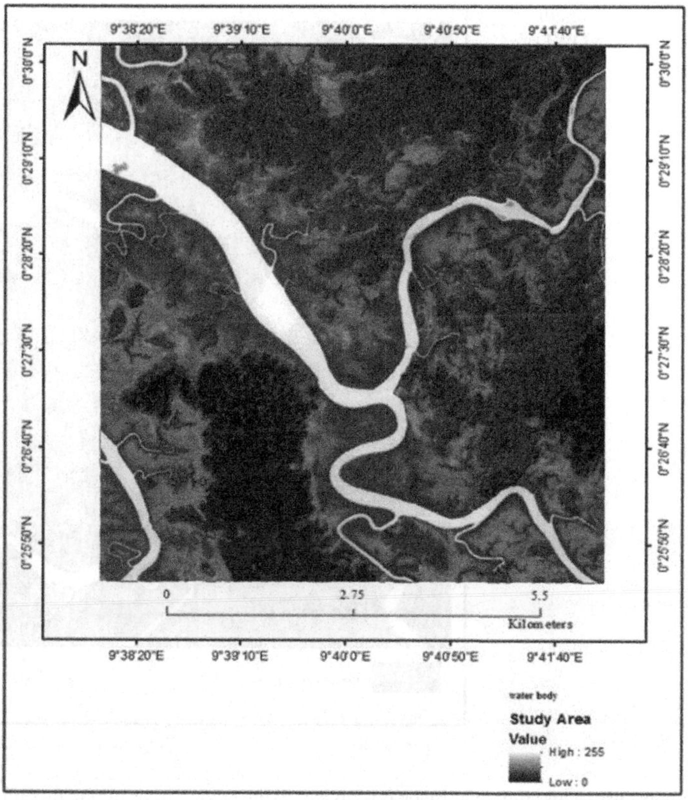

Fig. 7 NDWI of the region of interest using Sentinel-2 image

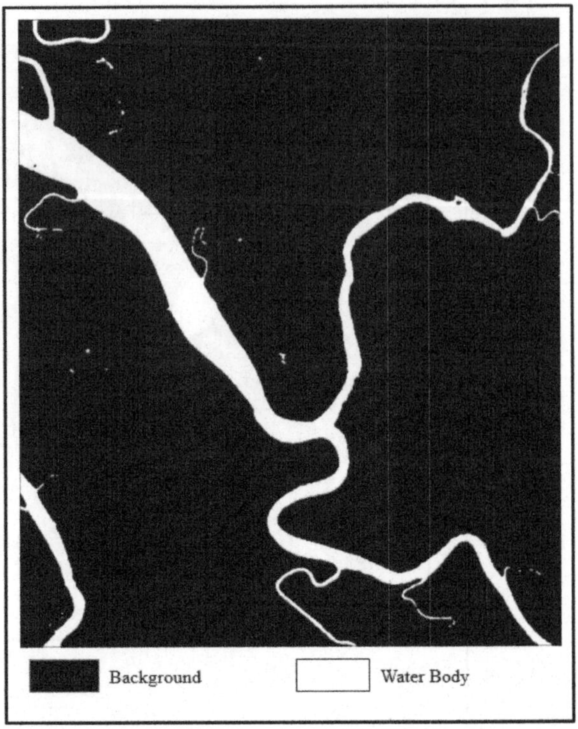

Fig. 8 Classification result of water body from NDWI of Sentinel-2 imagery

Acknowledgements This study was made possible with the resources of Indian Institute of Remote Sensing (IIRS), Indian Space Research Organization (ISRO) Dept. of Space, India. Thanks are given to JPL (NASA) and European Space Agency (ESA) for providing freely downloadable UAV SAR and Sentinel-2 remotely sensed data.

References

1. Waterbody|Definition of waterbody in English by Oxford Dictionaries (Online). Available: https://en.oxforddictionaries.com/definition/waterbody. Accessed 01 Mar 2019
2. Liu W, Yamazaki F, Gokon H, Koshimura SI (2013) Extraction of tsunami-flooded areas and damaged buildings in the 2011 Tohoku-oki earthquake from TerraSAR-X intensity images. Earthq Spectra 29(SUPPL):1
3. Jackson CR, Apel JR (2004) Chapter 0. Synthetic aperture radar marine user's manual synthetic aperture radar marine user's manual. Director, no
4. Aher SP, Ph D (2014) Synthetic aperture radar in Indian remote sensing. Int J Appl Inf Syst 7(2):2012–2015
5. Joy S, Lu Xx (2004) Application of remote sensing in flood management with special reference to monsoon Asia: a review. Nat Hazards, 283–301
6. Donga W, Pingb Q, A rapid method for water target extraction in SAR image. Isprs.Org, 237–242

7. Cheney M, Borden B (2009) Problems in synthetic-aperture radar imaging. Inverse Probl 25(12)
8. Sato A, Yamaguchi Y, Singh G, Park SE (2012) Four-component scattering power decomposition with extended volume scattering model. IEEE Geosci Remote Sens Lett 9(2):166–170
9. Du Y, Zhang Y, Ling F, Wang Q, Li W, Li X (2016) Water bodies' mapping from sentinel-2 imagery with modified normalized difference water index at 10-m spatial resolution produced by sharpening the swir band. Remote Sens 8(4)
10. Brisco B (2015) Mapping and monitoring surface water and wetlands with synthetic aperture radar. Remote Sens Wetl Appl Adv, no., pp 119–136
11. Brivio PA, Colombo R, Maggi R, Tomasoni R (2002) Integration of remote sensing data and GIS for accurate mapping of flooded areas. Int J Remote Sens 23(3):429–441
12. Liu C (2016) Analysis of sentinel-1 SAR data for mapping standing water in the Twente region. Univ, Twente - ITC
13. McFEETERS SK (1996) The use of the normalized difference water index (NDWI) in the delineation of open water features. Int J Remote Sens 17(7):1425–1432
14. Rokni K, Ahmad A, Selamat A, Hazini S (2014) Water feature extraction and change detection using multitemporal landsat imagery. Remote Sens 6(5):4173–4189
15. Wang Y, Ruan R, She Y, Yan M (2011) Extraction of water information based on RADARSAT SAR and Landsat ETM+. Procedia Environ Sci 10(PART C):2301–2306
16. Yommy AS, Liu R, Wu AS (2015) SAR image despeckling using refined lee filter. In: Proceedings—2015 7th international conference Intelligent human-machine system cybernetics. IHMSC 2015, vol 2, pp 260–265
17. White L, Brisco B, Dabboor M, Schmitt A, Pratt A (2015) A collection of SAR methodologies for monitoring wetlands. 7(6)

Comparative Study on Crop Type Classification Using Support Vector Machine on UAV Imagery

Vijaya Kumar Vasantha and Venkata Reddy Keesara

Abstract In Indian agricultural practices, single crop cultivation is rare and uncommon. This poses a real challenge for crop type classification using single date imagery. Site-specific information of crop type is required for agricultural management which includes technologies aiming at productivity and profit while practicing eco-friendly environment. Unmanned Air Vehicles (UAV) are effective image acquisition platforms for many agricultural applications. UAV's can acquire high levels of spatial details compared to standard remote sensing platforms. Single date RGB imagery of 5 cm spatial resolution obtained from processing the raw data was used for the classification of different types of crop. Traditional pixel-based analysis of remote sensing data results in inaccurate classification due to low spatial resolution, mixed pixels, and crop pattern variability. This can be overcome by using high-resolution UAV data and machine learning methods like Support Vector Machine (SVM). In the present study SVM kernel functions namely linear, sigmoid, radial basis and polynomial function are adopted and compared for mapping the crop types. The classification shows that the radial and sigmoid kernel functions give high accuracy when compared with the rest by performing the accuracy assessment for all four classifiers. These crop classifications are important for greenhouse gas modeling, agrarian policy, and agro-environmental studies.

Keywords Crop classification · Kernel · Support vector machine (SVM) · Unmanned air vehicle (UAV)

1 Introduction

Crop productivity and yield in India must be increased to overcome the current food scarcity and future challenges due to the vast increase in the population and income growth. Crop image acquisition is important to know the crop productivity

---

V. K. Vasantha · V. R. Keesara (✉)
Department of Civil Engineering, NIT Warangal, Warangal 506004, India
e-mail: kvreddy@nitw.ac.in

V. K. Vasantha
e-mail: vvijaya@student.nitw.ac.in

K. Jain et al. (eds.), *Proceedings of UASG 2019*, Lecture Notes in Civil Engineering 51,
https://doi.org/10.1007/978-3-030-37393-1_8

and to estimate crop yield. Misclassification of crop type leads to a major error in the estimation of yield, which is done due to mixed pixel and coarser resolution of remote sensing images. Misclassification is problematic where different types of crops with similar phenologies are grown together, which is common in India. An increased spectral and spatial resolution in remote sensing is a way to rectify misclassification errors.

Remote sensing images acquired by satellites or Unmanned Air Vehicle (UAV), whether they are low resolution or ultra-high resolution [1] have been applied for monitoring vegetation in different research areas for many years [2]. However, traditional satellite or aircraft images cannot capture plant-level details, high-resolution images such as UAV images have been widely applied in vegetation analysis due to the advantages of high resolution, high frequency, and easy operation [3]. UAV is becoming more popular in recent years because of its advantages as an economic and repeatable method with easily-controlled equipment. UAV can capture ultra high-resolution images that provide detailed features suitable for estimating the location of individual plants within several hectares [4]. As the flying altitude of UAV can be adjusted, the resolution of the acquired images ranges from meters to centimeters.

UAV's used for remote sensing and surveying usually involves a camera for high-resolution image acquisition. Commonly, cameras used in UAV's contain three visible bands (RGB bands) only. Due to the high resolution of UAV images, the texture is often used for crop classification. From a data availability point, it is not always possible to acquire time-series images of UAV due to restricted or definite accessibility to the area of interest. Thus, it is essential to improve the image classification performance when a single number of UAV images are available for crop classification [5]. Selection of appropriate classification methods is important to get accurate classification results. Since the 2000s, ANN (Artificial Neural Network) algorithms such as SVM and Random Forest (RF) were widely applied to crop classification with remote sensing data.

Scientists and researchers have made great progress in inventing and improving advanced classification algorithms to achieve higher classification accuracy. The classification algorithms include SVM, Maximum Likelihood (ML), RF and heterogeneous clustering approaches [6]. ML is one of the most commonly used supervised classification methods in remote sensing. ML-based on pixels assumes the data within each input feature is normally distributed, and the pixel that attains the highest probability is assigned to the class [7]. SVM is another supervised classification method that assigns each training sample into one of the two pre-defined classes [8].

1.1 Support Vector Machine

Machine learning techniques are commonly used for Classification of the given data. Vector data can be mapped into one of the various class labels in the classification procedure [9]. In supervised setting, it is done by looking at a set of input-output examples of the function. The finite input-output example data which is used for learning the classification function is called the training data. SVM is one of the successful supervised learning methods for this problem. They have strong theoretical foundations and have shown excellent empirical success in various fields. Support Vector Machines are trained so that the decision function would classify the unseen example data accurately. This ability to classify unseen example data accurately is referred to as generalization. High generalization capability is one of the main reasons for the success of SVMs. Let's see the basic idea behind the SVMs at first. Given a set of a n-dimensional vectors, a linear classifier tries to separate them with a 1-dimensional hyperplane. There are many hyperplanes that might classify the data. If we define "margin" as the distance between the nearest samples on both sides of the hyperplane, two classes are separated by a hyperplane which has the largest margin between them. we can change the linear classification it into non-linear classification by using non-linear kernels. SVM is a binary classifier, multiclassification is also possible using SVM by combining several binary classifiers. For multiclass classification, SVM uses pairwise classification method. For example, if there are three classes (vegetation, urban and water), SVM classifies vegetation against non-vegetation classes, water against non-water classes and remaining pixels it will assign to urban class.

A penalty parameter in SVM classifier allows a certain degree of misclassification. The non-linear SVM classifier is used to classify more than two classes by separating the hyperplane with maximum margin. Using kernel functions, we can compute the separating hyperplane without carrying the mapping into feature space. Equations (1), (2), (3), and (4) represents the mathematical form of the kernels linear, polynomial, radial basis, and sigmoid respectively [10, 11].

1. **Linear Kernel**: If we have a linearly separable data in the original input space, we need not map the input space into a high-dimensional space. We can use the linear kernel in such a situation, which is a dot product of the two vectors in the original space.

$$k(x, y) = (x, y) \tag{1}$$

2. **Polynomial Kernel**: Polynomial Kernels of degree p are given by

$$k(x, y) = ((x, y) + 1)^p \tag{2}$$

   where p is the polynomial degree
3. **Radial Basis Kernel**

$$k(x, y) = \exp \frac{-\|x - y\|^2}{2\sigma^2} \tag{3}$$

   where σ is the parameter which controls the radius and $\sigma > 0$.
4. **Sigmoid Kernel**

$$k(x, y) = \tanh(m(x, y) + c), \text{for some (not every)} m > 0 \text{ and } c > 0 \tag{4}$$

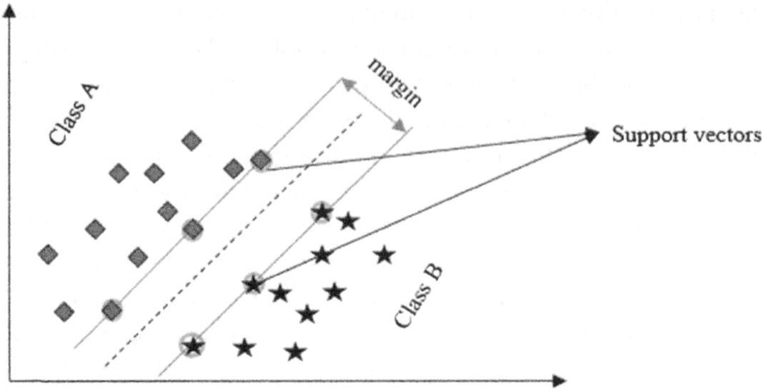

Fig. 1 Example of margin and support vectors in the case of linearly separable data

Figure 1 shows the linear SVM model in which there are two classes namely class A and class B they are separated by a maximum separable margin. Support vectors are the data points which are. Support vectors are the vector points used to train the dataset.

2 Study Area

The study area is Arale village, Satara district, Maharashtra. Figure 2 shows the geographical location of the study area. The geographic location of the study area is 17.7500° N, 74.0513° E which located in the western part of Maharashtra. Agriculture is an important activity in arale village and Satara district. More than 70% of the village population depends directly or indirectly on agriculture practices. The agricultural sector plays an important role in the development of the Satara district and arale village. The district is bounded by Solapur to east, Ratnagiri to west, Pune to north and Sangli to the south. Major portion of the vegetation depends on the rainfall and river system. The main rivers of Satara district are Krishna and Koyna. Approximately 170 km of the Krishna river flows through Satara district, Koyna is the largest tributary of the Krishna in the district. The average rainfall of the district is 1040 mm, the district is influenced by three monsoon seasons of rains. South-west monsoon (June–September), post-monsoon (October–December), pre-monsoon (March–May). The mean minimum and mean maximum temperatures are 12.50 °C and 38.50 °C respectively. Sugarcane, jowar, bajra, groundnuts, tomato, etc., are the major cultivated crops of the study area.

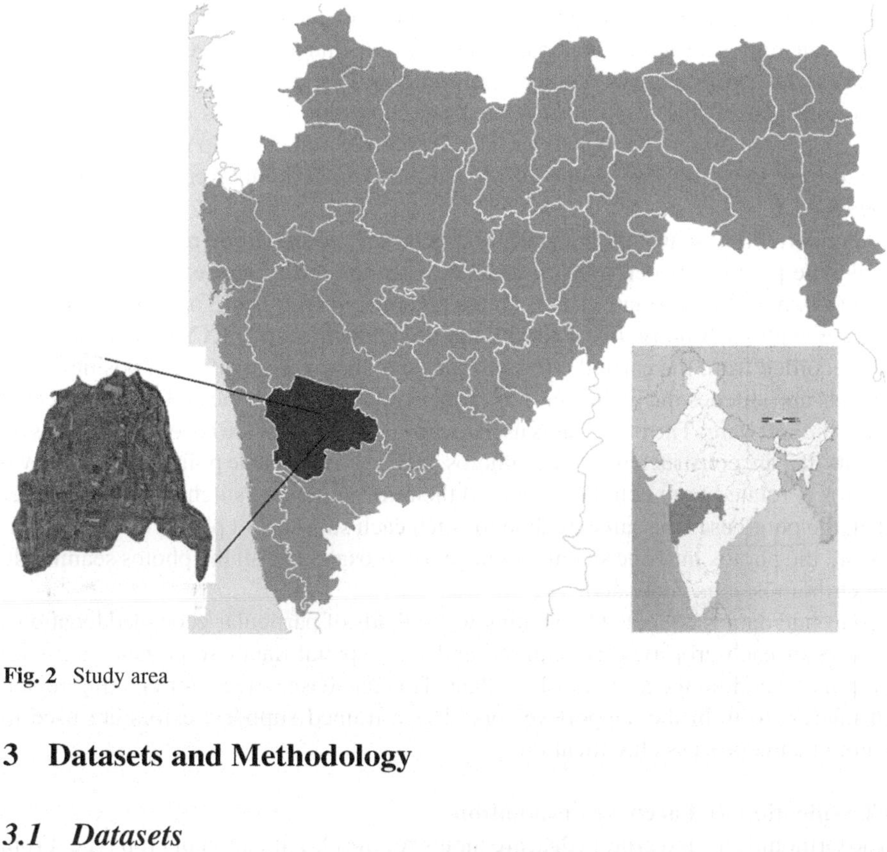

Fig. 2 Study area

3 Datasets and Methodology

3.1 Datasets

Digital photographs were collected using a UAV operated by Terra drone India in the month of August for the study area (Arale, Satara) this time period coincides with the Kharif season. Three-band (red, green, and blue) images were acquired using phantom four drones. The spatial resolution of 5 cm per pixel is obtained throughout the study area.

3.2 Methodology

Data collection and processing
The raw UAV images were processed in Agisoft Photoscan software to generate orthomosaic which is input to crop type classification purpose. The processing started with inputted individual UAV images and their GPS info, camera parameters and the exterior image orientation [12]. In the next step, key points were extracted based on the optimized SIFT algorithm. If two key points are found to be the same on two

different images, the two points are called one pair of tie points. The tie points in each pair represent the same location in adjacent images and are used to link two images. The software was automatically finding millions of tie points on images, and each pair of the tie point generated one 3D point. Millions of 3D points with known positions were generated into the 3D point cloud. The top points of the point cloud were used to generate the DSM, and the ground points were used to develop the DEM.

Agisoft Photoscan automatically extracts key points from individual images, builds tie points between images, and estimates the location of each feature, based on the UAV GPS information and camera parameters provided. The process matches images with each individual latitude, longitude, and altitude (XYZ) values, which are recorded from the carried GPS. Mosaics the photos from different positions and angles, and stitches the photos into a seamless image through the algorithms selected from the software. The procedures used to mosaic the photos are based on previous tie points that are constructed between photos. According to the tie points, one of the two photos is rotated to match the other, and then the photos are stitched together. After that, the photo uses the same method to stitch each subsequent photo, thereby stitching all the photos into one seamless image. By mosaicking all the photos seamlessly an Orthomosaic is generated.

Ground data is collected by visiting to the fields of particular geocoded locations, images of each crop type is acquired and a geospatial database is generated with crop name and image as an attribute data. This database is used for creating region of interest to train the support vectors. These trained support vectors are used to automate the process classification.

Classification and accuracy evaluation

The Orthomosaic and ground data are inputs for the classification of crop type. From the ground data training sites were created using an ROI (Region of Interest) tool. Seven different training classes were given as input vectors to the classifier. SVM classification is discussed in detail in Sect. 1.1. Linear kernel and non-linear kernels (Radial, Polynomial, and Sigmoid) are used to perform linear and non-linear classification. Pairwise classification strategy is used to perform multiclass classification. Classification output of SVM classifier is values of each pixel for each class, these values are used to calculate the probability which ranges from 0 to1 and probability values are stored as rule images. SVM classifier performs classification by selecting the highest probability. When the RBF kernel is used, the parameters of cost and gamma had been optimally determined. Large values of cost and gamma result in overfitting to the training data, yielding poor generalization ability of the classifier. For the polynomial kernel polynomial order of 2 is given as an input. The accuracy assessment is performed using the ground reference data collected for the study area. For each kernel function, classification accuracy assessment is performed to estimate overall accuracy (OA) and kappa coefficient. The OA determines how the kernel function is performing classification. The overall methodology of present study is shown in Fig. 3.

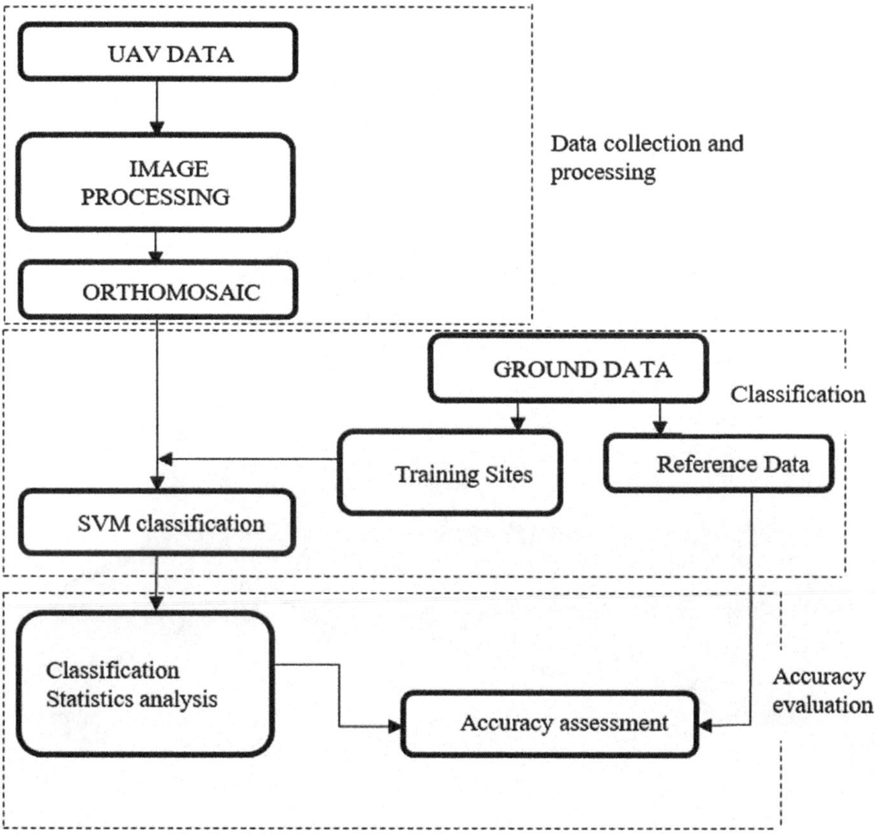

Fig. 3 Methodology flowchart

4 Results and Discussion

4.1 Orthomosaic

The Orthomosaic is generated from raw UAV images using Agisoft Photoscan (Fig. 4) with a spatial resolution of 5 cm. Orthomosaic is generated by merging individual ortho rectified images after interior orientation, exterior orientations and seamless transition between adjacent image.

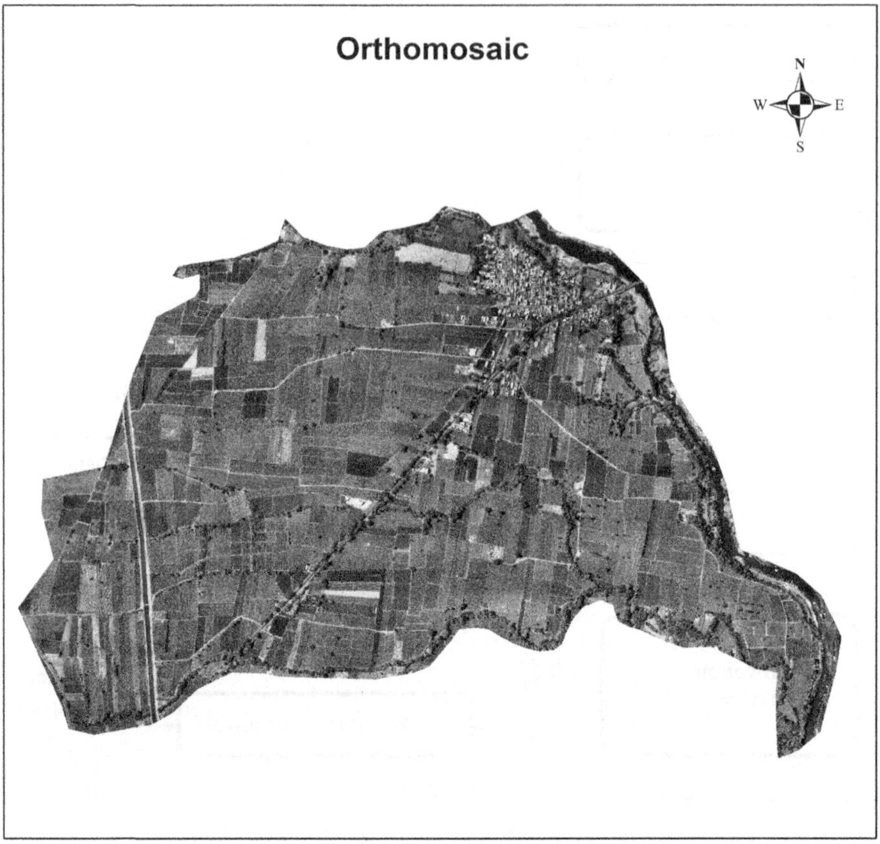

Fig. 4 Orthomosaic of the study area

4.2 Classification and Accuracy assessment

SVM is a commonly used machine learning algorithm in remote sensing data analysis. This study applied four kernel functions to extract crop type parcels. SVM classifier classified Seven types of classes namely sugarcane, jowar, groundnuts, bajra, Temporary barren (TB), vegetables and water which are selected based on the available training and reference data. Sugarcane is the major crop in the study area followed by jowar, groundnut, vegetables, etc., The classification results are shown in Fig. 5. At some sites, sigmoidal kernel classified jowar as TB and due to the moisture content in the soil, some of the sites are classified as water. Linear kernel gave very less accuracy in classifying the orthomosaic because of its capability to classify two classes at a time. Linear kernel classified some of the sugarcane sites as bajra and jowar.

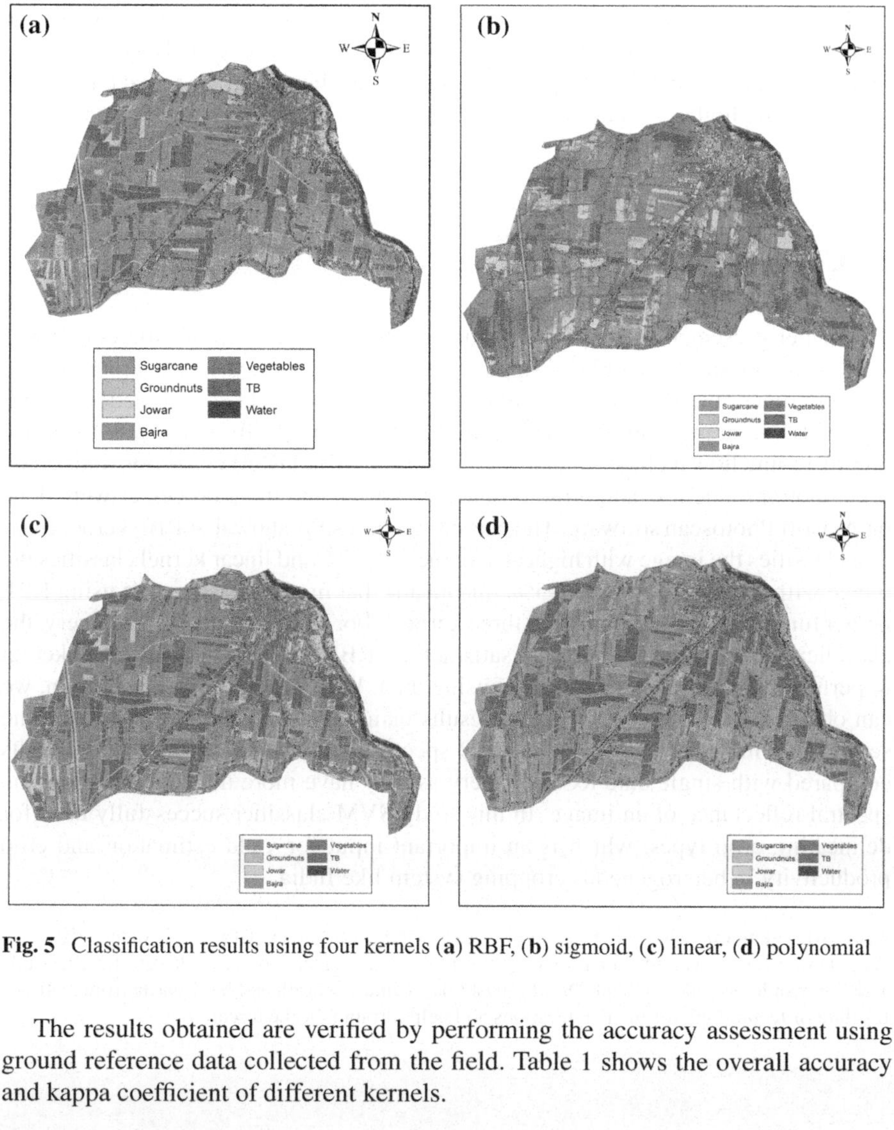

Fig. 5 Classification results using four kernels (**a**) RBF, (**b**) sigmoid, (**c**) linear, (**d**) polynomial

The results obtained are verified by performing the accuracy assessment using ground reference data collected from the field. Table 1 shows the overall accuracy and kappa coefficient of different kernels.

Table 1 Classification accuracy and Kapa coefficient

	RBF	Sigmoid	Polynomial	Linear
Overall accuracy (%)	87	84	80	75
Kappa coefficient	0.8453	0.795	0.743	0.691

RBF kernel classified the image with higher accuracy (87%) compared with other kernels. Whereas sigmoid, 3rd-degree polynomial and linear kernels gave an accuracy of 84%, 80%, and 75% respectively. The result shows that the radial basis function gives the highest overall accuracy, whereas linear classifier gives least overall accuracy.

5 Conclusion

This paper presented a comparative study on the performance of different SVM's kernels for classification of UAV single date RGB imagery in the agricultural region. For classification different SVM classifiers based on several kernel functions (RBF, Sigmoid, Polynomial, and Linear) are applied to classify RGB imagery. UAV image preprocessing flow includes image orientation, DEM and DSM extraction, individual orthophoto production, and orthomosaic generation, and these processes were done on Agisoft Photoscan software. The classification results show that RBF kernel function classifies the image with highest accuracy of 87% and linear kernel classifies the image with lowest accuracy of 75%, this means that image classification using RBF kernel function is better than other three kernels. For a single date RGB imagery, the classification results obtained were satisfactory. RBF kernel is the most used kernel to perform classification of remote sensing data. With higher spatial resolution, we can obtain satisfactory classification results using single RGB imagery. With time series UAV images or by including NIR spectral band we can obtain better results compared with single date RGB imagery as they have more information about the spectral reflectance of an image. In this study SVM classifier successfully used for delineating crop types, which is an important input for yield estimation and crop productivity in heterogeneous cropping system like India.

Acknowledgements The authors gratefully acknowledge to Prateek (CEO) and Anand (VP) from Terra Drone India Pvt Ltd., for providing the UAV images and ground truth data for the study area. We would also like to thank Dr. M. Shashi, K. Kumar, Sharath and Sai Charita from National Institute of technology for their suggestions and help throughout the research.

References

1. Pande-Chhetri R, Abd-elrahman A, Liu T, Morton J, Victor L (2017) Object-based classification of wetland vegetation using very high-resolution unmanned air system imagery. Eur J Remote Sens 50:564–576. https://doi.org/10.1080/22797254.2017.1373602
2. Melesse AM, Weng Q, Thenkabail PS, Senay GB (2007) Remote sensing sensors and applications in environmental resources mapping and modelling, 3209–3241. https://doi.org/10.3390/s7123209
3. Salamí E, Barrado C, Pastor E (2014) UAV flight experiments applied to the remote sensing of vegetated areas, 11051–11081. https://doi.org/10.3390/rs61111051

4. Colomina I, Molina P (2014) Unmanned aerial systems for photogrammetry and remote sensing: a review. ISPRS J Photogrammetry Remote Sens 92:79–97. https://doi.org/10.1016/j.isprsjprs.2014.02.013
5. Böhler JE, Schaepman ME (2018) Crop classification in a heterogeneous arable landscape using uncalibrated UAV data. Remote Sens. https://doi.org/10.3390/rs10081282
6. Ustuner M, Nutrition P (2014) Crop type classification using vegetation indices of Rapideye imagery. Int Arch Photogrammetry Remote Sens Spat Inf Sci, 195–198. https://doi.org/10.5194/isprsarchives-xl-7-195-2014
7. Otukei JR, Blaschke T (2010) Land cover change assessment using decision tree, support vector machines, and maximum likelihood classification algorithms. Int J Appl Earth Obs Geoinf, 27–31. https://doi.org/10.1016/j.jag.2009.11.002
8. Vapnik VN (1999) An overview of statistical learning theory. IEEE Trans Neural Netw 10:988–999
9. Archila Bustos M, Jirström M, Hall O, Dahlin S, Öborn I, Marstorp H (2018) Classification of Maize in complex smallholder farming systems using UAV imagery. Drones 2:22. https://doi.org/10.3390/drones2030022
10. Mankar NV, Khobragade A, Raghuwanshi MM (2016) Classification of remote sensing image using SVM kernels. In: IEEE WCTFTR 2016—proceedings of 2016 world conference on futuristic trends in research and innovation for social welfare, pp 1–5. https://doi.org/10.1109/startup.2016.7583977
11. Yekkehkhany B, Safari A, Homayouni S, Hasanlou M (2014) A comparison study of different kernel functions for SVM-based classification of multi-temporal polarimetry SAR data. https://doi.org/10.5194/isprsarchives-xl-2-w3-281-2014
12. Barazzetti L, Brumana R, Oreni D, Previtali M, Roncoroni F (2014) True-orthophoto generation from UAV images: implementation of a combined photogrammetric and computer vision approach. II:23–25. doi: 10.5194/isprsannals-II-5-57-2014

Drone-Based Sensing for Leaf Area Index Estimation of Citrus Canopy

Rahul Raj, Saurabh Suradhaniwar, Rohit Nandan, Adinarayana Jagarlapudi and Jeffrey Walker

Abstract Leaf Area Index (LAI) is an important parameter in the measuring of crop health. Temporal changes in the LAI provide important information about changes in the structure of the canopy and biomass over time. In this study, RGB images of the top of the canopy are collected by using a drone and through image processing; the coverage of green canopy is calculated from the images. Subsequently, by using the gap fraction, the LAI is estimated through the Beer-Lambert law. The data is collected from Warud taluka of Amravati district of Maharashtra, India. The area is severely under biotic and abiotic stresses. A multi-rotor quadcopter, which can carry a camera, is used to fly over the citrus farm on a predefined path. A camera that is mounted on the drone takes RGB images of the top of the canopy at a continuous interval with 70% frontal and 50% side overlap. These images are stitched together and an orthomosaic image layer is formed. Mathematical models are used to find the LAI from the images. Ground truth data is collected by a ceptometer within two hours of the flight of the drone. The two LAI datasets (LAI from the digital image and the LAI values from the LAI meter) are correlated, with R^2 equal to 0.73.

Keywords Leaf area index · Drone-based mapping · Precision agriculture · Citrus · Image processing

---

R. Raj (✉)
IITB-Monash Research Academy, Mumbai, India
e-mail: rahul.rahulraj@monash.edu

S. Suradhaniwar · R. Nandan · A. Jagarlapudi
Indian Institute of Technology, Bombay, India
e-mail: saurabh.suradh@gmail.com

R. Nandan
e-mail: rohitnandan@iitb.ac.in

A. Jagarlapudi
e-mail: adi@csre.iitb.ac.in

J. Walker
Monash University, Melbourne, Australia
e-mail: Jeff.Walker@monash.edu

K. Jain et al. (eds.), *Proceedings of UASG 2019*, Lecture Notes in Civil Engineering 51,
https://doi.org/10.1007/978-3-030-37393-1_9

1 Introduction

Good-quality, high-resolution farm data is very useful to biotech companies, government agencies, fertilizer manufacturers and commodity traders. However, due to the limited workforce and costly instruments/sensors, the collection of high spatial and temporal resolution farm data is a challenging and time-consuming process. Leaf area index (LAI) is an important crop physical parameter, which needs to be collected in order to quantify the light-use-efficiency of the vegetation [1]. The LAI is unit-less and is defined as the leaf area of half of the vegetation canopy (one-side of the leaf area) per unit ground surface area [2]. The LAI shows the amount of foliage area per unit ground surface area [3]. The definition holds good for broadleaved trees with flat leaves; however, if foliage elements are wrinkled, bent or not flat then vertical projection may not result in the highest value [4]. The LAI affects CO2 uptake of vegetation by affecting the effective stomatal area. Therefore, measuring temporal/seasonal variations of the LAI is important not only in the quantification of available biomass but also in the enhancing of the understanding of gaseous exchanges between the atmosphere and the canopies, so that crop growth models can be calibrated [5, 6]. The direct method of LAI measurement is a destructive method in which every green leaf of a plant is destructively sampled and one-sided area of each leaf is measured. The cumulative sum of the leaf area is then divided by the total ground area from which the LAI is to be calculated. Non-green leaves are not considered, because they do not contribute to photosynthesis [7]. The LAI values that are obtained using the direct method should be considered as a reference while comparing the LAI values that are obtained from an indirect method. In various researches that have been conducted at different palaces, it has been established that an indirect method instrument such as LAI-2000 or LAI-2200 is strongly correlated with the direct method LAI, with R^2 greater than 0.8 [8, 9].

In this study, instead of a destructive method, the LAI-2200 instrument is used to calculate the LAI. These LAI values are considered as ground truth and are compared with the LAI values that are estimated by using digital images of the top of the canopy, which are taken by a drone-based camera. A citrus farm (mandarin orange) is researched by placing an RGB camera on a quadcopter, which is flown at a height of 25 meters in the afternoon. Subsequently, within two hours, the LAI is collected through Licor's LAI-2200 instrument (ceptometer) from across the farm with high spatial resolution. The drone images are stitched and divided into smaller grids such that it represents the same area that is captured by the ceptometer. Each cropped image is converted to a grayscale image of greenness values, which is followed by the separating of the green and non-green pixels to calculate the gap fraction, by using the histogram method. The modified Beer-Lambert law is used to find the LAI through gap-fraction values. It is found that the LAI that is obtained through two different methods are correlated, with R^2 equaling 0.73. The framework of the study is shown in Fig. 1.

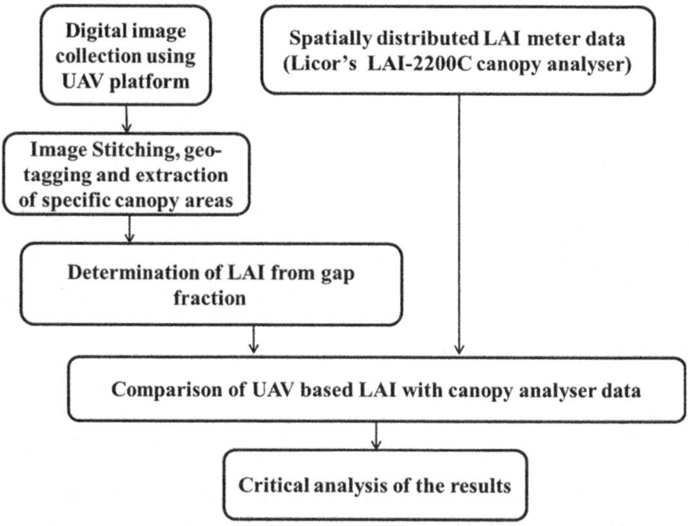

Fig. 1 The framework of the study

2 Materials and Methods

The overall approach of this proposed method in order to estimate the LAI from RGB images of the top of the canopy consists of five key steps:

(1) Acquisition of high spatial resolution RGB images of the top of the canopy of the citrus-farm by using a low-altitude quadcopter-camera system; (2) collection of in situ LAI values by using the ceptometer; (3) post-processing of Drone-based images to create an orthomosaic, to geo-reference and to extract the area of interest from the orthomosaic; (4) application of image processing techniques on images of the area of interest to quantify the gap fraction and to find the LAI; and (5) comparison and analysis of the extracted LAI by using drone images along with the LAI that is collected by using the ceptometer.

2.1 Site Description, Ground Truth LAI, and Acquisition of Images from UAV

The experiment was carried out on a 15-year-old citrus crop farmland that is located in Nagziri village of Warud taluka of Amravati district in Maharashtra, India. Amravati is in the north-east area of Maharashtra. Geographically, the study area lies approximately between 21026′54.4″ N - 21026′59.1″ N and 78009′13.1″ E - 78009′10.2″ E, as shown in Fig. 2a [10]. The drone-based RGB images of the top of the canopy were collected in the JPEG format on 24 Feb 2016 between 11:00 a.m. and 01:00

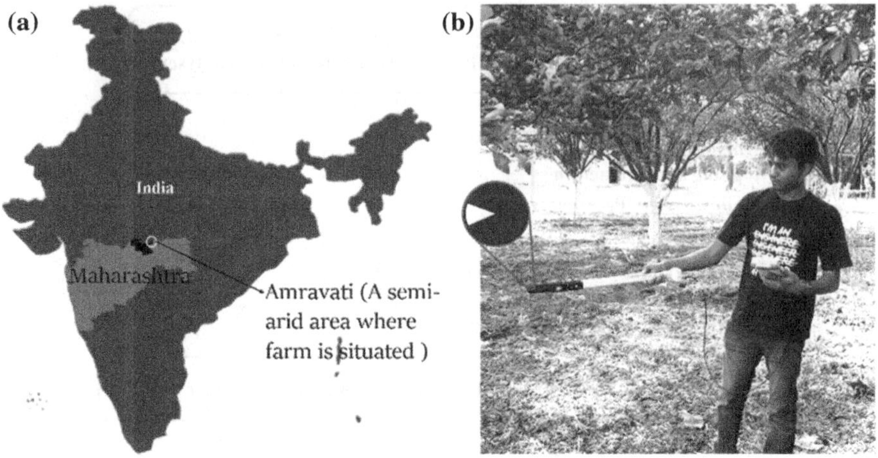

Fig. 2 **a** Geographical location of the farm that lies in the central part of India and falls under a semi-arid zone. **b** The orientation of 45° view angle cap of the ceptometer while collecting the LAI data

p.m. Immediately after the collection of data from the drone, Licor's LAI-2200C plant canopy analyzer was used to acquire ground truth LAI data from the farm. A view angle of 45° (315° masking) was used for the ceptometer. The masking cap of the ceptometer's lens was oriented in such a way that the view angle remained the farthest from the person who collected the data (Fig. 2b). For this study, the plot is virtually marked into 23 rows (R_0–R_{22}) (Fig. 3), and from those rows, ground truth LAI values are obtained using the ceptometer. In each row, certain LAI points were collected; however, rows 6, 15 and 22 were not considered, because it was subsequently found that the data from these rows were not acquired properly. Total 60 useful data points are collected from ceptometer by moving the ceptometer in a serpentine motion under the citrus canopy farms. After approx. every five metres of the walk under the canopy, ceptometer-based geotagged LAI data is collected.

Fig. 3 Rows of the farm for LAI data collection

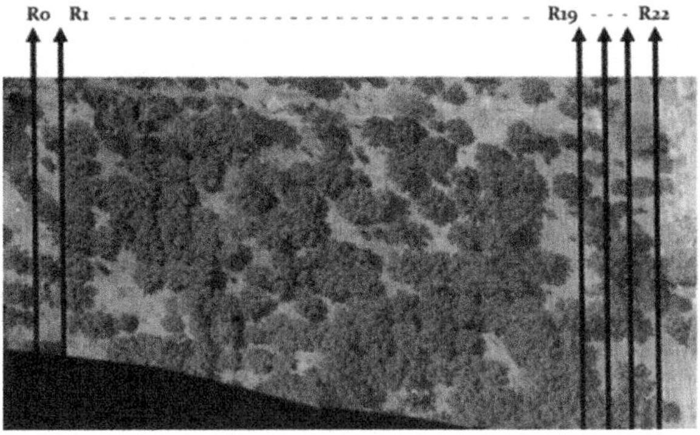

The FOV of the fisheye lens of ceptometer was 136°, and it was manually operated approximately at a height of 1 m from the ground. The estimated area from where the light was supposed to be captured by the fisheye lens of ceptometer was around 25 square metres at 2.5 metres from the ground, which is assumed to be the average height of the canopy when measured from the ground.

A digital camera (Canon IXUS 160 20MP) that was mounted on a multirotor UAV (Fig. 4) was programmed to capture images continuously at three-second intervals while the UAV flew along a predefined path over the citrus growing area. The camera had FoV of 55° × 50°. The sky was clear, and the day was sunny. The drone that was used in the experiment was a quadcopter with GPS, IMU and a DJI controller on it. The quadcopter had the ability to fly on autopilot along a predefined path that was defined by using open-source software, 'Mission planner'. The drone was flown at a height of 25 meters. In two consecutive images that were taken by the drone-mounted camera, 70% frontal and 50% side overlap of ground-pixels was maintained (Fig. 5). It is very important to maintain sufficient overlapping between consecutive and side images because otherwise an orthomosaic cannot be formed [11]. To maintain this minimum frontal and side overlap, the speed of the drone was maintained at 2.6 m/s. With these restrictions, the image capturing speed of the camera is fixed to one image per three seconds. The drone was flown in a serpentine motion, as shown in Fig. 5.

These images are stitched together by the PiX4D software and geotagged by the QGIS open-source software. The geotagged layer has a spatial resolution of around 3 cm $*$ 3 cm. From the ortho-mosaic image, those areas that are supposed to be seen through the ceptometer lens are manually cropped with the help of the geotagged LAI map and stored separately as a JPEG file. Sixty sub-images are cropped from the orthomosaic map. These cropped images are further processed and LAI values are calculated and correlated with Licor's LAI instrument data.

Fig. 4 Quadcopter used in the study

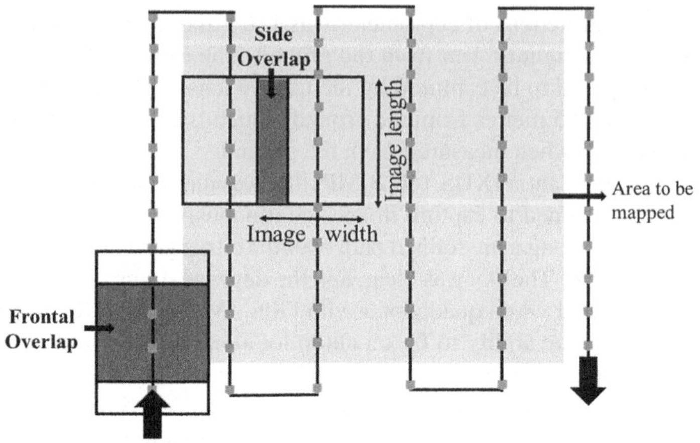

Fig. 5 Frontal and side overlap of images taken by the drone-mounted camera

2.2 Green Canopy Cover and LAI Estimation from Drone-Based Images

Various empirical methods can be employed to estimate the LAI from satellite images. In these techniques, ground measurements of the LAI are correlated with indices such as Normalized Difference Vegetation Index (NDVI), which are obtained from satellite images. However, it is seen that as the LAI increases above 3, NDVI values start to saturate; the NDVI values also do not show much sensitivity to canopies that have an LAI that is greater than 3 [12]. The technique used in this study does not use the NDVI index method to calculate the LAI. Here, the concept of light absorption by the canopy is used. If the density of the canopy in an image is likened to the concentration of solvent in a solution, then the light that is absorbed by the canopy can be likened to the light that is absorbed by the solution. Similarly, as a higher concentration of solvent leads to higher absorption of light, a higher density of leaves or a lower number of background pixels in an image can be likened to a higher LAI based on the Beer-Lambert–Bouguer law. In order to use this law to estimate the LAI, it is very important to quantify green canopy cover or gap fraction in an image. Green canopy cover is the representation of the density of the canopy over a ground area that is captured in that image. Gap fraction simply means the ratio of the total number of non-green pixels to the total number of pixels in the image. This gives us a quantification of the extent of the sparseness of the canopy. The process of estimating the LAI from the gap fraction can be linked to the 'Beer-Lambert–Bouguer' law [13, 14]. Sixty data points are used to compare the drone-based image results with ground truth values. To estimate the green canopy cover from a digital RGB image, it is recommended that the image be converted to the grayscale format by using a combination of red, green and blue bands [15]. For vegetation images, to retain the greenness of the image, Eq. (1) is the best combination, in which R, G and B are

Fig. 6 From the orthomosaic layer, an area of interest is cropped out. Red, Green and Blue layers are separated, and the greenness layer is formed by using the formula, 2G-R-B. This greenness image is converted to grayscale to plot the histogram

digital numbers of the pixel [16].

$$Greenness = 2G - R - B \quad (1)$$

The greenness image has only one layer, and it can be converted to a grayscale image (Fig. 6). This grayscale image should contain two classes, that is, canopy pixel (brighter colour representing green) and background pixel (darker colour representing non-green). To classify the two classes, different supervised and unsupervised techniques can be used. Here, Otsu's method, which is an unsupervised technique, is implemented on the image. Otsu's method of image classification performs better if the image has two classes [17]. After applying Otsu's method on an image, a histogram that contains two binomial distributions is plotted, as shown in Fig. 7. Since the two distributions usually overlap slightly, the threshold that separates the two classes needs to be found. The point, T_0, in Fig. 7 is the threshold point. All pixel values to the left of T_0 represent the background, while all pixel values to the right of T_0 represent the green canopy. The two peaks represent two classes, and the threshold point, T_0, separates the two classes. The generated histogram contains high-frequency noise, as shown in Fig. 7. This high-frequency noise should be smoothed with a low-pass filter in order to make the detection of T_0 automatic. Once T_0 is determined, the green canopy cover can be estimated by dividing the cumulative sum of pixels that are to the right of T_0 by the total number of pixels in the histogram. Conversely, the gap fraction of the image can be estimated by calculating the ratio that is obtained after dividing the cumulative sum of pixels that are to the left of T_0 by the total number of pixels in the histogram. By considering the random spatial distribution of the leaves, the gap fraction is related to the LAI of the area by a Poisson distribution [13], where the LAI is related to the gap fraction by a formula

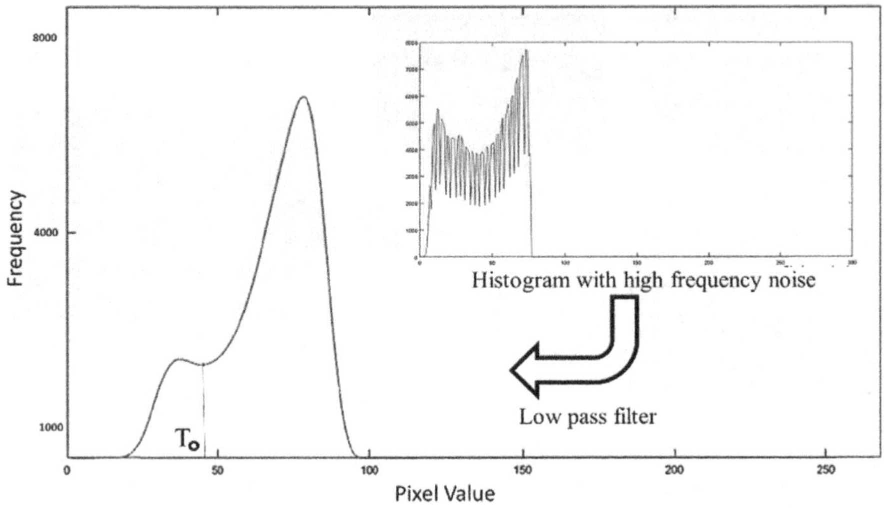

Fig. 7 A smooth histogram is obtained by applying a low-pass filter to the original histogram

that is given in Eq. (2).

$$LAI = -ln(Gapfraction)/k, \tag{2}$$

In Eq. (2), k is termed the canopy extinction coefficient, and its value is estimated at 0.5 [18]. Here, it should be noted that the gap fraction is calculated for images that were taken at a vertically downward angle when the solar zenith angle was also near zero. If solar zenith angle or camera view angle changes, the effect of these changes can be observed in the gap-fraction values, and formula x will need modifications to enable the incorporation of the changes [19]. The gap fraction and the LAI are calculated for all the 60 sub-images that are extracted from the orthomosaic layer.

3 Results and Discussion

3.1 Comparison of Estimated LAI with Its Ground Truth

LAI values that are estimated from drone images and those that are obtained from the ceptometer are compared by applying linear regression to obtain the coefficient of determination (R^2). Python open-source software is used to process the data. By implementing linear regression on all 60 data points, it is found that R^2 between Licor's LAI values and drone-based LAI values is 0.73 (Fig. 8a). The optimised value of canopy extinction coefficient shown in Eq. (2) is also calculated by checking the model accuracy by changing the value of 'k' in the range of 0.3–0.7. It is found

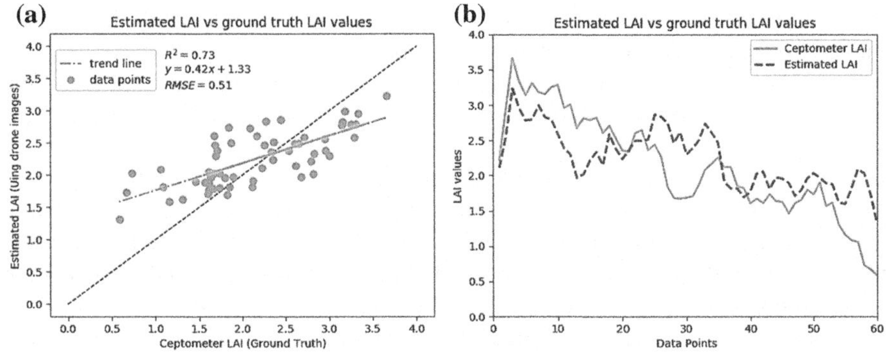

Fig. 8 **a** The scatter plot of LAI data obtained from ceptometer and estimated using drone images. The R^2 found is 0.73. **b** The visual trend analysis plot of LAI data obtained from ceptometer and estimated using drone images

that at k = 0.5 the two different platforms LAI values are best correlated i.e. $R^2 = 0.73$. The estimated root mean square error (RMSE) is 0.51. It is observed that the LAI of those images that have a lower canopy density (higher gap fraction) is slightly overestimated, while the LAI of those areas that have a denser canopy (lower gap fraction) is slightly underestimated when compared to the LAI values of the ceptometer. The underestimation or overestimation is because the ceptometer has fisheye lens; therefore, the effect of the lights that emanate from its sides affects it more along with the scattered light from those areas that are not in the view angle of the lens. However, in the case of drone images, the only deciding factor of the LAI is the gap fraction that is seen from the top of the canopy images; the areas that are outside that particular image do not affect the LAI values at all. In Fig. 8b, a comparison between the ground truth LAI and the estimated LAI can be observed. It can also be observed that when the ground truth LAI is greater than 3, the estimated LAI is underestimated, but the increase/decrease change trend remains the same (Fig. 8b). When the ground truth LAI is less than 1.5, that is, the canopy is open, the estimated LAI is overestimated, and even the increase/decrease change trend is not maintained. However, overall, both methods seem to yield well-correlated LAI values.

3.2 *Critical Analysis of the Limiting Factors of LAI Estimation*

The difference in LAI values between the same data points are also contributed to by the fact that the area of the cropped image (cropped from orthomosaic layer) will always be different from the area that is seen by the fisheye lens of the ceptometer;

this is because the ceptometer view area is in the shape of a triangle, while the cropped images are approximated to a rectangular shape.

The classification of the canopy and the non-canopy pixels are not infallible because any background pixel that is not a part of the canopy but green in color (for example, weed, green plastic and such others) can be classified as a green canopy pixel and will contribute to error. Otsu's method, which is used for the classification, works well when the image has two classes with sufficient areas of visibility of both the classes in the image. If the canopy is extremely dense or highly open, its histogram will contain only one peak, and the identification of the classification threshold point, T_0, will be impossible. In such cases, LAI cannot be quantified by using this technique. It is also difficult to fly the drone at a constant height with a constant speed and angle (due to changing wind conditions), which, sometimes, blurs the image pixels. To reduce this blur, the drone should fly at a low wind condition, and different flight paths and directions should be used while collecting the images from multiple flights.

Acknowledgements The authors would like to acknowledge Ms Mrunalini R. Badnakhe for her help in acquiring the necessary permissions to fly the drone. The help of Drona Aviation Company has also been critical in the collection of drone images.

Conflicts of Interest The authors declare no conflict of interest. The funders had no role in the design of the study; in the collection, analyses, or interpretation of data; and in the decision to publish the results.

Funding This research was funded by Information Technology Research Academy (ITRA), division of Department of Electronics and Information Technology (DeitY), Ministry of Communications and Information Technology (MCIT), Government of India.

References

1. Qu Y, Meng J, Wan H, Li Y (2016) Preliminary study on integrated wireless smart terminals for leaf area index measurement. Comput Electron Agric 129:56–65
2. Chen JM, Black TA (1992) Defining leaf area index for non-flat leaves. Plant Cell Environ 15(4):421–429
3. Chen JM, Cihlar J (1996) Retrieving leaf area index of boreal conifer forests using landsat TM images. Remote Sens Environ 55(2):153–162
4. Jonckheere I, Fleck S, Nackaerts K, Muys B, Coppin P, Weiss M (2004) Methods for leaf area index determination part I: theories, techniques and instruments. Area 121: 1–42
5. Bonan GB (1993) Importance of leaf area index and forest type when estimating photosynthesis in boreal forests. Remote Sens Environ 43(3):303–314
6. Running SW, Coughlan JC (1988) A general model of forest ecosystem processes for regional applications I. Hydrologic balance, canopy gas exchange and primary production processes. Ecol Model 42(2):125–154
7. Fang H, Li W, Wei S, Jiang C (2014) Seasonal variation of leaf area index (LAI) over paddy rice fields in NE China: Intercomparison of destructive sampling, LAI-2200, digital hemispherical photography (DHP), and AccuPAR methods. Agric For Meteorol 198:126–141
8. Stroppiana D, Boschetti M, Confalonieri R, Bocchi S, Brivio PA (2006) Evaluation of LAI-2000 for leaf area index monitoring in paddy rice. Field Crop Res 99(2–3):167–170

9. Liu J, Pattey E (2010) Retrieval of leaf area index from top-of-canopy digital photography over agricultural crops. Agric For Meteorol 150(11):1485–1490
10. GSDA. Available online: http://www.webcitation.org/70mdA0g4B Accessed on 9th July 2018
11. Pix4D-Before starting a project. Available online: http://www.webcitation.org/71olV0COP Accessed on 9th July 2018
12. Carlson TN, Ripley D (1997) On the relationship between NDVI, fractional vegetation cover, and leaf area index. Remote Sens Environ. 62: 241–252
13. Nilson T (1971) A theoretical analysis of the frequency of gaps in plant stands. Agric. Meteorol. 8, no. C: 25–38
14. Liu J, Pattey E, Admiral S (2013) Assessment of in situ crop LAI measurement using unidirectional view digital photography. Agric For Meteorol 169:25–34
15. Graham EA, Yuen EM, Robertson GF, Kaiser WJ, Hamilton MP, Rundel PW (2009) Budburst and leaf area expansion measured with a novel mo-bile camera system and simple color thresholding. Environ Exp Bot 65(2–3):238–244
16. Richardson D, Jenkins JP, Braswell BH, Hollinger DY, Ollinger SV, Smith ML (2007) Use of digital webcam images to track spring green-up in a deciduous broadleaf forest. Oecologia 152(2):323–334
17. Kittler J, Illingworth J (1985) On threshold selection using clustering crite-ria. Syst Man Cybern IEEE Trans 5:652–655
18. Goudriaan J (1988) The bare bones of leaf angle distribution in radiation models for canopy photosynthesis and energy exchange. Agric For Meteorol. 38:155–255
19. Baret F, de Solan B, Lopez-Lozano R, Ma K, Weiss M (2010) GAI estimates of row crops from downward looking digital photos taken perpendicular to rows at 57.5° zenith angle: theoretical considerations based on 3D architecture models and application to wheat crops. Agric For Meteorol 150(11):1393–1401

12. [illegible]
13. [illegible]
14. [illegible]
15. [illegible]
16. [illegible]
17. [illegible]
18. [illegible]
19. [illegible] Model [illegible] 1496

Dynamics of Target Detection Using Drone Based Hyperspectral Imagery

Sudhanshu Shekhar Jha and Rama Rao Nidamanuri

Abstract Imagery from advances in imaging and acquisition platforms, such as Unmanned Aerial Vehicles (UAV) having finer spatial resolution helped solve problems in identifying plant species, agricultural monitoring, rock characterization, albeit it poses plausible challenges arising from complex object-sensor dynamics. Reflectance spectra from close-range imaging are found to be significantly affected by the shape of the object, illumination angle, and light source position. Assessment of close-range hyperspectral imaging for target detection is useful across different application scenario for drone-based hyperspectral imaging. Over an experimental study site in Bangalore, drone-based hyperspectral imagery was acquired with the goal of detecting and identifying various artificial targets placed in a complex imaging geometry with different target-backgrounds. The acquisition platform used a compact hyperspectral imaging sensor mounted on a drone at a flying altitude of 95 m. Different metallic sheet targets, painted with green color, and natural metallic color are used. Green metallic target is placed and imaged in various target-background environments whereas the other metallic target is placed on open soil background. Target detection algorithms: spectral angle mapper (SAM), matched filter (MF), adaptive cosine estimator (ACE), constrained energy minimization (CEM) are evaluated using the statistical performance indicators using ROC curves. For a partially visible target, with a probability of detection at 99%, the probability of false alarm for ACE, CEM, MF, and SAM is found to be 63%, 96%, 29%, and 18% respectively. For nearly camouflaged target PFA is found to be 75, 94, 59 and 79% for the ACE, CEM, MF and SAM detector with a probability of detection at 99%. Results warrant further refinements in existing hyperspectral target detection algorithms for close-range hyperspectral imaging.

Keywords Close-range imaging · Target detection · Hyperspectral · Drone

S. S. Jha (✉) · R. R. Nidamanuri
Indian Institute of Space Science and Technology, Kerala 695547, India
e-mail: Sudhanshushekhar88@gmail.com

K. Jain et al. (eds.), *Proceedings of UASG 2019*, Lecture Notes in Civil Engineering 51,
https://doi.org/10.1007/978-3-030-37393-1_10

1 Introduction

Material classification and detection are the two prominent applications of hyperspectral remote sensing due to its capability to characterize materials based on their spectral properties [1]. Typically a target is defined as a material of interest in the imagery and its detection basically constitutes a hypothesis testing problem, where each pixel of the image is either a target or non-target/background pixel. Hyperspectral target detection has been one of the centers of research in recent times in applications of hyperspectral image analysis, especially for surveillance and mineral mapping. Briottet et al. [2] have evaluated the performance of signature-based spectral target detectors for numerous military targets and found encouraging results. Further, Yuen and Richardson [3] have explored the potential of various target detection algorithms that could be used for surveillance and security purposes, both, when a priori target information is not available and when there is sufficient knowledge about the target. Over the past few years, hyperspectral imaging techniques for remote sensing applications have significantly evolved from traditional imaging platforms, such as satellite and aircraft, to unmanned aerial vehicles (UAV) to drone-based acquisition [4]. With an increase in the spatio-spectral resolution of the hyperspectral imagers, a complex interaction between object and sensor has been observed. The reflectance spectra, in the case of close-range imaging, are found to be significantly affected by the shape of the object, illumination angle and light source position [5]. With these technological advancements, it is but natural that detection/classification of man-made materials and recording anomaly in natural phenomena will become the central theme of close-range based hyperspectral imaging applications. Although classification problems have been pursued in a relatively wider perspective, close-range hyperspectral target detection has not been reported to the best of our knowledge.

The aim of our study is to test the performance of some of the popular target detectors: spectral angle mapper (SAM), matched filter (MF), adaptive cosine estimator (ACE) and constrained energy minimization (CEM) on drone-based close-range sets of hyperspectral imagery. Specifically, we assess the robustness of target detectors in complex imaging geometry and their performance under multiple target-background combinations. The performance of various target detectors is evaluated on the basis of ROC curves computed between different target detectors.

2 Method and Materials

2.1 Target Detector Algorithms

Spectral Angle Mapper (SAM)

SAM is one of the most commonly used algorithms in hyperspectral remote sensing and used in many kinds of spectral material detection problems. Geometrically, it measures the similarity between two n-dimensional vectors based on the cosine of the angle between two vectors. It is easier to compute than the other target detectors. Mathematically, SAM score r is given by:

$$r_{SAM}(\mathbf{x}) = cos^{-1}\left[\frac{\mathbf{s}^T\mathbf{x}}{\sqrt{(\mathbf{s}^T\mathbf{s})(\mathbf{x}^T\mathbf{x})}}\right] \tag{1}$$

where **s** is the known target spectrum and **x** is the pixel under test.

Matched Filter (MF)

While SAM is limited by the assumption of a zero-mean and white background which is an unrealistic assumption for detecting materials in a real-world scenario, MF on the other hand, allows having a background with a normal distribution having finite mean and covariance. The detector assumes an additive model where background and target have equal covariance. Mathematically MF score r is given as:

$$r_{MF}(\mathbf{x}) = \frac{\mathbf{s}^T\mathbf{C}^{-1}\mathbf{x}}{\mathbf{s}^T\mathbf{C}^{-1}\mathbf{s}} \tag{2}$$

where **s** is the target spectrum, **C** is the background covariance matrix, and **x** is the pixel under test. It must be noted that mean is subtracted from the data cube and the target spectrum before applying the MF detector.

Constrained Energy Minimization (CEM)

Since hyperspectral imagery are characterized by high spectral resolution, there exists a high possibility of interference of background signal in the target signal. CEM algorithm suppresses the interference of undesired signals in the target signal, thereby accentuating the target signal. Mathematically, CEM score r is given as:

$$r_{\mathrm{CEM}}(\mathbf{x}) = \frac{\left(\mathbf{s}^{\mathrm{T}}\mathbf{R}_{\mathrm{LxL}}^{-1}\mathbf{s}\right)^{-1}}{\left(\mathbf{R}_{\mathrm{LxL}}^{-1}\mathbf{s}\right)^{\mathrm{T}}}\mathbf{x} \tag{3}$$

where **s** is the target spectrum, **R** is the background correlation/covariance matrix and **x** is the pixel under test.

Adaptive Cosine Estimator (ACE)

ACE is derived from the generalized likelihood ratio test (GLRT), which incorporates a substitution model for background, unlike the MF which assumes background noise to be additive in nature. Mathematically, ACE score r is given as:

$$r_{\mathrm{ACE}}(\mathbf{x}) = \frac{\left(\mathbf{s}^{\mathrm{T}}\mathbf{R}_{\mathrm{LxL}}^{-1}\mathbf{x}\right)^2}{(\mathbf{s}^{\mathrm{T}}\mathbf{R}_{\mathrm{LxL}}^{-1}\mathbf{s})\left(\mathbf{x}^{\mathrm{T}}\mathbf{R}_{\mathrm{LxL}}^{-1}\mathbf{x}\right)} \tag{4}$$

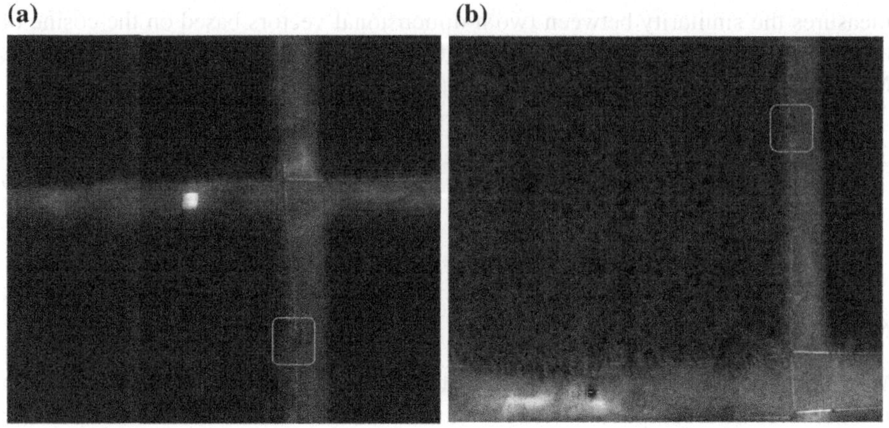

Fig. 1 Target-background setup for **a** target placed in a partial cover by background, **b** target placed in a near-full camouflage by background

where **s** is the target spectrum, **R** is the background correlation/covariance matrix and **x** is the pixel under test.

2.2 Experimental Design

The experimental setup for the present study is carried out in an agricultural field at University of Agriculture Science, Bangalore, India. For analyzing the dynamics of close-range target detection problems, we designed a controlled experimental setup, in which the target is placed next to an agricultural crop as shown in Fig. 1.

The target for this study is a 2 ft. × 2 ft. sized metallic sheet painted with green color. The green color provides a natural camouflage when placed with vegetation as the local background. The target is imaged using a drone-based Cubert hyperspectral sensor (altitude of ~95 m) having 138 spectral bands ranging from 450 to 998 nm with a spectral resolution of 4 nm and spatial resolution of 1–2 cm. Two different target-background environments: (1) target is placed such that half of the target is fully exposed to the imaging sensor and the other half is either covered or affected by the agricultural crop in the neighborhood (2) target is placed in such a way that the agricultural crop covers almost the full extent of the target, are realized for the evaluation of the target detection algorithms mentioned in the Sect. 2.1. The aim of the experimental design is to evaluate the performance of the detection algorithms for the close-range hyperspectral target detection in the presence of a cluttered background.

3 Results and Discussion

The performance comparison of various target detection algorithms for different cases mentioned in Sect. 2.2 is shown in Fig. 2. Table 1 shows the performance of target detection algorithms, known as the power of detection P_D, as the function of P_{FA} (Probability of false alarm) at various confidence level. Performance variation is observed across the detection rate as a function of the required accuracy. It is observed that as we increase the threshold of P_D, P_{FA} increases sharply in some scenarios such as jump of P_{FA} is approximately doubled in case of MF for Fig. 1a as we increase the threshold of P_D from 90 to 95%.

As evident from Fig. 1a, when the target is relatively visible and extent of background on the target is limited, SAM is able to achieve a $P_D \geq 99\%$ at $P_{FA} = 18\%$, which is better than the other mathematically complex detectors such as ACE and CEM. As the target-background environment becomes complex for the second case, the performance of all the target detector decreases substantially.

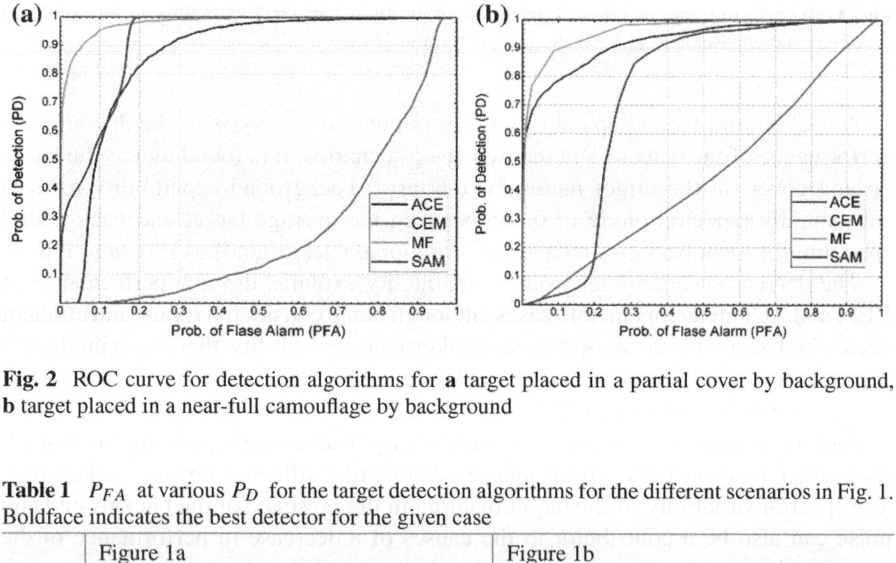

Fig. 2 ROC curve for detection algorithms for **a** target placed in a partial cover by background, **b** target placed in a near-full camouflage by background

Table 1 P_{FA} at various P_D for the target detection algorithms for the different scenarios in Fig. 1. Boldface indicates the best detector for the given case

	Figure 1a				Figure 1b			
P_{FA}					P_{FA}			
P_D	ACE	CEM	MF	SAM	ACE	CEM	MF	SAM
80%	0.186	0.921	**0.019**	0.162	0.117	0.781	**0.036**	0.285
85%	0.220	0.934	**0.030**	0.167	0.184	0.853	**0.058**	0.309
90%	0.275	0.941	**0.058**	0.169	0.243	0.891	**0.087**	0.387
95%	0.364	0.948	**0.097**	0.176	0.428	0.941	**0.181**	0.505
99%	0.631	0.962	0.290	**0.187**	0.753	0.980	**0.278**	0.792

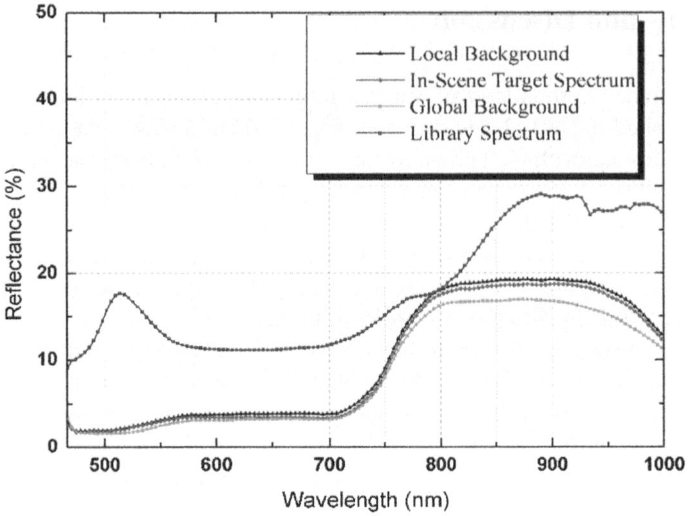

Fig. 3 Average in-scene target-background separability and spectral variability of target library spectrum in comparison to in-scene target spectrum

A background-target separability was computed for analysis of the downgraded performance of the detectors in the two given scenarios. It is found that as the background cover on the target increases, the target-background separability reduces substantially (spectral angle of 0.18° between the average target and background spectrum for local background and 1.6° for global background) as shown in Fig. 3.

The decreased target-background separability explains the low performance of CEM and ACE detectors in both cases, although being considered robust and efficient detectors. Due to this decreased target-background separability, there is an inefficient suppression/separation of the target-background spectrum which eventually lowers the performance of ACE and CEM. In addition, we find that while the library spectrum for the target signature is different from the average background spectrum (see Fig. 3), the average in-scene target spectrum is predominantly influenced by the background. The spectral variability of the target material in the presence of the background and noise can also be a contributor to the causes of a decrease in performance of the considered target detectors in the present study.

4 Conclusion

With the present designed experimental setup, an attempt has been made to study the detection performance of various popular target detection algorithms when the targets are positioned in a complex environment and imaged at close range. We have evaluated the performance of target detectors using ROC metric and found that while better results are obtained for a target positioned relatively in open background condition, the same is not true when the dynamics of target positioning becomes complex. Moreover, we also find that target-background spectral separability and variability has a substantial effect on the performance of target detectors. While it may seem trivial to detect a target at close range due to its apparent clear view, a close examination of the detection results suggests that the automatized spectral discrimination of different materials at high spatial resolution is seemingly challenging and existing detection algorithms require significant refinements.

Acknowledgements We would like to express our sincere gratitude to the Department of Biotechnology, (DBT), Government of India for funding this research (Grant Number: DBT/IN/German/DFG/14/BVCR/2016) as part of Indo-German consortium of DFG Research Unit FOR2432/1. We also thank the University of Agricultural Sciences, Bangalore, India for facilitating the conduct of experiments. We acknowledge all the Indo-German research scholars participating in the spectral imaging data campaign.

References

1. Manolakis D, Shaw G (2002) Detection algorithms for hyperspectral imaging applications. IEEE Signal Process Mag 19(1):29–43
2. Briottet X, Boucher Y, Dimmeler A, Malaplate A, Cini A, Diani M, Bekman HHPT, Schwering P, Skauli T, Kasen I, Renhorn I (2006) Military applications of hyperspectral imagery. In: Targets and backgrounds XII: characterization and representation, vol 6239, p 62390B, International Society for Optics and Photonics
3. Yuen PW, Richardson M (2010) An introduction to hyperspectral imaging and its application for security, surveillance and target acquisition. Imaging Sci J 58(5):241–253
4. Adão T, Hruška J, Pádua L, Bessa J, Peres E, Morais R, Sousa JJ (2017) Hyperspectral imaging: a review on UAV-based sensors, data processing and applications for agriculture and forestry. Remote Sens 9(11):1110
5. Robles-Kelly A, Simpson-Young B (2013) Imaging spectroscopy for scene analysis: challenges and opportunities. IET Comput Vis 7(6):467–477

Blockchain and UAV: Security, Challenges and Research Issues

Renu, Sanjeev Sharma and Sandeep Saxena

Abstract UAV is advance air vehicles with sensors and software to fly without a human pilot. This is an emerging technology with a tremendous potential to revolutionize warfare and to enable new civilian applications. There is a need to ensure that no malicious entity can disturb the transmission of data during UAV Ad hoc Network (UAANET) communications. Currently, UAV having various vulnerabilities and hacked by a different set of skilled persons. In this paper, we outline security requirements for UAANET along with the existing attacks. Before giving any security solution, we first deeply understand common possible Attacks and their impact on society. We presented UAANET general security limitations and challenges with the existing solutions by Blockchain Technology. The Blockchain technology uses the combination of cryptography, a consensus algorithm, and a distributed ledger to create a decentralized trustworthy platform. In addition, open research issues are also discussed.

Keywords Blockchain · UAV security · UAV · UAANET · MANET

1 Introduction

UAV Ad hoc Networks (UAANETs)can be said as the subset of the well-known mobile ad hoc network (MANET). It consists of multiple small Unmanned Aerial Vehicles (UAVs) and Ground Control Stations (GCS). These UAVs collaborate in order to relay data (command and control traffic and remotely sensed data) between each other and to the Ground Control Station (GCS). Compared to other types of ad hoc networks, UAANETs have some unique features and bring several major challenges to the research community. UAVs are controlled remotely and can fly autonomously as perfixed and predetermined Flightplan. For a UAV it is required to communicate with the controller sending back some information about its position

Renu (✉) · S. Sharma
RGPV, Bhopal, India

Renu · S. Saxena
Galgotias College of Engineering and Technology, Greater Noida, India

K. Jain et al. (eds.), *Proceedings of UASG 2019*, Lecture Notes in Civil Engineering 51,
https://doi.org/10.1007/978-3-030-37393-1_11

and speed etc. periodically so that the operator allows him to modify the flight settings accordingly if required. In the context of the UAANET, Blockchain technology can be applied to ensure the successful processing of multiple transactions, the tracking, and coordination of various UAVs. UAVs and CGS can communicate securely by having an encrypted chip. The UAV's encrypted chip communicates with a chip reader on a control point (GCS). The chip reader verifies the chip's cryptographic signature and checks its identity on the blockchain. Once permission is confirmed, the communication can be completed. Since blockchain technology is decentralized, there is no central authority or specific security administrator is necessary. Blockchain technology is based on cryptography algorithms that are designed to ensure the prevention of data distortion and ensure high security.

The rest of the paper is organized as follows: Sect. 2 justifies the objective of the paper, Sect. 3 gives the overview of Blockchain and UAANET with their general architectures, Sect. 4 is based on the current security requirement of UAANET with its vulnerabilities and attacks Sect. 5 presents research issues and current projects in this area. At the last Sect. 6 concludes the paper.

2 Scope and Objective

The paper analyzes the working principles of Blockchain, framework, and benefits of Blockchain, as well as identifies the scope of integration of blockchain with UAANET as two emerging technologies. In [1], UAANET architecture is compared with other ad hoc architecture by giving their benefits and UAANET (UAV Ad-hot network) is declared as the best architecture. They also discussed the security challenges that must be faced and gave suggestions to improve the security of UAANET. In 2016 L Gupta etc. covered important issues related to UAV communication networks [2]. Also compare MANETs, VANETs and UAV networks with two possible ways of connection Infrastructure-based or ad hoc considering several issues like power consumption, mobility, and life of the UAV. From the above surveys, which focused on the establishment and communication links between UAVs. As per our acknowledge there is no study focused on the current technology and secured communication aspects between UAV and GCS. To fulfill the gap, the main objective of this paper is to analyze blockchain architecture for providing reliable and secured communication between the UAV and the GCS. We look at the combination of blockchain and the UAANET as it's used in insurance and will increasingly be, moving beyond the traditional security solution. The paper projects that blockchain could be better security solution with ongoing attacks and vulnerabilities in UAANET.

3 Overview of Blockchain and UAANET

To secure this sensitive and dynamic environment, the integration of blockchain technology and the UAANET may be projected as an integrated emerging trend.

3.1 Architecture of Block-Chain

The Blockchain is seen as the next huge revolutionizing technology. It changed the way we work in Blockchain was first implemented as digital cryptocurrency known as *bitcoin*. We are living in a society with information exchange through the Internet, where each transaction involved a trusted third party. These parties are accountable for secure exchange or breaches. Two manage this party is neither easy nor efficient. Blockchain eliminates the need for a trusted third party by using the decentralized public ledger. Instead of relying on an intermediary they follow *consensus algorithm* for maintaining mutual trust. Thus, the building blocks of a blockchain-based are:

- Participating parties or devices
- Consensus protocol
- Poof of work
- Cryptographic hashes
- Digital signatures.

A lot of industries need efficient architecture which increased adoption of growing blockchain technology. The diagram of the structure of blockchain is given (Fig. 1).

3.2 UAV Communication Architectures

The UAV is an acronym for Unmanned Aerial Vehicle, which is an aircraft with no pilot on board. UAVs can be remote-controlled aircraft by a system embedded on Ground Control Station (GCS). UAANET is a sub-category of the well-known magnet in which nodes communicates with each other without the need for a fixed infrastructure. Each UAV acts as an end system. All UAVs are assumed to cooperate and relay information. The ad hoc architecture means constant changing topology as UAV mobility. In UAANETs, the GCS is considered as a regular end node which can have a fixed or mobile geographical position. It communicates with the nearest UAV which acts as a gateway. Communication may be UAV-to-UAV, UAV-to-GCS, and GCS-to-UAV (Fig. 2).

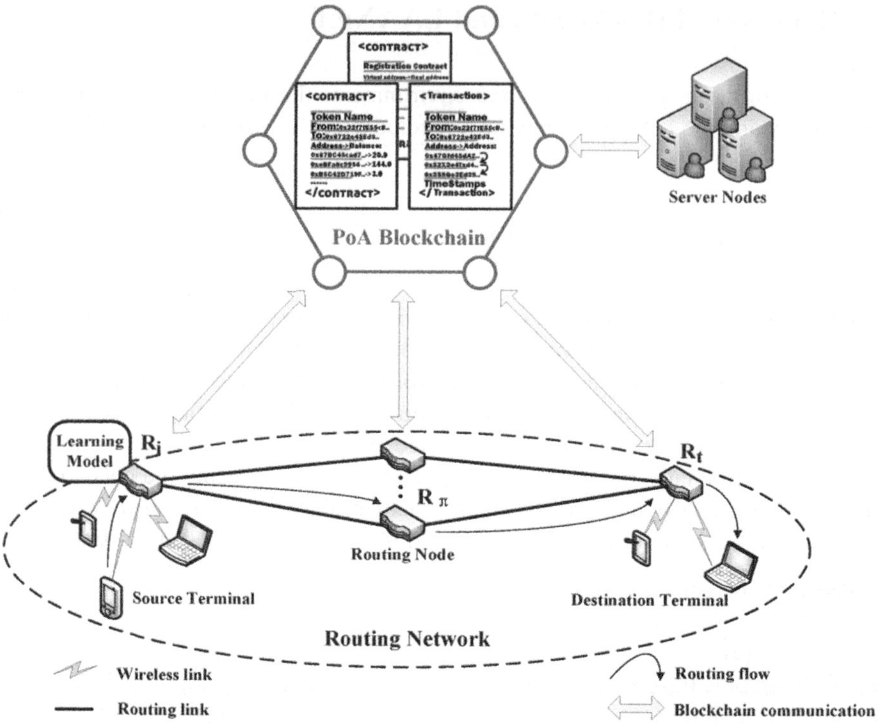

Fig. 1 Structure of blockchain for secure routing in MANET

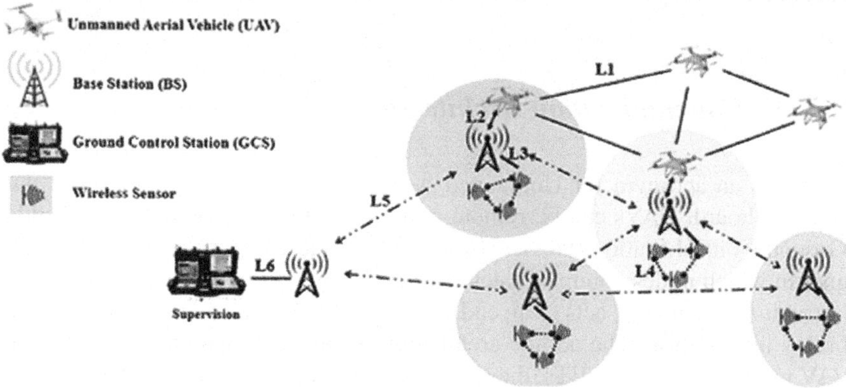

Fig. 2 Structure of UAANET as a subset of MANET

3.3 UAANET as a Subset of MANET

Ad hoc networks have been largely investigated by the research community for a set of mobile systems air vehicles [3, 4]. UAANET is also an infrastructure-less network with multiple nodes to forward data packets and having self-organized abilities, distributed information management with communications and cooperation between nodes to perform data delivery [5–7]. However, it also has some different feature from MANET:

1. UAV can cover mission area because of relatively high speed, the so large number of UAVs are not generally required.
2. Network connectivity is more critical in UAANET as compared to MANETs or VANETs. The disturbance in the communication could be caused by link quality fluctuates.
3. UAANETs nodes are usually assumed to have enough energy and computing power compared to nodes in MANET.
4. UAV mobility is three dimensional and different from other this brings challenges in misbehavior detection which was based on node position.
5. UAANETs are deployed for real-time applications so the control traffic should have a small delay.

3.4 Model Algorithm to Implement Blockchain in UAANET

1. *Start*
2. *Node (UAV) requests for a transaction execution in the network*
3. *A query is passed to the cloud layer*
4. *A connection is created between node and blockchain*
5. *A service provider is chosen from the list of SPs*
6. *A block is created with UAV's ID, SP's ID and time stamp*
7. *The newly created block is broadcasted to all the users in a blockchain network for the authentication of a newly created block.*
8. *All the nodes in the network certified the newly created block*
9. *If a block is certified*

   Then Block will be added to the end of a blockchain
   Else block will be rejected due to unauthorization

10. *The incentive will be given to CSP.*

4 UAANET Security Requirement and Challenges

UAANET is all about getting and transmitting large amounts of data through systems and linkages on real-time. Since any business needs privacy and security, the major task here is to ensure the protection of all data and communication [8]. Apart from privacy, it is important to ensure safe data delivery to the right place, form and time.

4.1 Vulnerabilities in UAANETs

In this section, we examine various causes for making UAANETs vulnerable to attack.

i. UAANETs use a wireless medium which is open for misuse and replay of massages which allows getting sensitive information, thus violating data confidentiality. In such case pilot does not current information from UAV, the UAV will not get any command for a time meanwhile an attacker to capture a UAV.
ii. UAANETs like traditional MANET suffers from the scarcity of central authority attacker can enter inside as well as outside the network because of no centralized key management system.
iii. In UAANET due to high mobility, it is a challenging task to differentiate between a truly misbehaving node and a legitimate node that appears to be misbehaving.
iv. UAANETs require the cooperation from all nodes participate in the routing process but there is no guarantee that a path between two nodes is secured.

4.2 Existing Attacks in UAANET

UAANET is exposed to different types of threats and attacks. This is caused by constraint characteristics came as hierarchic from the parent class of MANET. The purpose of attack targeting the network consists of absorbing and controlling network traffic, disrupting the routing function and injecting malicious nodes. A diversity of attacks in a MANET environment have been extensively described in the literature. Several classifications have has been proposed in the literature for MANETs. Instead, to classify threats and attacks based on basic routing functions, we illustrate the attack on the type of traffic between UAVs. We can categorize attacks as one which could harm traffic control, second is related to control beacon messages and the third for real-time traffic.

(A) *In the Blackhole attack*, an attacker attempts to advertise that it has a fresh route. By generating forge control packets, the adversary node may succeed in becoming part of the network route. Then, once chosen as the intermediate node, the attacker drops the packets instead of processing them. This prompts the GCS to choose the route involving the adversary node as the better one. To avoid detection from its neighbors, the attacker can selectively forward some packets and discard others and listed as a grey hole attack [9].

(B) *DOS attack* in UAANETs allows an adversary to behave as legitimate nodes in the network. The purpose of this attack is to exclude a UAV from the network by suspending its communication with other nodes. The attacker drains their energy by continuously generating data packets and sends them to target UAVs to. This can be extremely harmful to small size UAVs with limited storage and capacity. This causes the UAVs either not to respond to route request or to execute forged packets inserted by an attacker.

(C) *The wormhole attack* involves two attackers one attacker records packets at a particular location and replays them to another attacker by using a high-speed private network using tunnel Once the wormhole tunnel is selected as a path, the malicious attackers can discard, or modify data traffic.

(D) *Sybil attack* on UAANET could be used to disable all links related to a UAV or a set of UAVs. Here an attacker can attempt to send multiple control packets using different identities. The adversary node could use random identities or the identity of another node to create confusion in the routing process.

(E) In Byzantine attack, a compromised intermediate node or a set of compromised intermediate nodes carry out attacks such as creating routing loops, forwarding packets through non-optimal paths, or selectively dropping packets. These actions result in the disruption or degradation of routing services.

Implementing blockchain technology and distributing the contents to a large number of nodes and making it nearly impossible for hackers to attack. Blockchain significantly reduces the chance of data changed by unauthorized parties [10]. By using blockchains to protect the data, a system can ensure that it's invulnerable to hackers unless every single node is simultaneously wiped clean. Some companies are already implementing blockchain in this area to prevent attacks from occurring. To destroy or corrupt a blockchain, a hacker would have to destroy the data stored on every UAV in the network, each one having a copy of some or all the data. It is not possible that hacker downs an entire network simultaneously.

5 Research Issues and Current Projects

The technology behind blockchain is versatile and incredibly useful for the future of the Internet, allowing users to better secure their data. Innovative uses for blockchain technology are already becoming popular in different fields by implementing rigorous encryption and data distribution protocols on a network, any business can ensure that their information will remain safely intact and out of the reach of hackers [11, 12]. Various researches and startups are moving in this area for securing the various type of network using blockchain.

5.1 UAV and Blockchain Together: Projects Around World

The list of projects is growing day by day for blockchain in drone applications around the world. In the initial stages for both blockchain and drone technologies, there are a limited number of projects in stages of development. In [13] basic requirements for air networking and communications are explored with mobility, coordination, and control and also found the need for certification issues on multiple UAVs to deploy in airspaces. Here are examples of some popular projects as the merging of blockchain and drone technologies.

Drone Package Delivery: Chronicled is working to empower the security of drones connected via the blockchain. They gave a model for safe package delivery via drone with microchips that give delivery drones a unique identity on the blockchain, which UAV applications use to give. If permission is confirmed, the door opens, and delivery can be completed.

Drone Delivery: The Dorado Platform is serving as a food delivery platform of Foodoutgroup.com. Dorado's launched a prototype beyond food delivery to deliver anything the customer wants and transform the way goods move around cities by enabling anyone to have anything delivered on-demand.

Drone Flight Right-of-Way Grants: (Avigation Easements Rights and Ownership) ("AERO") proposed a distributed ledger system for making navigation easy.

Package Tracking System: Walmart uses blockchain technology to track packages delivered by unmanned drones titled "Unmanned Aerial Delivery to Secure Location."

6 Conclusion

UAVs and blockchain both are getting popular very fast. To destroy or corrupt a blockchain, a hacker would have to destroy the data stored on every UAV in the network, each one having a copy of some or all the data. It is not possible that hacker downs an entire network simultaneously. Use of blockchain technology in UAANET over the cloud and distributing the contents to a large number of nodes and making it nearly impossible for hackers to attack. Blockchain significantly reduces the chance of data changed by unauthorized parties. In this paper, we studied the overview and architecture for securing hostile areas UAVs. We explored various attacks in UAANET and examined various advantages of blockchain that can be used to improve the communication between the flying vehicles and the ground terminal. This paper is focused on secure data exchanged between the components of the UAANET. There are still some loopholes for the security of blockchain itself that will be treated in our future works.

References

1. Maxa JA, Mahmoud MSB, Larrieu N (2017) Survey on UAANET routing protocols and network security challenges. Ad Hoc and Sensor Wireless Network
2. Gupta L, Jain R, Vaszkun G (2016) Survey of important issues in UAV communication networks. IEEE Commun Surv Tutor 18(2):1123–1152
3. Vey Q, Pirovano A, Radzik J, Garcia F (2014) Aeronautical ad hoc network for civil aviation. In: communication technologies for vehicles, Springer, 2014
4. Zyskind G, Nathan O, Pentland A (2015) Decentralizing privacy: using blockchain to protect personal data. In: 2015 IEEE security and privacy workshop, San Jose, CA, 180–184 2015
5. Dorri A, Kanhere SS, Jurdak R, Gauravaram P (2017) Blockchain for IoT security and privacy: the case study of a smart home. In: IEEE Percom workshop on security privacy and trust in the internet of thing
6. Ekblaw A, Azaria A, Halamka JD, Lippman A (2016) A case study for blockchain in healthcare: the metric prototype for electronic health records and medical research data
7. Hartenstein H, Laberteaux KP (2008) A tutorial survey on vehicular ad hoc networks. Commun Mag IEEE 46(6):164–171
8. Maxa JA, Mahmoud MSB, Larrieu N Survey on UAANET routing protocols and network security challenges. https://hal-enac.archives-ouvertes.fr/hal-01465993
9. Li X, Jiang P, Chen T, Luo X, Wen Q A survey on the security of blockchain systems. ScienceDirect Futur Gener Comput Syst
10. Puthal D, Malik N, Mohanty SP, Kougianos E, Yang C The blockchain as a decentralized security framework. In: White paper
11. Salman T, Zolanvari M, Erbad A Security services using blockchains: a state of the art survey, IEEE
12. Dinh TTA, Wang J, Chen G, Liu R, Ooi BC, Tan K-L, Block Bench: a framework for analyzing private blockchains. In: ACM international conference on management of data (SIGMOD'17), ACM, New York, NY, USA, 1085–1100 2017
13. Namuduri K, Wan Y, Gomathisankaran M Mobile ad how networks in the sky: state of the art, opportunities, and challenges. In: Proceedings of the second ACM MobiHoc workshop on Airborne networks and communications, ACM, 25–28 2013
14. Lin IC, Liao TC (2017) A survey of blockchain security issues and challenges. Int J Netw Secur 195(5):653–659

REFERENCES

Placement Optimization of Surveillance Cameras: Visibility Analysis

Divyasree Gaju and Deva Pratap

Abstract India's focus on smart urban environments opens an opportunity to evaluate the applicability of technological advancements. This also however brings new challenges home, with safety being the foremost. There is dire need to monitor common public spaces in urban setup with surveillance cameras to address safety concerns. One of the main issue that is faced in camera surveillance is choosing locations to ensure the maximum possible coverage. This was achieved through visibility analysis of urban environment. UAV data provides higher spatial resolution and hence detailed information is obtained from it. Such a data is especially useful in urban areas where there is a need for precise analysis of the urban landscape for various applications. Better inclusion of vegetation is expected with elevation models and point cloud data which is generally neglected in the crude 3D models. Dense point cloud which was obtained from the UAV data and its surface reconstruction was used for analysing visible spaces. Analysis of visibility maps has been done from the isovists generated for each vantage point (camera placement) with ray tracing method.

Keywords UAV · Visibility analysis · Point cloud · Surveillance camera · Isovists · Ray tracing

1 Introduction

1.1 UAV

Unmanned Aerial Vehicles have made technological advancements in almost all the arenas of the engineering like archeological documentation, hazard monitoring, asset mapping, earth observation and particularly 3D surveying [1] and its applications.

D. Gaju (✉) · D. Pratap
NIT Warangal, Warangal, India
e-mail: divyasreegaju@student.nitw.ac.in

D. Pratap
e-mail: deva@nitw.ac.in

K. Jain et al. (eds.), *Proceedings of UASG 2019*, Lecture Notes in Civil Engineering 51,
https://doi.org/10.1007/978-3-030-37393-1_12

Urban monitoring is one such application where precise information needs to be obtained and analyzed for the efficient functioning of any administrative unit. Data acquired by a UAV provides such accuracy because of the high-resolution sensors that could be carried on it [2].

1.2 Visibility Analysis

Visibility analysis gives information about spatial relationships between any two points in either two dimensional or three-dimensional space based on Line of Sight (LoS) concept [3]. It is generally performed on a city model i.e., a geometric model that was constructed from two dimensional and elevation data. The analysis tests if an LoS (ray) intersects any obstacle present in its way [4]. Precise 3D city modeling generally uses the LiDAR data. Despite the accuracy that is obtained from such a data, the city models are crude because of their exclusion of many details like the vegetation, temporary obstructions etc. And hence such a model developed is far from reality [5]. Often, obtaining LiDAR data is not economic and hence the aerial photographs obtained can be used to generate the elevation models using photogrammetric methods. Such models offer far simpler alternative to the active point clouds obtained from the LiDAR sensors [6].

1.3 Camera Surveillance

With fast growing urban landscapes across India, it is very important to look for economic ways to monitor and cater to the needs of such areas. Urban safety is one such aspect which needs constant surveillance of urban environments [7]. Camera surveillance is one of the best options and for this purpose a thorough understanding of the landscape is necessary. This not only is difficult, considering the number of towns and cities that exist in the country, but is also costly. And hence choosing the optimum number of cameras and their locations becomes important [8].

1.4 Isovists in Visibility Analysis

An isovist is the visible space volume from any given point in space (2D/3D). Visibility maps generated using such isovists for all the probable viewpoints will help in analyzing the visibility of an area. Volume of view plays a major role in deciding the significance of the vantage point chosen, in this case, camera location. Locations should be chosen such that there is minimum overlapping visible space for any two given points [7].

2 Study Area and Datasets

2.1 Study Area

The study area chosen for this project is a small area of an urban local body called Umred from Nagpur district of Maharashtra, India. It is a densely populated town and has high density urban settlements. It is located at 20° 51' 14 N and 17° 19' 29E, at an elevation of 290 meters above sea level (Figs. 1 and 2).

Fig. 1 Political map of Maharashtra showing Umred

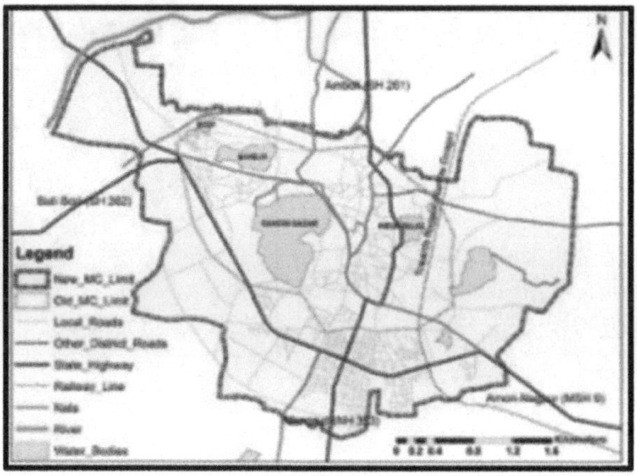

Fig. 2 Administrative map of Umred

3 Methodology and Data Preparation

Based on objectives and literature review, the following methodology has been adopted for present work as shown in the flowchart (Fig. 3). A detailed explanation is given in the section below.

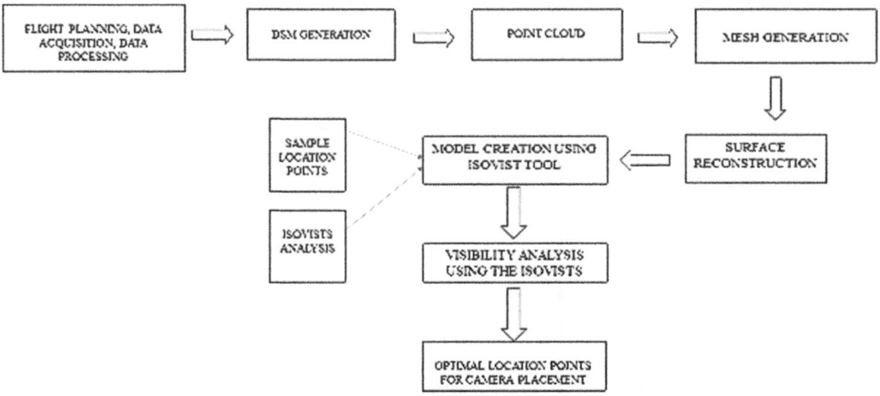

Fig. 3 Flowchart of methodology followed in the present study

Fig. 4 Orthomosaic of the sample area

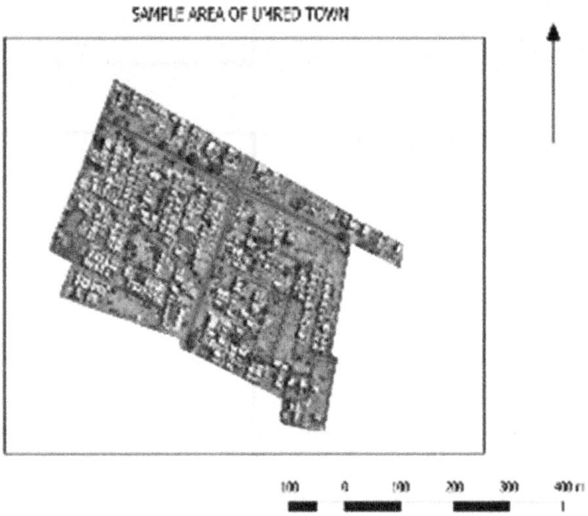

3.1 Flight Planning and Image Acquisition

Flight planning consists of the preparation of a navigation map for the area of choice along with the parameters like camera specifications, side lap, end lap, scale, flying height, yaw, pitch and roll. UAV is then flown and geotagged images are acquired with 6 cm resolution.

3.2 Image Processing

The images so obtained are processed using an image processing software. The first step requires the removal of unwanted oblique images. The locational data of every image in a csv format is obtained after flight. The images are then renamed as per in the csv file. The next step requires for the images to be referenced to the location. For this, the photos are added, the csv is imported, camera calibration values are given as input, and the projection is specified. The workflow involves aligning photos, and batch processing the set of operations that are to be performed.

3.3 DSM and Orthomosaic Generation

A dense point cloud which has high quality and mild depth filtering mode is generated using the tie points. Digital elevation model and subsequently Digital Surface model is generated using the dense point cloud. And finally the Point cloud, DSM and Orthomosaic are exported. Digital Surface Model obtained from the stereo images of a region depicts the near reality of the terrain details. The passive cloud allows us to construct a mesh and reconstruct the surface after that.

3.4 Data Preparation for the Sample Area

A small portion of the town has been chosen because of the computational constraints. The Orthomosaic (Fig. 4) and Digital Surface Model (Fig. 5) were trimmed and a point cloud simplification with a sample size down to 1/3rd of the original point cloud size was done (Fig. 6). For the resampled points, normals are computed (Figs. 7 and 8).

With the Screened Poisson Surface Reconstruction method [9], mesh of octree depth 12 and its surface was reconstructed (Figs. 9 and 10).

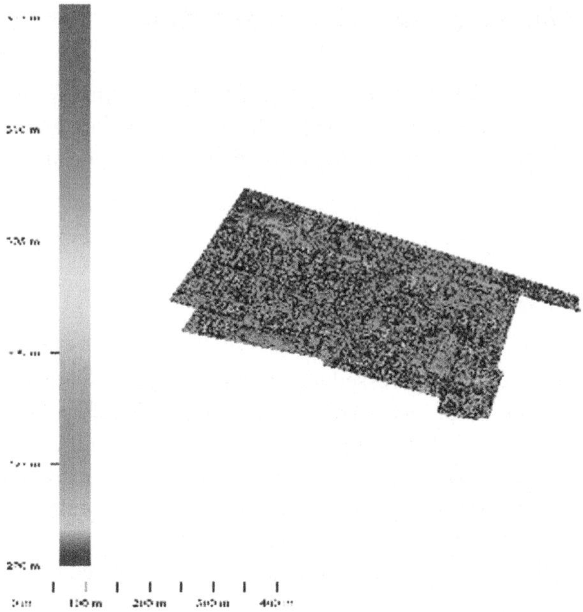

Fig. 5 DSM of the sample area

Fig. 6 Point cloud of the Crossroads in the sample area

3.5 Sample Camera Locations

Cameras locations are given at crucial locations like road junctions, street corners, public places, building corners etc. These were manually chosen from the orthomosaic of the sample area and were created in QGIS as a layer with height, id, description and coordinates as attributes. A total of 158 locations were proposed at a spacing

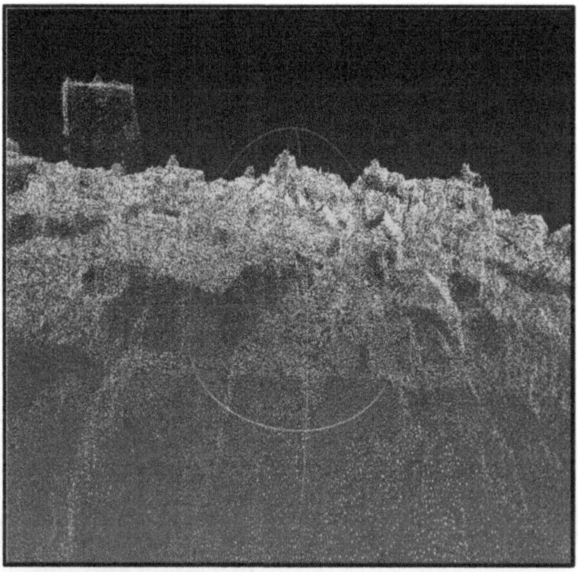

Fig. 7 Simplified and noise-free point cloud

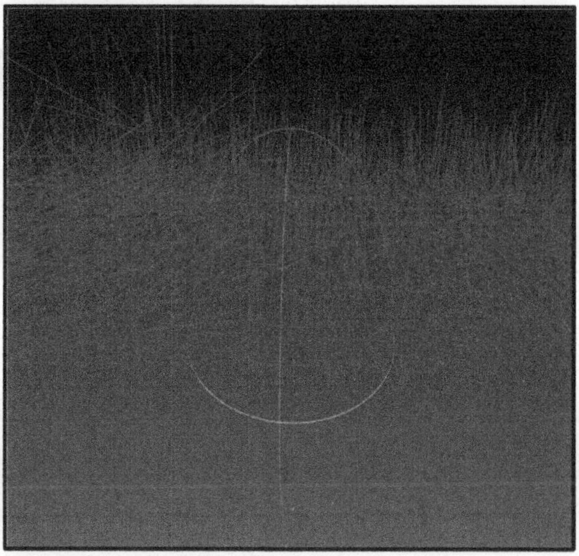

Fig. 8 Normals of the simplified point data

of approximately 30 m according to the camera specifications. Most of the outdoor surveillance cameras have 30–40 m range, with 100–140° wide angle (Fig. 11).

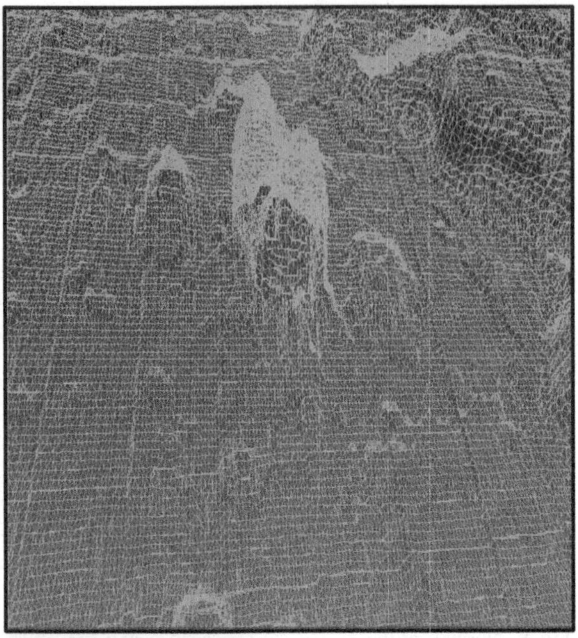

Fig. 9 Mesh with octree depth 12

Fig. 10 Surface generated from the mesh

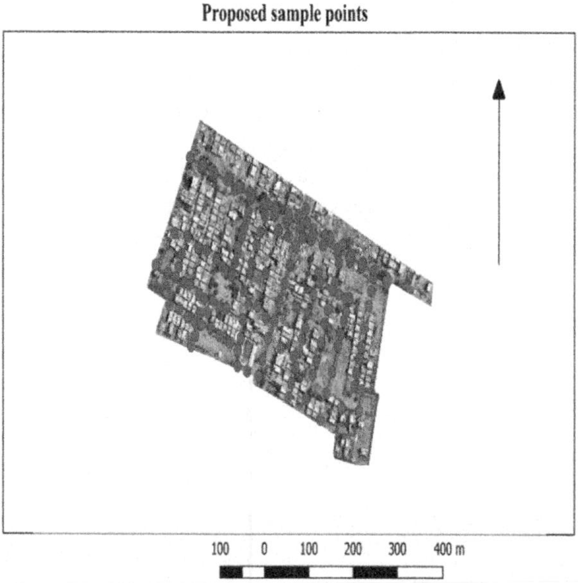

Fig. 11 Proposed sample locations for camera placement

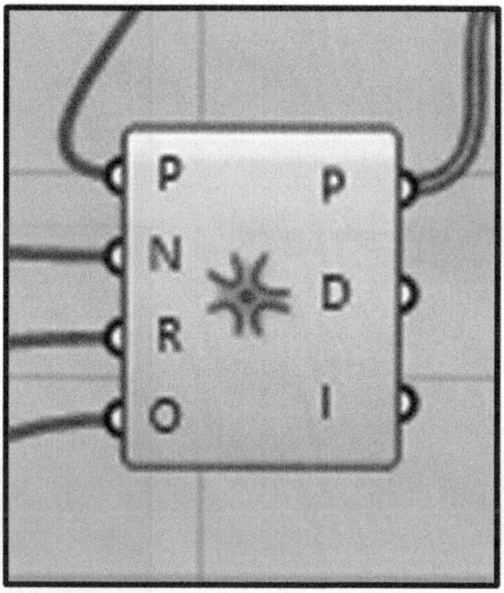

Fig. 12 Isovist tool from Grasshopper

3.6 *Model Creation and Application*

Most of the research so far in this arena uses complicated algorithms to solve the camera placement problem using visibility analysis. Isovists are visible areas from a vantage point. Proposed camera locations are the vantage points that were chosen and isovists for each point was generated using a model created in the Grasshopper interface of Rhino software. Rhino is a high end CAD software and Grasshopper is a virtual scripting language interface which is embedded in rhino. An inbuilt tool called isovists (Fig. 12) is available in Grasshopper using which the model was created. This tool is based on the concept of ray-tracing. From the vantage point, a user defined number of rays are projected and they are intersected by the input obstacle. The points where the rays intersect are noted as locations as a list (Fig. 13).

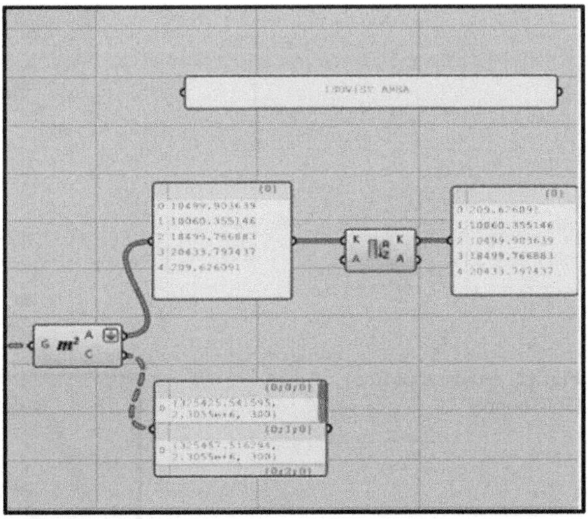

Fig. 13 Areas as a list

Fig. 14 Boundary surface for a point

Fig. 15 Proposed camera locations at crossroads in the sample area

3.7 Visibility Map Generation

Boundary surfaces (Fig. 14) for each vantage point are converted to shapefiles. These shapefiles which are representing the Visibility maps that consist coverage areas for all the proposed camera locations were imported to QGIS. Intersection was performed for those coverage area layers, this adds a new field in the attribute table. This is later queried for the areas less than 10% of the input coverage areas. This results in those Visibility map shapefile layers having less than the threshold. The corresponding camera locations are finalized.

4 Results

Figures 15 and 16 show the proposed and optimized locations at a crossroads in the chosen area. Out of 20 proposed locations in this particular crossroads junction, 8 locations have been identified as optimal for the surveillance camera placement. The final optimized number of locations are 111 (Fig. 17) out of 158 locations (Fig. 12). The visibility maps for the final camera locations are shown in Fig. 18.

5 Conclusion

In this study, the application of isovists in visibility analysis for surveillance camera placement has been reviewed. An investigation of geometric permutations of isovists that result from different combinations of isovist origins and targets was done. The preparation of target is one of the main steps of this project which determines the

Fig. 16 Optimized camera locations at crossroads in the sample area

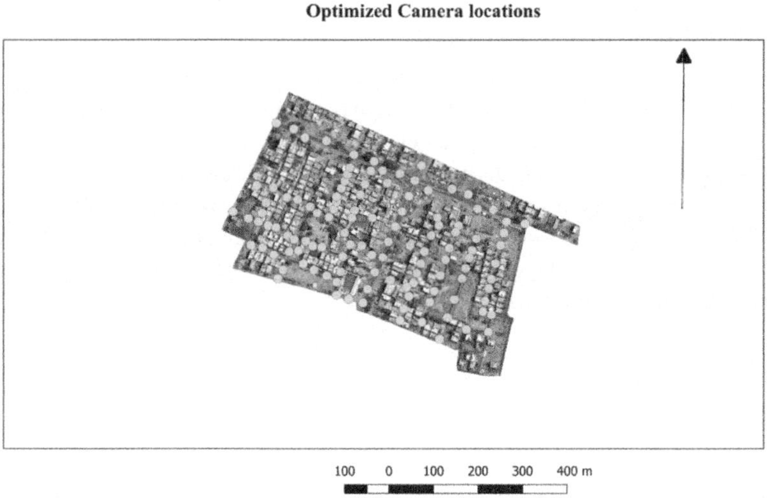

Fig. 17 Final optimized camera locations

accuracy of the Visibility maps. The results obtained for the sample area taken show that from 158 proposed locations, the number of optimized locations obtained were 111 and hence there has been a **reduction of 29% in number of camera locations**. This study has been done in the hope of optimizing the locations of camera placement for reducing the economic burden on the urban administration. The work proves to reduce manual work and hence saves time and workforce.

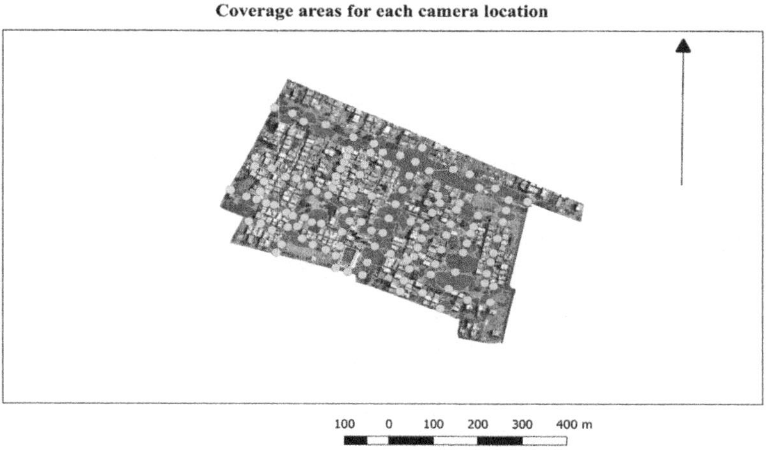

Fig. 18 Visibility maps for the final optimized camera locations

Acknowledgements The authors would like to express their gratitude to Terra Drone Ind pvt ltd for providing required input for completion of this project. The authors would like to thank Dr. Peter Bus whose valuable lessons available on the open platforms online helped us in understanding important concepts. These important concepts helped us in adopting the methodology used in the work. The authors also thank all the faculty members and fellow people for their valuable suggestions.

References

1. Shakhatreh H, Sawalmeh A, Al-Fuqaha A, Dou Z, Almaita E, Khalil I, Othman NS, Khreishah A, Guizani M (2018) Unmanned aerial vehicles: a survey on civil applications and key research challenges, pp 1–58
2. Koeva M, Muneza M, Gevaert C, Gerke M, Nex F (2018) Using UAVs for map creation and updating. A case study in Rwanda. Surv Rev 50: 312–325. https://doi.org/10.1080/00396265.2016.1268756
3. Hagstrom S, Messinger D (2011) Line-of-sight analysis using voxelized discrete lidar. Laser radar technology application XVI, vol 8037, p 80370B. https://doi.org/10.1117/12.884049
4. Koltsova A, Tunçer B, Schmitt G (2013) Visibility analysis for 3D urban environments. Model Comput Hum Factors-Vol. 2-Comput Perform eCAADe 31(2): 375–384
5. Czyńska, K (2015) Application of Lidar data and 3D-city models in visual impact simulations of tall buildings. In: The international archives of photogrammetry, remote sensing and spatial information sciences, vol 40, ISPRS Arch, 1359–1366. https://doi.org/10.5194/isprsarchives-xl-7-w3-1359-2015
6. Romero LF, Cervilla AR, Mérida M, Vías J, Tabik S (2016) Total 3D-viewshed map: quantifying the visible volume in digital elevation models. Trans GIS 21:591–607. https://doi.org/10.1111/tgis.12216
7. Lonergan C, Hedley N (2016) Unpacking isovists: a framework for 3D spatial visibility analysis. Cartogr Geogr Inf Sci 43:87–102. https://doi.org/10.1080/15230406.2015.1065761
8. Yabuta K, Kitazawa H (2008) Optimum camera placement considering camera specification for security monitoring. In: IEEE international symposium on circuits and systems, vol 2. 2114–2117. https://doi.org/10.1109/iscas.2008.4541867
9. Kazhdan M, Hoppe H (2013) Screened poisson surface reconstruction. ACM Trans Graph 32:1–13. https://doi.org/10.1145/2487228.2487237

References

Application of Unmanned Aerial Vehicle (UAV) for Damage Assessment of a Cultural Heritage Monument

Ekta Baranwal, Poonam Seth, Hina Pande, S. Raghavendra and S. K. P. Kushwaha

Abstract Disaster is the immense disruption of growing affair of the world, whether it is natural or due to human factors. Earthquake is the most demolished natural disasters which affect mankind in all aspects. One of the censorious disputes after an earthquake is assessing the damage of the area for the salvage of humankind, civilization, and restoration of the loss. Cultural heritages are the ones which is significant to history, gregarious, architecture, and science, and arises the need to preserve. In the context of conservation of architecture and heritage application, geoinformatics system is the decisive and efficient tool for monitoring, detecting, and assessment of architectural health of any built heritage. UAV photogrammetry is one of the recent techniques in image based remote sensing to do these tasks because of its various advantages over Terrestrial Close-Range Photogrammetry (CRP), aerial photogrammetry, and sometimes over non-image-based techniques of remote sensing. This paper presents the post-earthquake damage assessment of "Sulamani Pagoda" from Bagan, Myanmar which is an ancient city of Myanmar with archaeological belt or zone of 13 km × 8 km area and settled in rustling earthquake zone. In the year 2016, on 24th August, a hefty earthquake hit central Myanmar and did extensive damage in Bagan where about 400 temples were eradicated, in which Myauk Guni (North Guni), and Sulamani temple were severely damaged. After the

E. Baranwal (✉)
Department of Remote Sensing, Banasthali Vidyapith, Rajasthan 304022, India

P. Seth
Geoweb Services, IT and Distance Learning Department, Indian Institute of Remote Sensing, ISRO, Dehradun, Uttarakhand 248001, India
e-mail: poonam@iirs.gov.in

H. Pande · S. Raghavendra · S. K. P. Kushwaha
Photogrammetry and Remote Sensing Department, Indian Institute of Remote Sensing, ISRO, Dehradun, Uttarakhand 248001, India
e-mail: hina@iirs.gov.in

S. Raghavendra
e-mail: raghav@iirs.gov.in

K. Jain et al. (eds.), *Proceedings of UASG 2019*, Lecture Notes in Civil Engineering 51,
https://doi.org/10.1007/978-3-030-37393-1_13

earthquake 15 m tall tower portion of Sulamani Pagoda was crushed through the top of the east passageway. Unwrapping and classification of the point cloud is used to discriminate the damaged portions of the Pagoda and also to analyse and quantify the amount of damage occurred on the architectural heritage site.

Keywords UAV photogrammetry · Cultural heritage health · Damage assessment · Point cloud · Classification

1 Introduction

The prerequisite for detection and preservation of cultural heritage is well understood and professionals endeavour is to exploit any possible technique to consummate this aim. Several Literatures and documents are published related to the cultural heritage documentation. Unmanned Aerial Vehicle (UAV) are being utilized in may applications due to many reasons like its accessibilities, cost efficiency, high resolution and operational safely. A review paper focused on the significant issues for applying UAV images successfully [1, 2]. Another review paper highlighting the efficient application of drone in disaster mitigation. The author in this paper has presented his experiences for a rapid damage assessment for pre, post and after the earthquake scenario [3].

A research work was carried out on medieval S. Agostino church of central Italy for damage documentation and construction which was hit with three big earthquakes. In this work 3D aerial and terrestrial models were obtained using multi temporal perspective. Finite Element Modelling (FEM) was done to analyse the structure [4]. 3D reconstruction of the structures post-earthquake can be performed to actually access the damage portion [5]. Efficient point cloud classification can be done to identify the damaged portions [6]. A research carried out to assess the damage due to kumamoto earthquake, which occurred in japan. Using Lidar, Optical and SAR data and comparing with the GNSS observation, ground surveys and seismic records of the area [7] Road damage extraction is also done for post-earthquake damage assessment [8]. Damage due to Hurricanes can also be assessed with the help of UAVs [9]. Earthquakes generally cause structural damages and produces debris and rubble. Debris may include sediments, vegetation or relocated property. HOG was proved to be more efficient for textural analysis to identify the damage [10]. A paper discusses the application of satellite images and high-resolution drone images disaster reduction and relief responses. A disaster which occurred in Sichuan province caused extensive damage to roads and housing, which resulted in landslides and mountain collapse images were interpreted for quantification of the damage caused by the earthquake [11]. A research to rapidly develop 3D structures post wildfire and to document and assess the damage was done in [12]. Similar few researches have been carried out for 3D documentation of the cultural heritage structures [13], 3D documentation using lidar and photogrammetry was done for a cultural heritage using CAD [14].

UAV nadir images have limitations of high façade information. Object based image analysis is combined with 3D point cloud assessment of facades and roofs [15]. For damage assessment of taller buildings and the top portion are easily assessed with the help of UAV application post disaster [16, 17].

Documentation of any cultural heritage is usually done by recording the state of an object and area, before and after an event. Damage assessment of the cultural heritage is the part of this post-event documentation where the changes in the physical state of that object and area are analyzed to make various decisions regarding its conservation.

Literature shows numerous works based on damage assessment through photogrammetry as well as from Lidar data. These techniques for damage assessment are categorized as:

- Image based
  - Photogrammetry
  - IR Camera
- Non-image based
  - Terrestrial survey (Traditional)
  - Laser Scanner
- Combinative Methods
  - Photo-laser scanner
  - Structured Light.

While assessing the damage of ancient monuments, their significance and importance should be considered. Because it's the significance and importance of the monument which helps us to decide that, at which level it should be taken care of. Assessing the damage of these moments defines the class of its damage which helps to make decision for its conservation, whether it can be reconstructed or not. If it can't be reconstructed, then at least we can conserve it in the form of 3-D model or movie.

The main objective of this work was to evaluate damage of a cultural monument using post-earthquake UAV based images which is achieved through unwrapping and classification of the point cloud of the monument.

2 Study Area and Data Set

In this paper, the study area is "Sulamani Pagoda" from Bagan, Myanmar. Bagan or Burmese, is the ancient city of Myanmar whose archaeological belt or zone is spread over the 13 km × 8 km area. 10,000 plus Buddhist temples, pagodas (stupas) and monasteries were built and shaped in the Burmese plains in the midst of 11th and thirteenth centuries. Mostly temples are situated in old Bagan, also known as Temple City, one of the among world's most noteworthy archaeological sites which is also

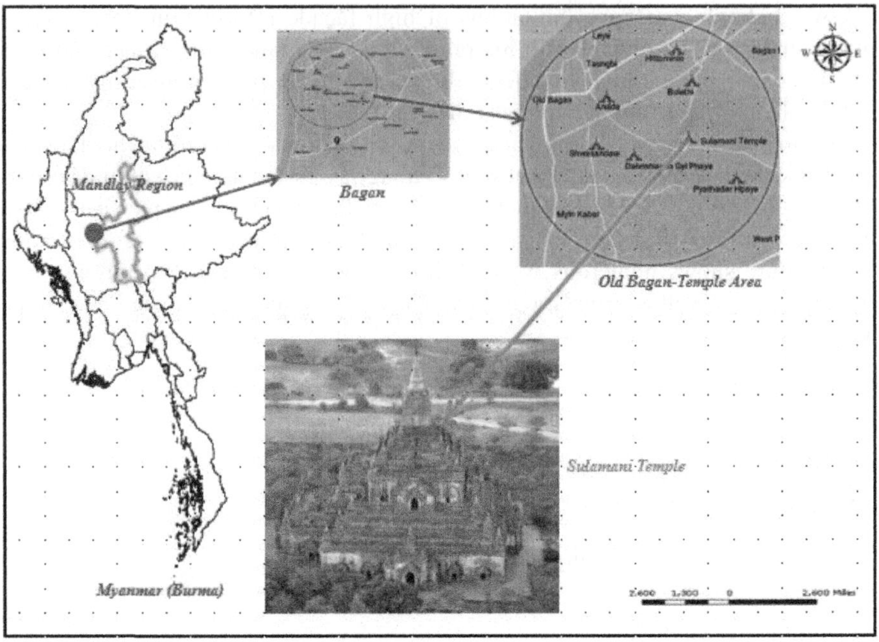

Fig. 1 Study area location

settled in rustling earthquake zone, and had endured through several seismism over the dimension of time. The location of the study area is shown in the Fig. 1.

In the year 2016, on 24th August, a hefty earthquake hit central Myanmar again and did extensive damage in Bagan. About 400 temples were eradicated, in which Myauk Guni (North Guni), and Sulamani temple were severely damaged due to this earthquake. After the earthquake 15 m tall tower portion of Sulamani Pagoda was crushed through the top of the east passageway as shown in Fig. 2.

Total 150 Unmanned Aerial Vehicle (UAV) RGB images were acquired at 90 m of flying altitude. Which were used to produce ultra-dense point cloud on which further analysis of damage assessment for the temple was done.

3 Methodology

When the data is collected, elementary processing of the data took place. This process results the products which are used for the next level of processing to achieve the purpose of the project. Processing of UAV images is not easy even processed through a computer program. Since these images are very high in resolution, they took much time to process and this processing time also depends on the density of the generated point cloud i.e. how dense the point cloud we required for our work. The primary

Fig. 2 Ground image of Sulamani temple: **a** Pre earthquake **b** Post-earthquake. *Source* Google images

output obtained after processing UAV images is the point cloud but in most of the applications, this point cloud is not used directly for further analysis instead 3-D model, DEM, and orthomosaic photo-generated through it, are utilized for analysis. The complete workflow is represented thru the flow diagram (Fig. 3). Initial part of

Fig. 3 Flow chart

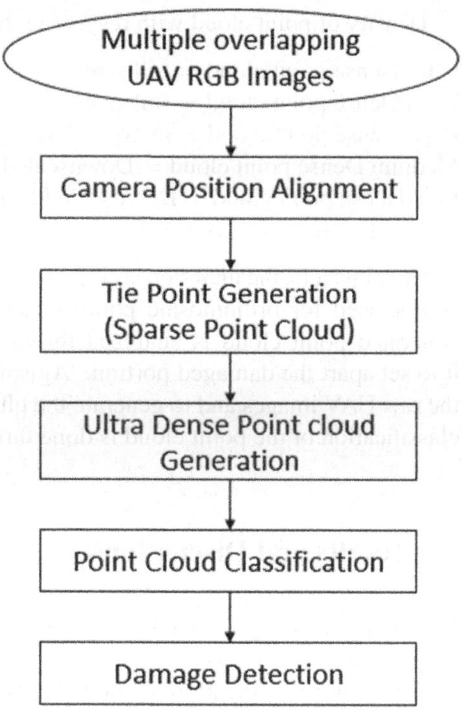

this methodology was by the processing of UAV images and here 150 UAV images are processed to obtain ultra-dense point cloud of the monument.

In general, 3D modernization and grayscale thresholding are 2 promising techniques that can process the images to make them much more facile to interpret. 3D reconstruction consists of utilizing a program to stitch together numerous images and the algorithm comprises of many phases which are as follows:- Firstly, the images prerequisite to be examined for key points. These are facets of the image with a discrete structure or feature contrasting anything on the image, conventionally an edge. This will avail the algorithm attain a logic of in what manner the images overlap (60% of side overlap + 80% of forward overlap is required). This is gone through each image by examining the pixel colors and culling distinct groupings. Next, these key points are compeer, and then sorts through the culled key points of each image and lined up with matching images. Error evaluation is performed for each match to ascertain the points that are identically tantamount, and any erroneous matches are rejected. This efficaciously matches all of the images together; though, the product is still in 2D. Then, the pictures are initially analyzed for deepness predicated on focal length. Images will have diverse calibres of pellucidity or focus predicated on the settings of the camera. By examining the pellucidity of apiece image, a comparative depth can be specified to each position on the image. This engenders a rough outline of the deepness across the entire surface. This comparative depth is then calculated horizontally which engenders a more precise model of the complexity and point cloud of images relative to the edifice holistically.

Density of point cloud with respect to the size of raw images used to create it, is:

Ultra Dense point cloud = full size
Ultra Dense point cloud = full size
High Dense point cloud = Downscaled to 50%
Medium Dense point cloud = Downscaled to 25%
Low Dense point cloud = Downscaled to 12.5%
Lowest Dense point cloud = Downscaled to 6.75%.

Conclusively, the final step is creating triangulation or mesh which is then textured and/or used for orthomosaic photo generation and/or for DEM generation. This generated point cloud is analyzed through unrolling it and afterward classifying it to set apart the damaged portion. 'Agisoft Photoscan' software is used to process the raw UAV images and to generate the ultra-dense point cloud while unrolling and classification of the point cloud is done through software 'Cloud Compare'.

4 Results and Discussions

UAV images were processed and generated point cloud of the monument is shown in Fig. 4.

Automatic classification of point cloud comprises of two stages. Firstly, the dense cloud is isolated into specific size cells. In every cell the absolute bottom point is

Fig. 4 Ultra-dense point cloud of Sulamani temple post-earthquake

recognized whose triangulation gives the primary estimation of the 3D model of terrain. Secondly, new points are added to this ground class, such as it fulfils two conditions: laying inside a specific length of space from the landscape model and the angle of terrain model with the associated line of this new point and a ground class point should be less than a specific angle.

The result of this automatic classification may or may not be acceptable for different applications. The alternate classification of dense point cloud can be done manually for the better result according to the need of various applications. Same procedure is also opted for this research. The unwrapped point cloud of the Pagoda (Fig. 5a) is classified manually to separate the damaged part of it (Fig. 5b). Different types of point cloud can be unrolled on the basis of geometry, either cylinder or cone. The unrolling can be done on different axis of orientation of the point cloud i.e. x, y and z axis, along the center of the axis. Even the radius along which the unwrapping is to done and the point on the axis along which the unwrapping is to be done or the software generates the axis along the point cloud gravity center can be specified. The result of this classification says that approximately 40% of the points from the entire point cloud of the monument are classified as damage.

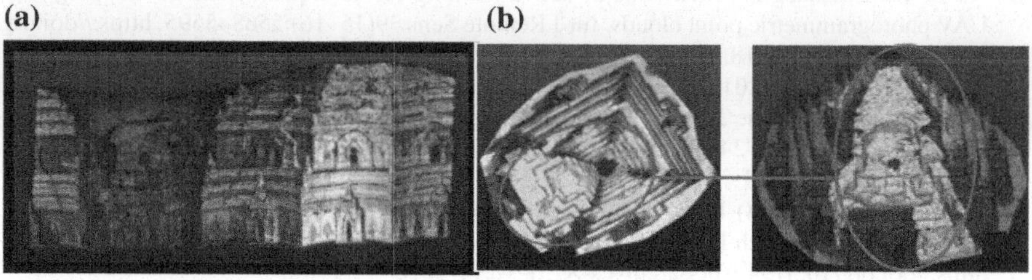

Fig. 5 **a** Unwrapped point cloud of Sulamani temple post-earthquake, **b** Classified point cloud of Sulamani temple post-earthquake: yellow portions are damaged part of the temple and debris

5 Conclusion

Through this research the authors would like to convey that remote sensing techniques can be very much efficient in quantification of any damaged portions. Which are not easily assessable by other means without near physical contact. The point cloud data produced from UAV images has provided with very precise actual geometry information of the cultural heritage site post-earthquake. Which has helped to identify the damaged portion distinguish. Authors believe that having 3D documentation of all the present valuable cultural heritage site can help in restoration of the structure post any disaster or calamity.

Recommendation and Future Scope With the help of temporal point cloud data of any cultural heritage site can prove to be a crucial asset in precise quantification of damaged portion.

With of help of pre and post disaster or calamity datasets, automatic quantisation approaches can be made more reliable than manual segmentation of damaged portions.

References

1. Shahbazi M, Théau J, Ménard P (2014) Recent applications of unmanned aerial imagery in natural resource management. GIScience Remote Sens 51(4):339–365. https://doi.org/10.1080/15481603.2014.926650
2. Calantropio A, Chiabrando F, Sammartano G, Spanò A, Losè LT (2018) UAV strategies validation and remote sensing data for damage assessment in post-disaster scenarios. Int Arch Photogramm, Remote Sens Spat Inf Sci-ISPRS Arch 42(3W4):121–128. https://doi.org/10.5194/isprs-archives-XLII-3-W4-121-2018
3. Restas A (2015) Drone applications for supporting disaster management. World J Eng Technol 03(03):316–321. https://doi.org/10.4236/wjet.2015.33c047
4. Grazzini A, Chiabrando F, Foti S, Lingua AM, Spano AT (2018) Damage assessment and seismic vulnerability analysis of S. Agostino Church in Amatrice. May 1–12. Retrieved from: https://www.researchgate.net/publication/325302055
5. Cusicanqui J, Kerle N, Nex F (2018) Usability of aerial video footage for 3-D scene reconstruction and structural damage assessment. Nat Hazards Earth Syst Sci 18(6):1583–1598. https://doi.org/10.5194/nhess-18-1583-2018
6. Xu Z, Wu L, Zhang Z (2018) Use of active learning for earthquake damage mapping from UAV photogrammetric point clouds. Int J Remote Sens 39(15–16):5568–5595. https://doi.org/10.1080/01431161.2018.1466083
7. Yamazaki, F, Liu W (2016) Remote sensing technologies for post-earthquake damage assessment: a case study on the 2016 Kumamoto earthquake. In: ASIA conference on earthquake engineering (6ACEE), 22–24. Retrieved from https://www.researchgate.net/profile/F_Yamazaki/publication/307351403
8. Chen Z, Dou A (2018) Road damage extraction from post-earthquake uav images assisted by vector data. Int Arch Photogramm Remote Sens Spat Inf Sci-ISPRS Arch 42(3):211–216. https://doi.org/10.5194/isprs-archives-XLII-3-211-2018
9. Adams S, Friedland C, Levitan M (2010) Unmanned aerial vehicle data acquisition for damage assessment in hurricane events. In: 8th international workshop on remote sensing for disaster management, 1–7 Sep 2014. Retrieved from http://www.enveng.titech.ac.jp/midorikawa/rsdm2010_pdf/13_adams_paper.pdf

10. Ghaffarian S, Kerle N (2019) Towards post-disaster debris identification for precise damage and recovery assessments from UAV and satellite images. XLII(June):10–14
11. Li C, Yuan X, Zhang J, Dulin Mi P, Li Z (2018) Earthquake damage monitoring and assessment based on high-resolution remote sensing images-take Lushun earthquake as an example. In: International conference on geoinformatics, 1–4 June 2018. https://doi.org/10.1109/GEOINFORMATICS.2018.8557152
12. Macchiarella ND, Robbins J, Cashdollar D (2019). Rapid virtual object development using photogrammetric imagery obtained with small unmanned aircraft systems -applications for disaster assessment and cultural heritage preservation. (January). https://doi.org/10.2514/6.2019-1974
13. Hassani F (2015) Documentation of cultural heritage techniques, potentials and constraints. Int Arch Photogramm Remote Sens Spat Inf Sci-ISPRS Arch 40(5W7):207–214. https://doi.org/10.5194/isprsarchives-XL-5-W7-207-2015
14. Kushwaha SKP et al. (2020) 3D Digital documentation of a cultural heritage site using terrestrial laser scanner—a case study. In: Ghosh J, da Silva I (eds) Applications of geomatics in civil engineering. Lecture notes in civil engineering, vol 33. Springer, Singapore
15. Fernandez Galarreta J, Kerle N, Gerke M (2015) UAV-based urban structural damage assessment using object-based image analysis and semantic reasoning. Nat Hazards Earth Syst Sci 15(6):1087–1101. https://doi.org/10.5194/nhess-15-1087-2015
16. Eudossiana V, Planning C (2008) Uav application in post—seismic environment. XL 4–6 Sep 2013
17. Dominici D, Alicandro M, Massimi V (2017) UAV photogrammetry in the post-earthquake scenario: case studies in L'Aquila. GeomatS, Nat Hazards Risk 8(1):87–103. https://doi.org/10.1080/19475705.2016.1176605

Conceptual Design and Comparative CFD Analyses on Unmanned Amphibious Vehicle for Crack Detection

R. Vijayanandh, M. Senthil Kumar, S. Rahul, E. Thamizhanbu and M. Durai Isaac Jafferson

Abstract Nowadays, Unmanned Amphibious Vehicles (UAVs) is being suggested for critical applications like fish detection, difficult mapping and atmosphere studies. One such filed are surface cracks inspection and in specific a big structure like dams might suffer a lot due to external structural cracks. As dams are being considered the cracks not only exist over the water surface it might also exist underwater, to inspect such cracks an UAV carrying with sensors is required. The major unique feature applied in this underwater vehicle is a common material for both the environments with the help of Kevlar composite. Generally, impact load withstanding capability is high in the Kevlar, which shortlisted Kevlar as suitable for an amphibious vehicle. For the purpose of inspection, image processing photogrammetric method is followed, in which the external surface of the dam is captured by a high definition camera which can take pictures in both air and in water is used. The pictures are then stitched together to get a complete pictorial view of the external surface of the dam. From the surface is further inspected to detect cracks and other structural defects. The entire image processing has been analyzed with the help of MATLAB, in addition to that the photogrammetric method is also suggested for crack detection. For the theoretical design approach of a UAV has been carried out for the purpose of components selection and thereby the conceptual design of UAV is modeled with the help of CATIA V5. The integrated working environment have been analyzed with the help of ANSYS Workbench 16.2.

Keywords Inspection · Propeller · Amphibious vehicle · Underwater · Photogrammetric · Cracks detection

R. Vijayanandh (✉) · M. Senthil Kumar · S. Rahul · E. Thamizhanbu · M. Durai Isaac Jafferson
Department of Aeronautical Engineering, Kumaraguru College of Technology, Coimbatore, Tamil Nadu, India
e-mail: vijayanandh.raja@gmail.com

K. Jain et al. (eds.), *Proceedings of UASG 2019*, Lecture Notes in Civil Engineering 51,
https://doi.org/10.1007/978-3-030-37393-1_14

1 Introduction

The fundamental study and research about Unmanned Vehicles (UVs) are emerging today due to its huge implementation in critical applications such as quick transportation, surveillance, explosive detection, crack detection, etc. Finalization of applications/working environments is the primary task involved in the UVs, which provide the view of Unmanned Aircraft configuration, control methodologies, selection of UVs components. Generally, three types of environmental working conditions based UVs are available in Aerospace industry to execute the critical applications, in which the three types are Ground environment based UVs, Aerial working conditions based UVs and water-based UVs. Each and every Unmanned Aircraft has its unique maneuvering in order to complete the mission in a successful manner [1]. In this paper proposes a design followed by numerical simulation of an UAV that can be flying on the air like the helicopter, move through the water in the forward motion. Determination UAV's requirement, Specifications of UAV, Conceptual design of UAV and its numerical simulation are the main four steps involved here for successful completion. The output of the final numerical results will provide the details about the withstanding capability of UAV for both water and air environments.

1.1 Unmanned Amphibious Vehicle (UAV)

The UAV is a type of advanced Unmanned Aircraft, which can able to maneuver both air and water in order to execute integrated applications such as crack detection on a dam, fish detection, etc. The amphibious environment is quite difficult due to its different working environments so the research on the design and analysis of UAV is emerged in everywhere. One of the important design considerations in UAV is to tackle both the environments without affecting the performance of a UAV also the major drawback in the UAV is unable to provide suitable material for both water and air environments, but this problem is to be solved with the help of implementation of Kevlar composite [2].

Determination of multirotor performance's parameter is one of the important in the literature survey, this will guide the entire process in the right manner and thereby the output may have high efficiency. Vehicle's Speed, total weight, range, endurance, configurations, and payload are the important performance parameters analyzed in the literature section. Nowadays, the general design approach of a UAV is completed with the help of a trial and error method. Trial and error method of UAV design processes is consisting of fixed mass or fixed speed. In this paper, the components of a UAV and its dimensions is approximately finalized by fixed speed condition as well as fixed mass condition. The maximum speed of a UAV is fixed as 10 m/s and the mass of UAV component has been fixed as 2 kg. The working performance and its dimensions have been derived with the inclusion of fixed mass (2 kg) and fixed speed (10 m/s) also working fluids [3].

1.2 Objective

The main objective of this paper is to design the safe and secure Unmanned Amphibious Vehicle in order to survive both water and air environments for the purpose of critical surveillance. In this paper, the numerical simulation of flow analysis using Ansys Workbench 16.2 and structural health monitoring system using image processing methodologies have been used in order to provide safety and security to the UAV, which can able to survive at both environments.

2 Design Methodologies Involved in UAV

2.1 Introduction

The important step by step procedures involved in the design of a UAVs is design requirements, optimized functional allocation, design synthesis and evaluation, and finally validation. An optimized UAV, with a minimum of undesirable side effects, requires the application of an integrated life-cycle oriented "system" approach [4]. Design methodologies involved in this paper are mainly depended on conceptual design, preliminary design and detail design of a UAV based on systems engineering approach.

Preliminary Design

UAV maximum take-off weight, propulsion system, propeller dimensions and its thrust producing capability are the primary parameters estimated in the preliminary design phase. In preliminary design, also comprises of few other non-important UAV parameters such as UAV size, UAV zero-lift drag coefficient and UAV maximum lift coefficient [4].

Conceptual Design

The main aim of conceptual design is to build up and characterize the activities of specific design to requirements for the UAV. The output of such activities is integrated and included in a UAV specification. The primary tool in this stage of design is the "selection" [4]. In conceptual design, problem definition and its clear definition support to start the design of the UAV in a good way. Also, with the help of design requirements, the configurations of UAV is generated in which different types of evaluation and analysis are taking place in this process with very low amount calculation. The 3-D view of UAV is the consolidated output of conceptual design, which must contain the fundamental components.

Detailed Design

Major components of UAVs and its subordinate parameters are evaluated in the detailed design. Generally, most of the operations involved in this stage are executed with the help of iterative methodology. Performance evaluation, stability evaluation, controllability analysis and flight simulation are the four criticism plays a vital role in the detailed design stage of a UAV. Determination of the zero-lift drag coefficient of UAV is the primary process must be included in the UAV performance evaluation. The requirements of the stability analysis are the estimation of all the components weight of the UAV and its cumulative center of gravity. For the successful mission, the UAV has to perform well in the perspective of maneuvering, which perfectly designed in the controllability analysis operational stage [4].

2.2 UAV Requirements—Important Parameters for UAV

High stiffness to weight ratio, lightweight, covert, user-friendly, good in electric connection, secure and easy operable in maneuvering are the primary characteristic need to be considered for integrated missions. For successful UAV's construction, the study about aircraft layout is an important one, which might comprise of aerodynamics, hydrodynamics, propulsion, flight control, image processing, autonomy, payload and system integration. The estimation of individual components weight and thereby the overall weight of the UAV are the primary requirements which play a major role in the selection of a propulsive system [5]. One of the difficult tasks involved in the UAV's construction is the selection of components because of its direct linkage with UAV's mission completion with high efficiency. Each and every component of UAV have internally linked with each other so in order to make a decision about the final configuration of each UAV, the designer must be fully aware of the function of each component.

2.3 Design of an Advanced UAV

The design process can able to provide the easiest representation of all the components of UAV in the optimized perspective. In general, the structure of aircraft have more complex than other prototypes so design plays a difficult role in this paper also the process of the model completion is executed with the help of CATIA, which is a unique tool for aerospace. UAV has to efficiently operate for both environments, in which the efficiency is depending upon the structural design and the relationship between fluid mediums with UAV's outer boundary. The outer design of the UAV's outer boundary intentionally constructed as streamlining body shape for drag reduction purpose hence another design considering the spline operation and structural stress study are completed using CATIA [6]. Figure 1 show the typical (top and bottom view) design modeler images of UAV in the isometric projection.

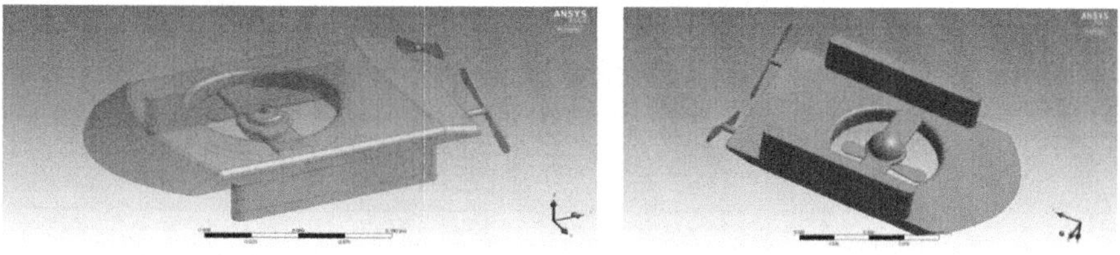

Fig. 1 Typical isometric view of UAV

2.4 Composite Material

In order to withstand both the environments, the suggested UAV has to overcome impact load, which is imposed by external fluids such as air and water. Generally, impact load has more dangerous than normal external load so ordinary materials fail to withstand at extreme conditions. Hence composite material is the only solution for this problem; in which Kevlar fiber mingled with epoxy resin have the capability to withstand high impact load [7]. The major drawback in the amphibious vehicle is common material selection for both the working environments. In order to solve the material based issue, in this paper suggested Kevlar composite which has the primary property of high impact load withstanding capability. Hence Kevlar is best to alternate for an amphibious type of vehicle, especially while in the critical water environments. The overall mechanical properties of composite material, which is Kevlar, the density of 1470 (kg/m^3), Poisson's ratio of 0.33, Young's Modulus (GPa) of 75, Bulk Modulus (GPa) of 73.529 and Shear Modulus (GPa) of 28.195.

3 Computational Analysis Results

The major aim of this numerical simulation is to predict the drag force and side forces on the UAV in order to check the suitability of the design for both the cases. Overall work is comprised of two cases, which are Vertical Take-off and Landing (VTOL) and forward motion for both water and air. Ansys Workbench 16.2 is a primary tool worked for engineering parameters estimation with the help of advanced work stations. The numerical simulations are mainly depending on the perfect conversion process of a physical model into a finite element model, so in this work 3-D tetrahedral is selected because of its unique coverage [8]. The entire physical model of UAV is captured by more than 5 lakhs tetrahedral elements for the purpose of the efficient outcome.

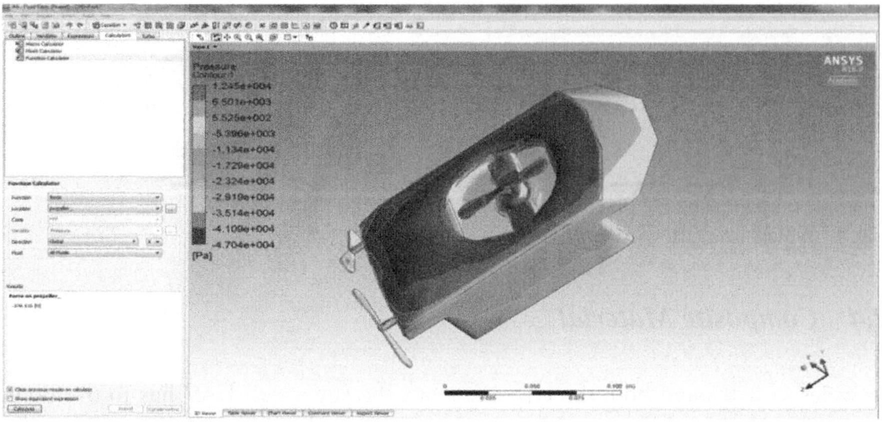

Fig. 2 Drag force on unmanned amphibious vehicle

3.1 Hydrodynamic Analysis

Numerical analysis of UAV to predict the behavior of hydrodynamic is carried out for two different cases, which are 5 and 10 m/s. The maximum rate of climb of UAV is fixed as 10 m/s so input fluid velocity have been finalized, in order to study low-speed behavior roughly 5 m/s also considered as one of the input velocities.

5 m/s in Vertical Take-Off and Landing (VTOL)

The drag force on UAV is predicted, which is revealed in Fig. 2. The construction material planned for this paper is Kevlar so estimated drag value is the acceptable limit.

Figure 3 shows the typical velocity variations on the UAV and the Fig. 4 reveals the pressure distribution on UAV for the velocity of 5 m/s.

10 m/s in VTOL

Figure 5 shows the typical velocity variations on the UAV and Fig. 6 reveals the pressure distribution on UAV for the velocity of 10 m/s.

3.2 Aerodynamic Analysis

Comparatively, the opposing force is quite low in case of air than water so the maximum working velocity of VTOL maneuvering of UAV is fixed as 15 m/s in the aerodynamics conditions. A comparative study will provide more confidence in the analysis so in aerodynamic analysis 5 and 10 m/s also considered as velocity inputs [9, 10].

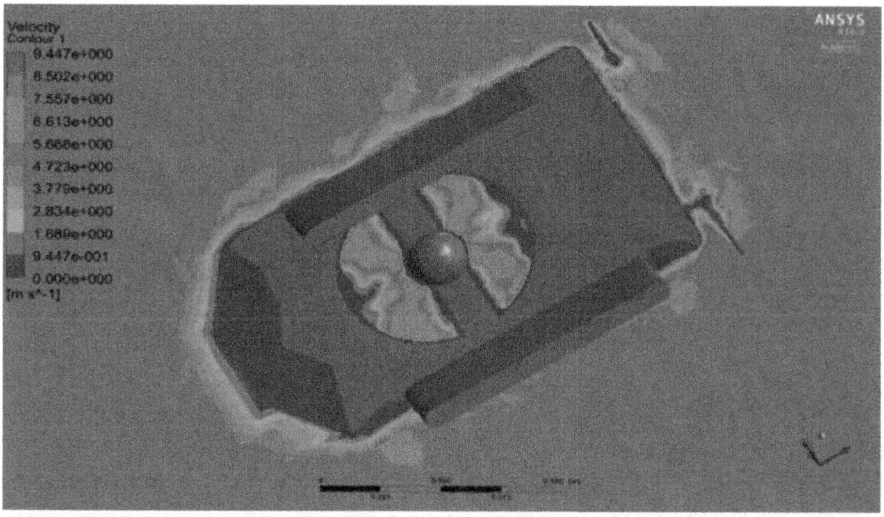

Fig. 3 Typical view of velocity variation

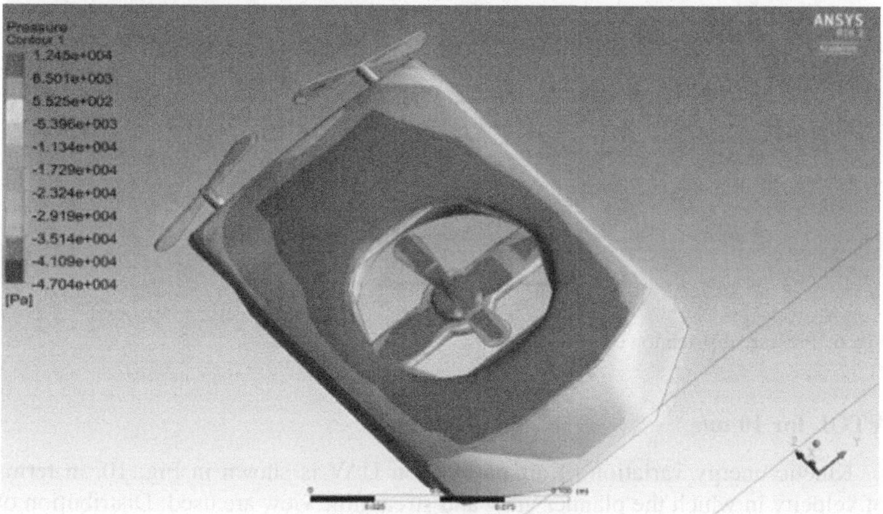

Fig. 4 Typical view of pressure distribution

VTOL for 5 m/s

Typical view of velocity representation on UAV is shown in Fig. 7 and the pressure distribution of UAV is revealed in Fig. 8.

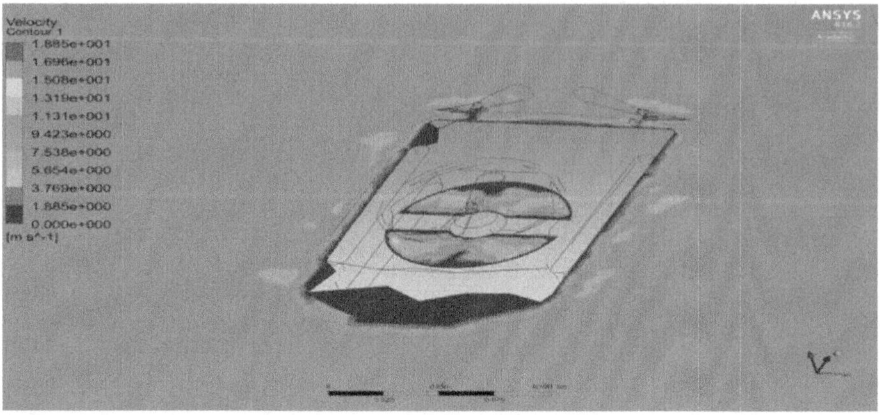

Fig. 5 Velocity variation on UAV

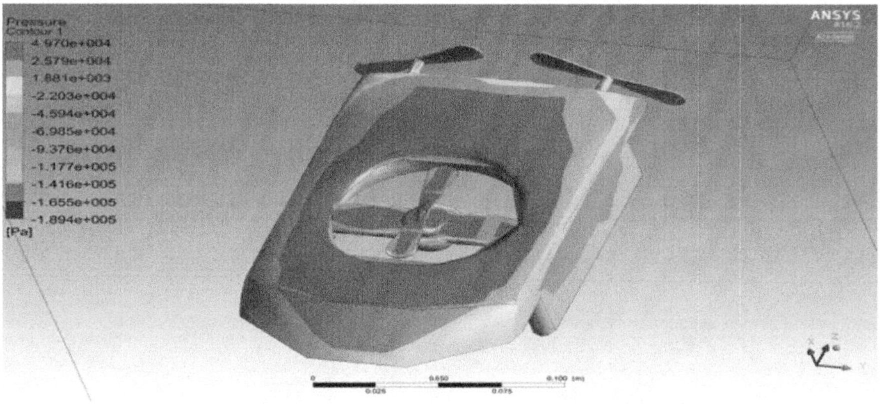

Fig. 6 Pressure distribution on UAV

VTOL for 10 m/s

Kinetic energy variation of air particle on UAV is shown in Fig. 10, in terms of velocity in which the planner view and streamline view are used. Distribution of potential energy on UAV with the respect to given input conditions and working fluid is shown in Fig. 9. The opposing force will finalize the performance of a vehicle so study about drag force is mandatory because of this mandatory, this paper also deals the drag estimation which is shown in Fig. 11.

VTOL 15 m/s

Typical view of velocity representation on UAV is shown in Fig. 12 and the pressure distribution of UAV is revealed in Fig. 13.

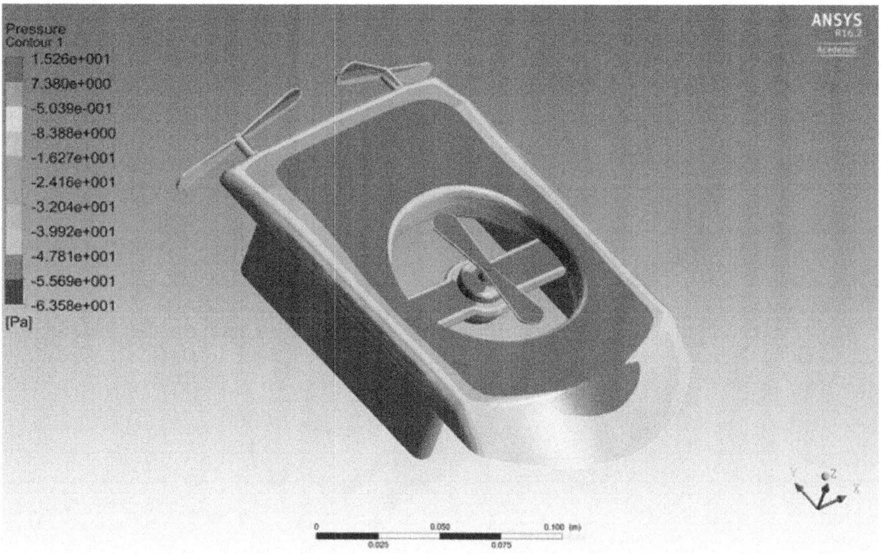

Fig. 7 Pressure distribution on UAV

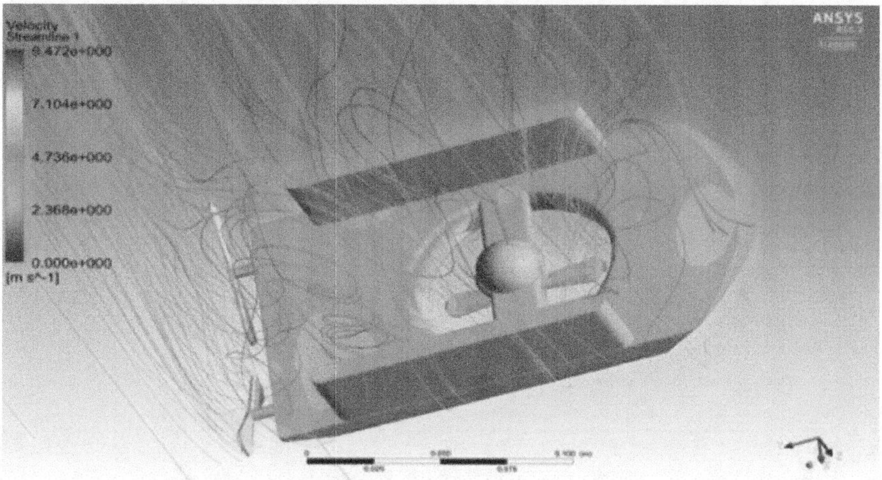

Fig. 8 Velocity streamlines

3.3 Comparative Analysis

The forces acting on the UAV are noted for both environments, which are listed in Table 1 and 2. The pressure distributions, velocity variations and force estimation for different velocities of water medium are listed in Table 1. Table 2 contains the

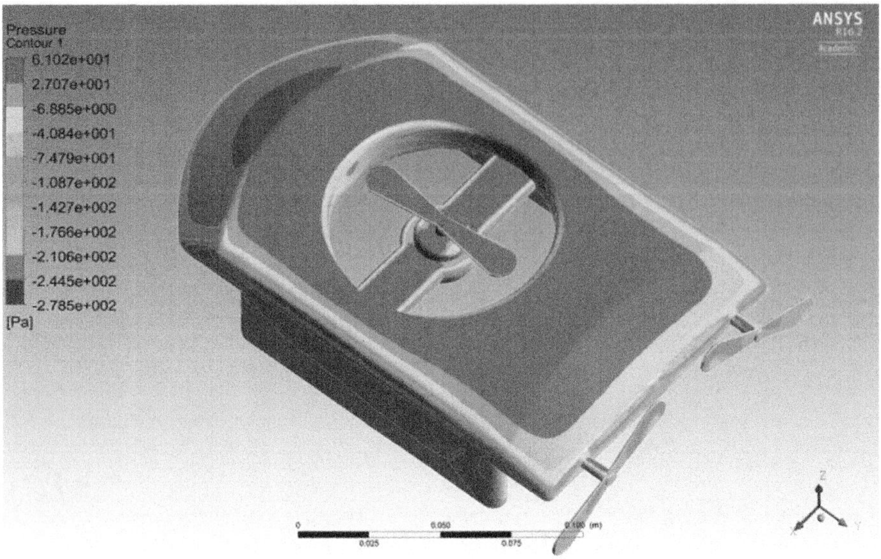

Fig. 9 Pressure distribution on UAV

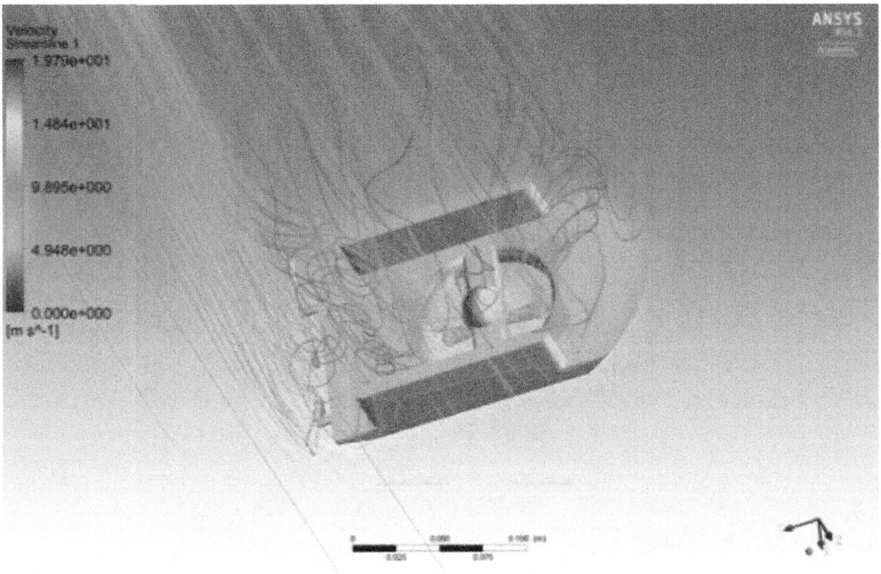

Fig. 10 A typical velocity variation view on UAV

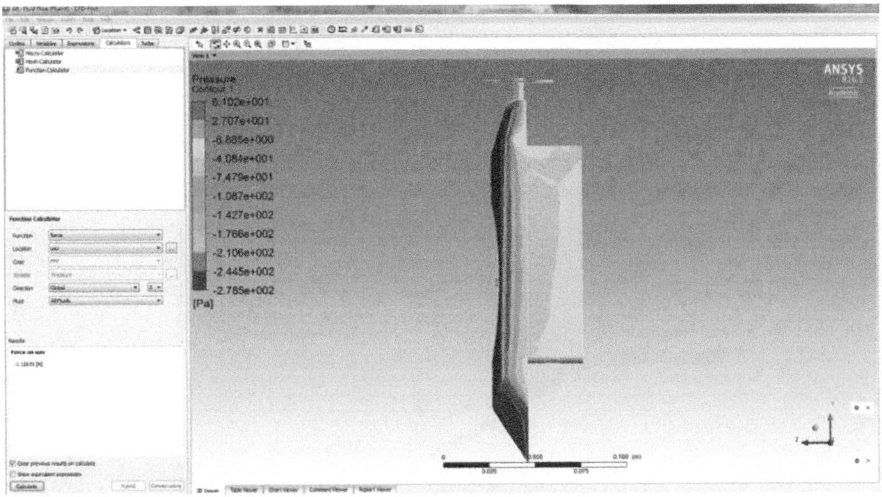

Fig. 11 Drag force on unmanned amphibious vehicle

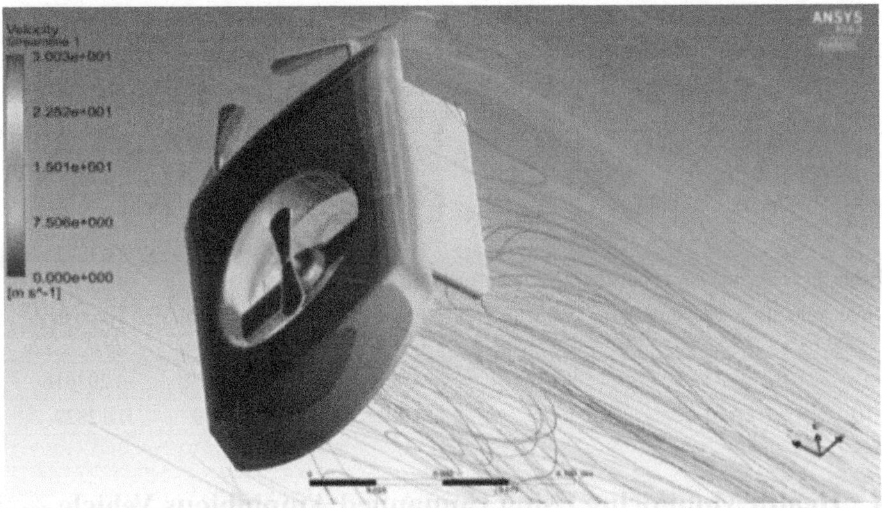

Fig. 12 A typical velocity variation view on UAV

flow properties and side forces of air medium on UAV while underwent different velocities such as 5, 10 and 15 m/s.

Fig. 13 Pressure distribution on UAV

Table 1 Flow analysis results of UAV in water medium

Inlet Velocity (m/s)	Pressure (Pa)	Velocity (m/s)	Force in X (N)	Force in Y (N)	Force in Z (N)
5	12,450	9.45	278.516	1.90	26.5316
10	49,700	18.85	1113.71	6.0980	104.787

Table 2 Flow analysis results of UAV in air medium

Inlet velocity (m/s)	Pressure (Pa)	Velocity (m/s)	Force in X (N)	Force in Y (N)	Force in Z (N)
5	15.26	9.472	0.00187615	0.0375478	0.293616
10	61.02	19.79	0.00152683	0.168849	1.12829

4 Health Monitoring Using Unmanned Amphibious Vehicle

4.1 Crack Detection on Dam Using Image Processing

Nowadays smaller advanced integrated UAVs are more comfortable than conventional UAVs for critical applications. With the help of dynamic detection techniques, intelligent video surveillance templates matching, the complex fault detection at an extremely critical location such as crack at the dam, crack at a wind turbine, etc. are estimated using advanced integrated UAVs. After a literature study, the suitable method is shortlisted for crack detection on the dam using UAV, which is nothing but cracks detection based on image processing. The flow chart of the crack detection algorithm used for this work is shown in Fig. 14. Also, MATLAB is used here for image processing in which the reference and input images are internet based sources.

Case 1: Result for crack detection

Case 1 discuss the image processing results and their prediction capture of a 2D surface with crack and reference images, which are taken from the internet (Figs. 15 and 16) for algorithm verification. This image processing includes more advanced crack detection algorithm to solve real-time crack detection with the inclusion of curved object detection. Advanced crack detection algorithm has been computed with the help of MATLAB.

Input Images

Result image

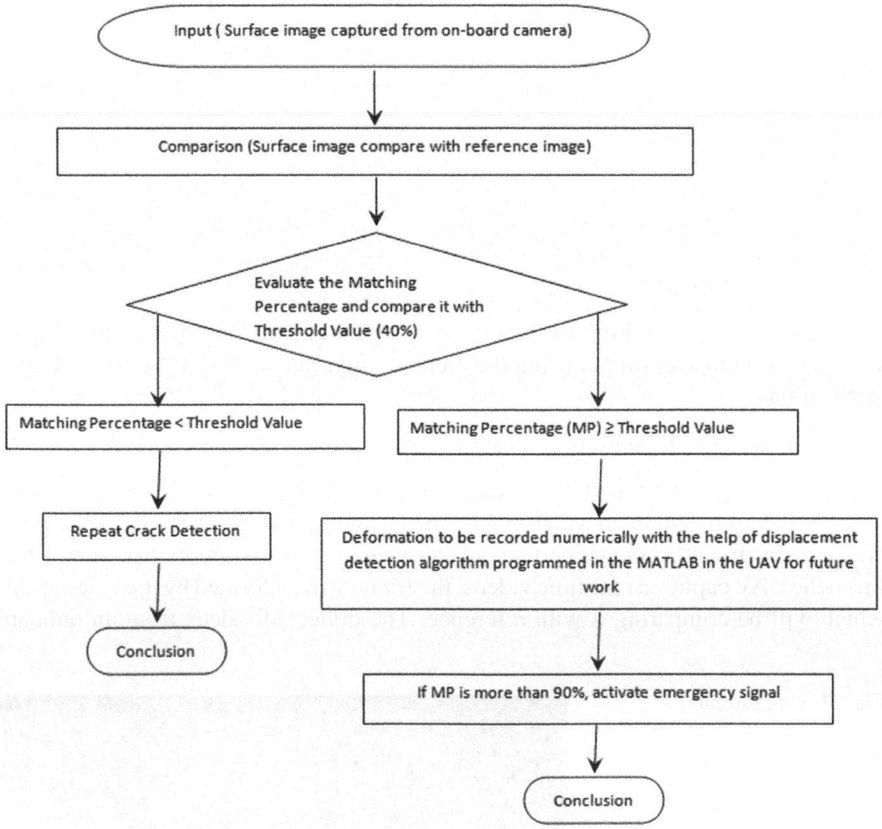

Fig. 14 Flow chart of a detection algorithm

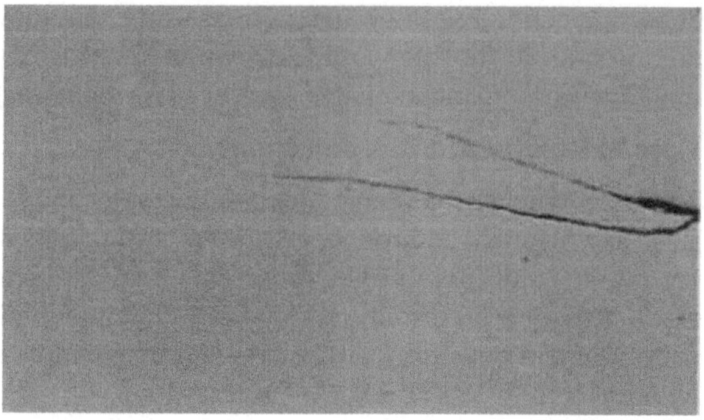

Fig. 15 Surface defect image

Fig. 16 Surface reference image

Figure 17 is the resulting image of crack detection for a different surface image with and without crack input using the method grayscale in MATLAB for real-time applications.

Case 2: Result for video importing crack detection

This work also proposed that the cracks are effectively detected by the vision-based navigation system, in which crack sensing methodology, the recording process, data transferring process and perfect detection algorithm are important components. From the UAV captured real-time videos, the frame will be formed by the video splits, which will be comparing it with reference. The collected videos from an onboard

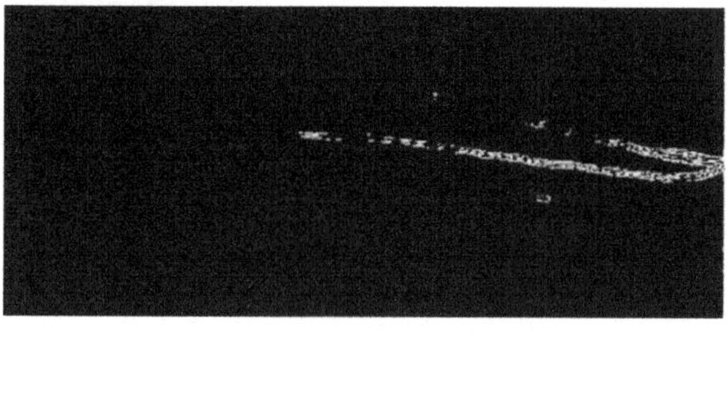

Fig. 17 Crack accuracy result

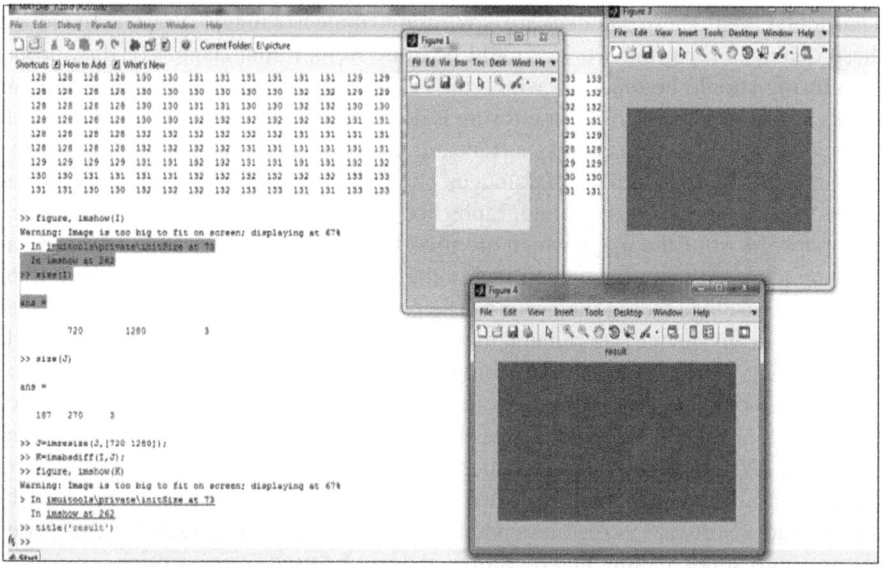

Fig. 18 Video importing crack detection

camera will be collected in a ground controller station and which will be image processed by Higher level image processing algorithm in MATLAB [11].

Case 2 is an advanced section, which deals with the vision-based navigation technique. In this case, crack on the surface is captured from the camera, which is transmitted to the algorithm to detect the modification on the boundary of the surface. All the inputs and the results are combinable shown in Fig. 18.

Case 3: Crack Detection on Dam using Photogrammetry

Dams are the big structures which are built across freshwater sources to store water, in which the size of a dam depends upon the water source's width, depth, water flow speed, storage requirements these vary the size and design of dams. A basic function of a dam is to withstand the hydrostatic pressure distributed over its structure by the river water. The design of a dam is made is such a way that the reaction force of the dam opposes the pressure from the water. To withstand a high pressure a dam must possess large reaction forces which are obtained by having a good structural strength. In general, the side that faces the water is known as the upstream side and the opposite side as the downstream side, in which the pressure act continuously over the walls of the upstream side [12]. Due to such external loads the structure of a dam experience damages, which leads a structural failure. The structural damage of a dam is catastrophic to prevent its regular monitoring of the external structure should be done. Most of the structural damage in dams could be seen externally. The damages won't suddenly result in a structural collapse. First, the damage will be in the form of a crack or a rupture of the outer surface. If these damages are not considered and left without repairing it result in bigger cracks of breakage of a particular area. These damages are slowly contribute to a bigger disaster. To prevent these damages repairing works should be done in its initial stage. For this, proper monitoring should be made by which the cracks and damages can be detected and works are done as per it. The monitoring is done mostly made by a photogrammetric method in which pictures of the surface are taken and then inspection is done by a human to locate the cracks and damages [13]. Photogrammetry is a process where inputs are submitted as the photographs and later on processed as the 3D models and maps. Most of the maps which are used by us nowadays are captured using an airplane attached with the camera facing the ground. The camera is attached to the wings and fuselage facing the ground. The photos which are given as the inputs can be clicked in any device like camera, mobile phones and can be Go pro. In these work, we are going to use Go pro as it is going to click pictures both in the atmosphere as well as in the water. Also, we are going to integrate the Photogrammetry as crack detection tool by capturing photos of the dam surface with overlap camera placed over the UAV. After the photos are clicked then combined using software's used for Photogrammetry, finally data inspected using visual by checking the changes in the dam surface. Software is recommended to use is Pix4D can be used for converting pictures into single picture and digitalize and can be used for inspection.

5 Conclusion

The conceptual design of an advanced UAV is modeled with the help of CATIA with as much as low dimensions in order to execute all the critical applications. The engineering parameters are estimated using Ansys Workbench 16.2 for UAV model for forwarding speeds of 5, 10 and 15 m/s of air medium. Similarly the hydrodynamic analysis also executed for the input velocities of 5 and 10 m/s with the help of Ansys Fluent 16.2. The results are plotted and compared for both case environments for VTOL operation. From the pressure variations graph, side forces result understood that the values are within the limit in the Kevlar UAV for both the environmental conditions. The material optimized UAV has planned to equip with an innovative system for health monitoring of different structures, which is based on computational crack detection measurements. With the help of CATIA, ANSYS and MATLAB, the numerical study on crack detection of different structures have been carried out successfully.

References

1. Raj Kumar G et al (2017) Conceptual design and structural analysis of integrated composite micro aerial vehicle. J Adv Res Dyn Control Syst 9((Sp–14)):857–881
2. Michelson RC et al (2010) Overview of micro air vehicle system design and integration issues, Georgia Tech Research Institute, Georgia Institute of Technology, Atlanta, GA, USA, pp 1–12.

https://doi.org/10.1002/9780470686652.eae401
3. Esakki B et al (2018) Design of amphibious vehicle for unmanned 3 mission in water quality monitoring using IoT. Sensors 18:3318. http://dx.doi.org/10.3390/s18103318
4. Sadraey M (2010) A systems engineering approach to unmanned aerial vehicle design. In: 10th AIAA aviation technology, integration, and operations (ATIO) conference, 13–15 Sep 2010, Fort Worth, Texas
5. Collins KA (1993) A concept of unmanned aerial vehicles in amphibious operations. Doctoral dissertation, Monterey, California. Naval Postgraduate School
6. Vijayanandh R et al (2017) Design, fabrication and simulation of hexacopter for forest surveillance. ARPN J Eng Appl Sci. 12(12):3879–3884. ISSN: 1819-6608
7. Pisanich G, Morris S (2002) Fielding an amphibious UAV: development, results, and lessons learned. In: Digital avionics systems conference, 2002. Proceedings. The 21st IEEE, vol 2, pp 8C4–8C4
8. Vijayanandh R et al (2017) Numerical study on structural health monitoring for unmanned aerial vehicle. J Adv Res Dyn Control Syst 9(6):1937–1958
9. Wang Z-J et al (2018) Conceptual design of a water-air amphibious unmanned vehicle. In: 2nd international conference on modeling, simulation and optimization technologies and applications (MSOTA 2018) ISBN: 978-1-60595-594-0
10. Frejek M, Nokleby S (2008) Design of a small-scale autonomous amphibious vehicle. In: Electrical and computer engineering, 2008. CCECE 2008. Canadian conference on IEEE, pp 000781–000786
11. Wei D et al (2018) Foreign maritime unmanned vehicle development status and trend. J Ordnance Equip Eng 39(7):33–35
12. Li X, Wang Z (2003) Unmanned underwater vehicle technologies. Mar Electr Technol 6:12–15
13. Liu F (2014) Conceptual design and sailing stability study on water air amphibious unmanned ship, Master degree thesis of Jiangsu University of Science and Technology

Conceptual Design and Optimization of Flexible Landing Gear for Tilt-Hexacopter Using CFD

R. Vijayanandh, P. Kiran, S. Indira Prasanth, G. Raj Kumar and S. Balaji

Abstract The lifetime issue due to structural failure in the advanced multi-rotor Unmanned Aerial Vehicle (UAV) is one of the prime factors towards the reduction of UAV's usages. The scope of this work is to safeguard the hexacopter components that are working in adverse conditions, in which retractable landing gear plays a vital role in the safety production. Implementation of retractable landing gear provides coverage to the hexacopter propellers in the flight mode also helps hexacopter to achieve a successful landing. The design construction of hexacopter and their subordinate components are fixed mass method, in which initially 1.5 kg is assumed as the overall weight with the inclusion of payload and retractable landing gear. Computational Fluid Dynamics (CFD) simulations of with and without retractable landing gear implemented in hexacopter have been carried out to study the aerodynamic effect of retractable implementation. In this work, CATIA used as a modeling tool for both cases and Ansys Workbench used as CFD solver tool.

Keywords CFD · Optimization · Landing gear · Safeguard · Tilt

1 Advanced Multi-rotor UAV

UAV plays an important role and it is the key for our modern technological world's surveillance. In UAV, a rotary wing aircraft with more than two rotors is termed as multi-rotor UAV. Due to its reliability, stability, control, compact, redundancy and low maintenance, multi-rotor emerged as an efficient UAV for complex applications. By the rotor construction, it is categorized as bicopter, tricopter, quadcopter, pentacopter, hexacopter [1]. The impact on flight execution occurred by the number of motors and its configuration. The fundamental working platform of this article is Tilt-Hexacopter, which can able to tilt for efficiency purpose. Thus but the drastic problem is difficult to fix suitable landing gear. Major problems in the Hexacopter have been analyzed in which the prime problems are it does not provide safety to its parts, the life cycle of the components is less, and it does not have safe landing and takeoff. To overcome

R. Vijayanandh (✉) · P. Kiran · S. Indira Prasanth · G. Raj Kumar · S. Balaji
Aeronautical Engineering, Kumaraguru College of Technology, Coimbatore, Tamil Nadu, India
e-mail: vijayanandh.raja@gmail.com

K. Jain et al. (eds.), *Proceedings of UASG 2019*, Lecture Notes in Civil Engineering 51,
https://doi.org/10.1007/978-3-030-37393-1_15

these problems the Hexacopter with retractable landing gear has been proposed and it's having the following advantages compared to hexacopter without landing gear. The advantages are safety provided through additional retractable landing gear, it can able to perform safe landing with lifetime [2]. The occurrence of lifetime failure in advanced multi-rotor UAV is the prime issue, which reduces the usage of UAVs. To overcome failure issue, this paper suggested the retractable landing gear, which provides safe landing as well as propeller safeguard while in the flight mode.

2 Conceptual Design Study

2.1 Design Tool

CATIA provides tools to complete product definition, including functional tolerances as well as kinematics definition. CATIA used in a wide range of applications for tooling design, for both generic tooling and mold & die. In the case of Aerospace engineering, an additional module named the aerospace sheet metal design offers the user combine the capabilities of generative sheet metal design and generative surface design Tilt-Hexacopter.

2.2 Tilt-Hexacopter Without Landing Gear

Figure 1 represents the Vertical Take-Off and Landing (VTOL) mode of Tilt-Hexacopter without landing gear and the forward speed mode is revealed in Fig. 2.

Fig. 1 VTOL motion

Fig. 2 Forward motion

2.3 Tilt-Hexacopter with Model-1 Landing Gear

Figure 3 represents the Tilt-Hexacopter with landing gear model-1 (normal mode) for VTOL case. Figure 4 represents the execution of forward speed operation of Tilt-Hexacopter with landing gear model-1. Figure 5 shows the combined view of tilt hexacopter with retract landing gear.

Fig. 3 VTOL mode

Fig. 4 Forward mode

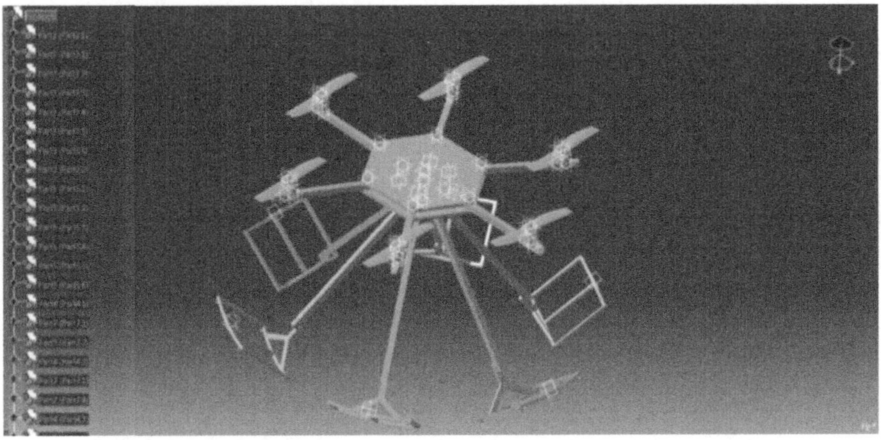

Fig. 5 Combined view of hexacopter with retract landing gear

2.4 Tilt-Hexacopter with Model-2 Landing Gear

Figure 6 represents the sectional view of Tilt-Hexacopter with Anti crash landing gear version-2 for VTOL operation. Figure 7 reveals the combined view of Tilt-Hexacopter with Anti crash landing gear version-2 for forward movement. A typical view of Tilt-Hexacopter with retractable landing gear is shown in Fig. 8, in which it's purposively modeled the landing approach in CATIA.

Fig. 6 VTOL operation

Fig. 7 Forward speed operation

3 Numerical Simulation

3.1 Introduction

A chosen system or control volume is analyzing virtually for obtaining the theoretical results (as compared with the experimental) with the required mathematical complex equation via numerical simulation. In this numerical simulation, CFD (Computational Fluid Dynamics) plays a vital role for generating the solutions for flowing of the fluid with or without the application of solids [3, 4]. It can be analyzed

Fig. 8 Anti crash landing gear hexacopter version 2 retract position

with the flow properties of velocity, pressure, temperature, and density with the help of Navier Stokes equation supported with numerical approaches. In this process, CFD first attributes the Navier Stokes equation simplified into Reynolds Averaged Navier Stoke (RANS) equation and then the control volume (as system) splits into the meshes or nodes. Then the RANS equation leads to governing the meshes and solving the required numerical equation with the help of boundary conditions [5, 6].

3.2 Boundary Condition

In the numerical simulation, boundary conditions are capable to initiate the whole process. In this project pressure based solver has been used because of the tilt hexacopter working environment, in which tilt-hexacopter has comes under the low-speed incompressible operation. The absolute system has been selected in order to cover the whole system with the global coordinate system [7, 8]. Rotating propellers plays a vital role in multi-rotor UAV so the effect due to propellers is much more important especially turbulence generation around the propellers is the focal role in the UAV aerodynamics. In this paper k-epsilon turbulence mode has been used, which have the capacity to capture the turbulence, flow separation with the help of its two equations such as kinetic energy of the turbulence and dissipation rate of turbulence. Since the multi-rotor UAV has limitation in the forward speed, which is coming under the low-speed operation so the velocity inlet is preferable for this simulation also pressure outlet has been suggested and used [9, 10]. After the assignment of artificial boundary conditions, the process of physical boundary conditions has been completed in which the no-slip conditions are given on the solid-fluid interaction surface on the tilt-hexacopter and free slip conditions have been to the external domain exterior part. 101,325 Pa has been given as operating pressure so the value of gauge pressure value is given as 0 Pa. Least square solution methodologies are selected because of the unstructured mesh generation also higher order equation has been selected in order to get high accrued results. Flow analysis over the Hexacopter is complicated analysis, which may affect the numerical results and thereby unstable variation takes place in the output residual so relaxation factor has been reduced 20% [11].

3.3 Result

In this article suggest the advanced numerical simulation is the only way to simulate the aerodynamics behavior of Hexacopter while undergoing critical maneuvering in the critical environment.

Hexacopter with landing Gear: Wireframe Model—VTOL: Velocity 30 m/s

Figure 9 represents the integrated view of Tilt-Hexacopter with landing gear, in which the pressure and velocity are predicted. Figure 10 shows the planar view of the velocity variation of VTOL mode at the velocity of 30 m/s.

Figure 11 represents the prediction of drag force of the Tilt-Hexacopter with landing gear, while in the VTOL mode as a wireframe model for 30 m/s.

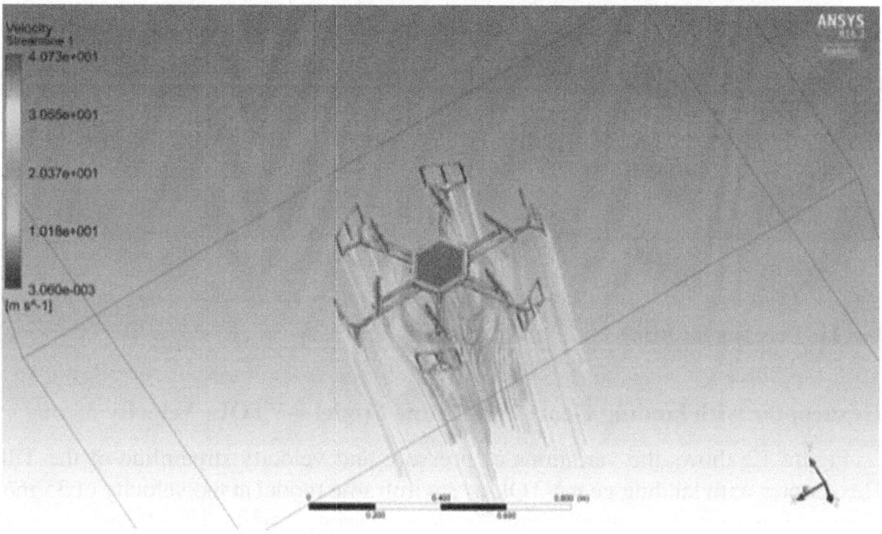

Fig. 9 Pressure and velocity streamline

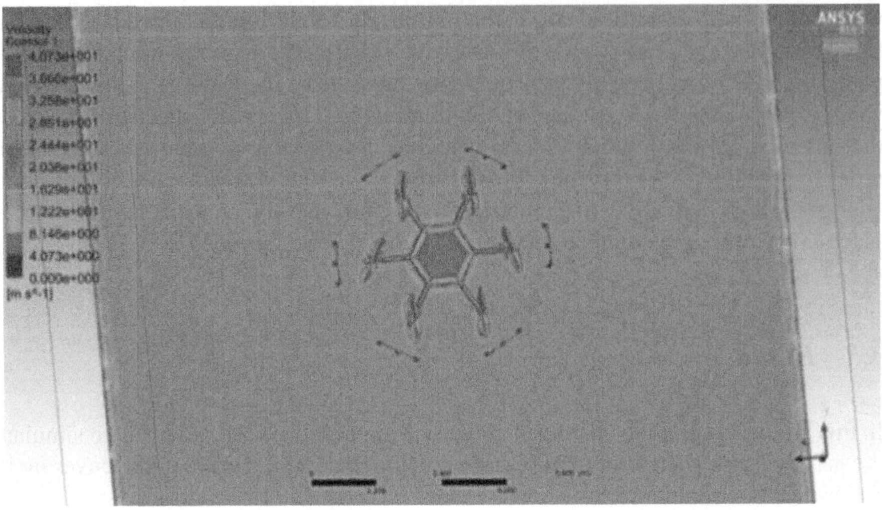

Fig. 10 Planar view of velocity varition

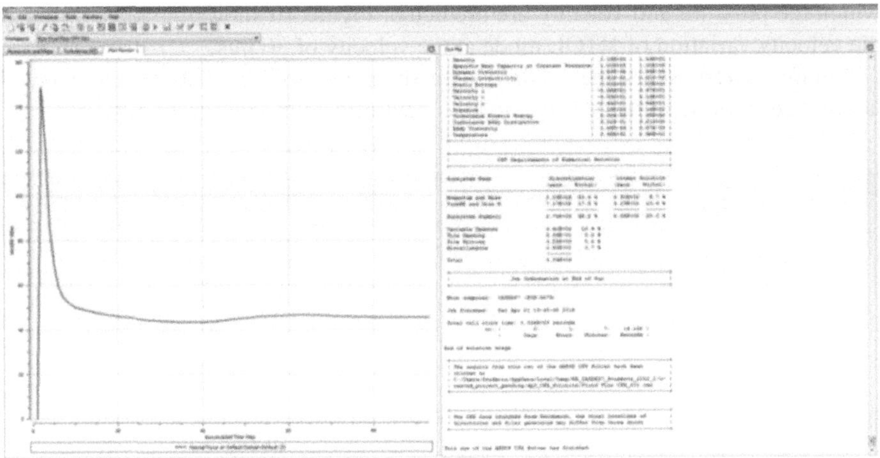

Fig. 11 Drag plot for 30 m/s

Hexacopter with landing Gear: Wireframe Model—VTOL: Velocity 35 m/s

Figure 12 shows the variations of pressure and velocity streamline of the Tilt-Hexacopter with landing gear VTOL as a wireframe model at the velocity of 35 m/s. Figure 13 represents the planar view of velocity variation of the Tilt-Hexacopter with landing gear for the given input velocity of 35 m/s.

Fig. 12 Pressure and velocity streamline

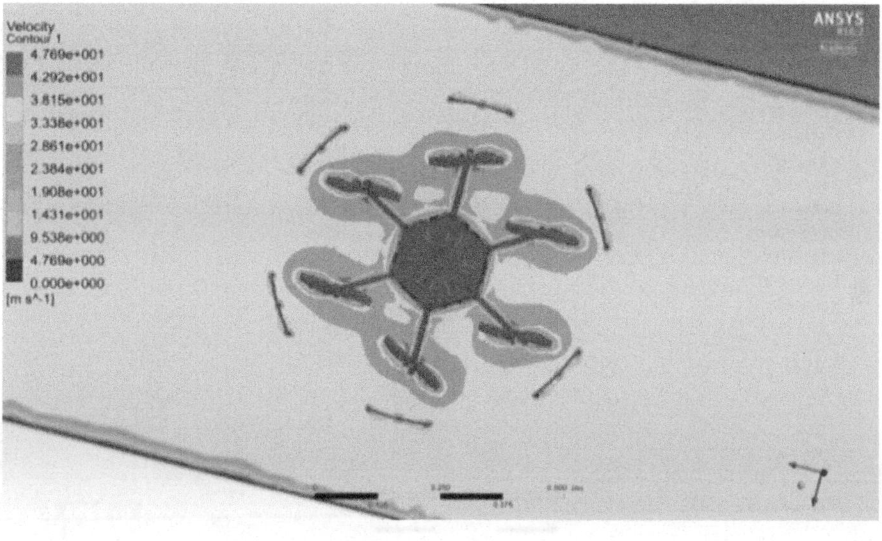

Fig. 13 Planar view of velocity varition

Hexacopter with Landing Gear: Simplified Wireframe Model—VTOL: Velocity 30 m/s

Figure 14 represents the pressure and velocity streamlines of the VTOL mode of the Tilt-Hexacopter with landing gear as a simplified wireframe model for the given boundary conditions. Figure 15 represents a typical view of velocity variations on the Hexacopter.

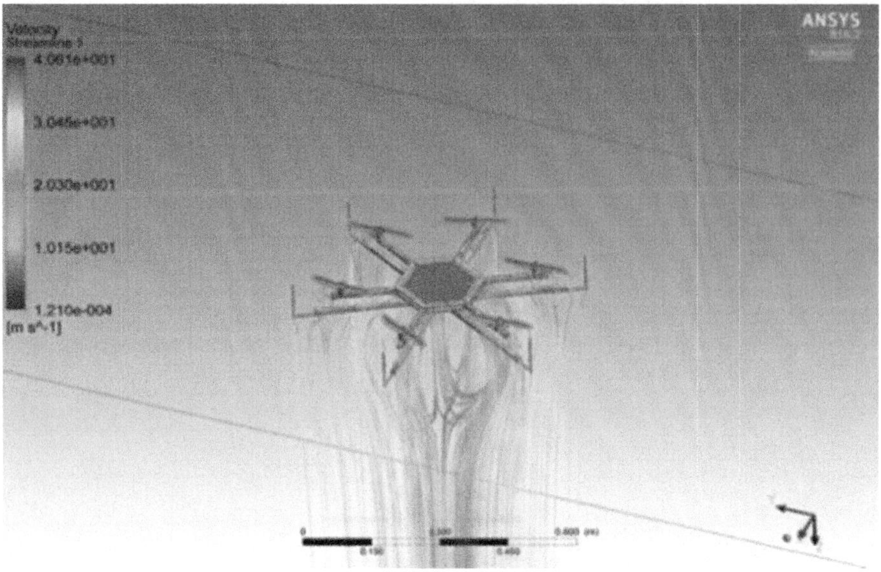

Fig. 14 Pressure and velocity streamline

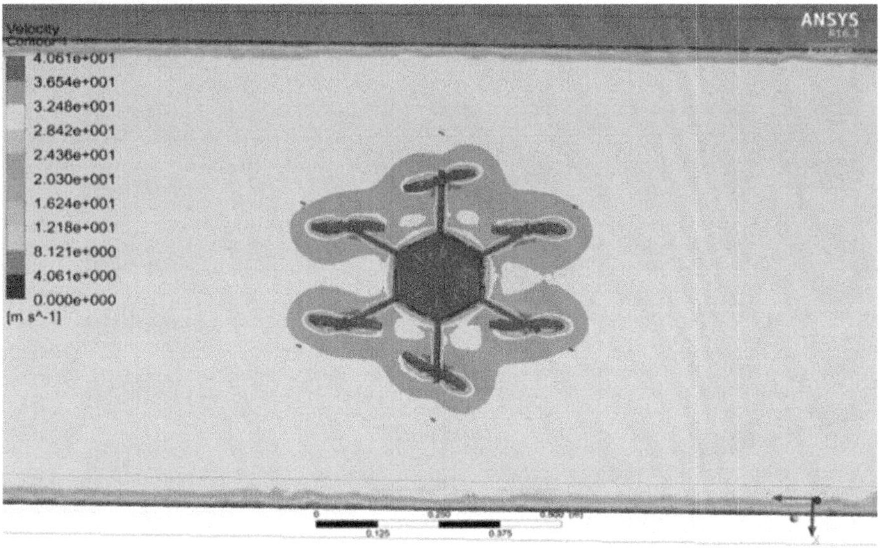

Fig. 15 Planar view of velocity variation

Hexacopter with Landing Gear: Simplified Wireframe Model—VTOL: Velocity 35 m/s

Figure 16 represents the combined plot of the Tilt-Hexacopter with landing gear, in which the red color on the UAV represents the pressure variations at the stagnation point and velocity around the UAV is projected in the streamlined mode. Figure 17 represents the planar view of velocity variations of Hexacopter with landing gear at the velocity of 35 m/s.

Tilt-Hexacopter with Landing Gear—Forward Motion: Velocity 30 m/s

Figure 18 represents the pressure and velocity streamline of the Tilt-Hexacopter with landing gear at the velocity of 30 m/s. Figure 19 represents the planar view of velocity variation of the tilt hexacopter with landing gear at the velocity of 30 m/s.

Tilt-Hexacopter with Landing Gear—Forward Motion: Velocity 35 m/s

The transformation of fluid energy into potential energy and kinetic energy are revealed in Fig. 20, in which the maximum red color represents the stagnation point. A typical planar representation of velocity variations around Tilt-Hexacopter is shown in Fig. 21.

Tilt-Hexacopter with Landing Gear—HTOL: Velocity 40 m/s

For the given input velocity of 40 m/s, the pressure and velocity streamlines on Tilt-Hexacopter are predicted, which are combined shown in the Fig. 22. Changes of air velocity around Tilt-Hexacopter is represented in the view of planer in the Fig. 23.

Fig. 16 Pressure and velocity streamline

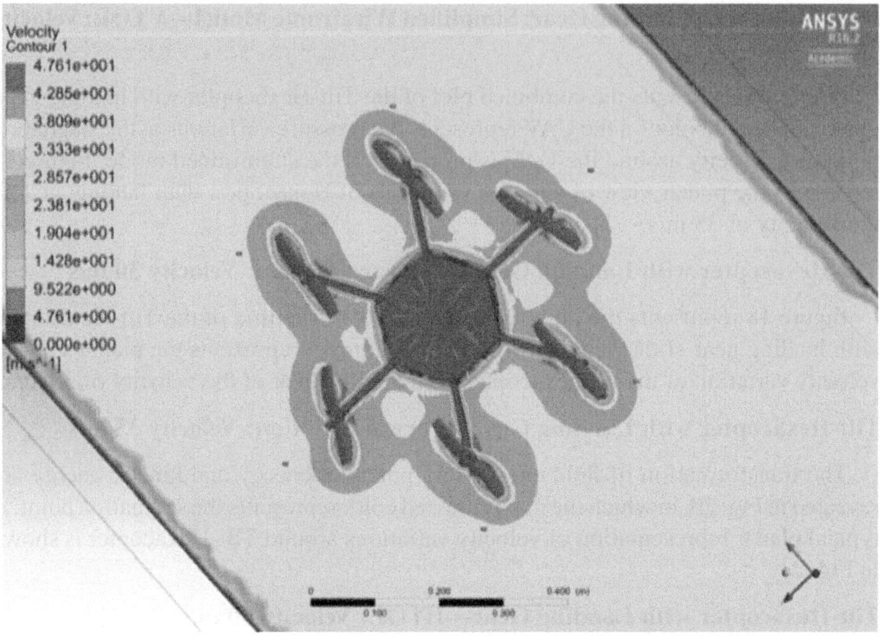

Fig. 17 Planar view of velocity variation

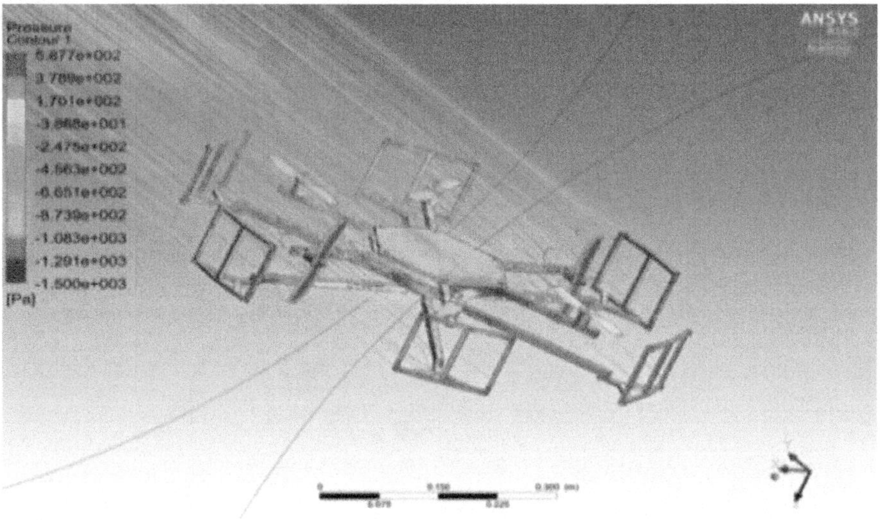

Fig. 18 Pressure and velocity streamline

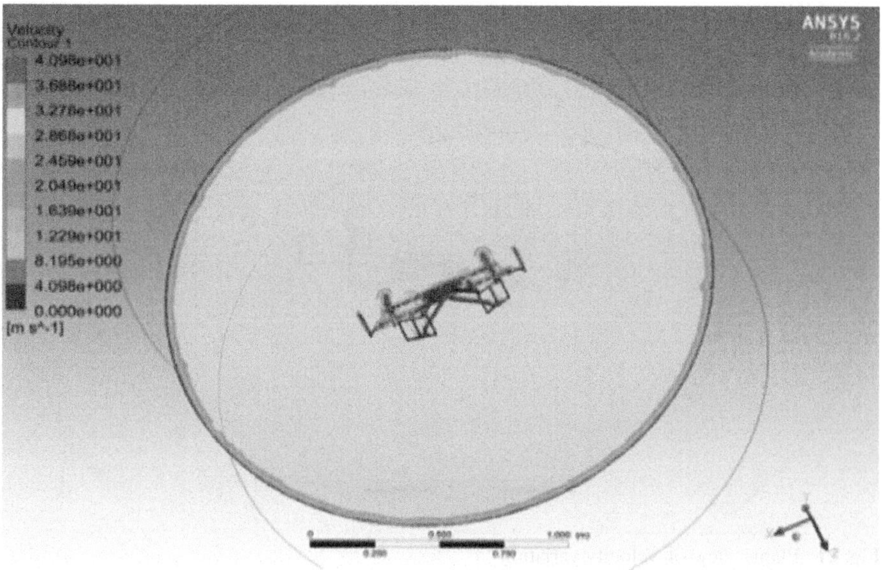

Fig. 19 Velocity variations

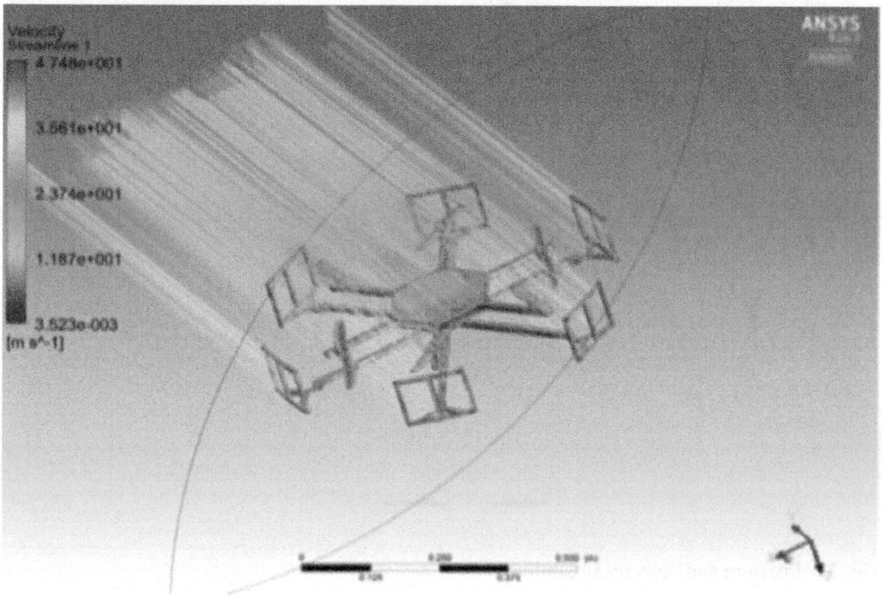

Fig. 20 Pressure and velocity streamline

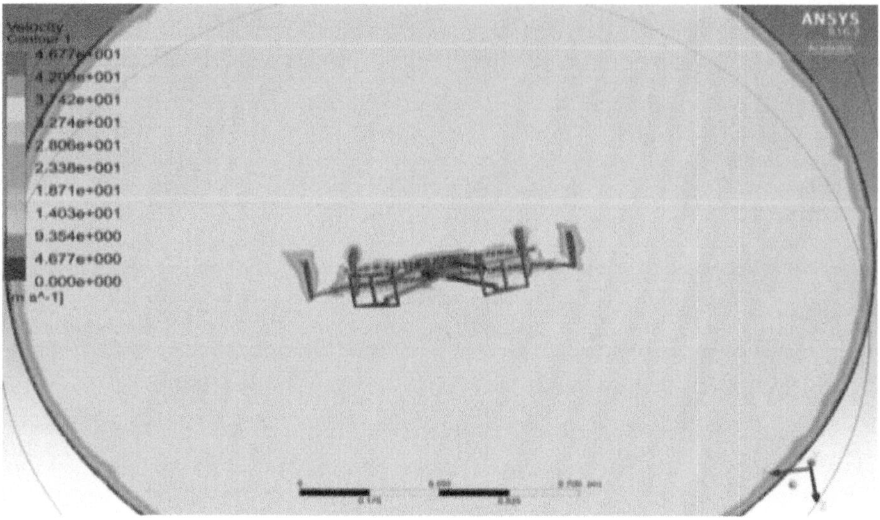

Fig. 21 Planar view of Velocity variation

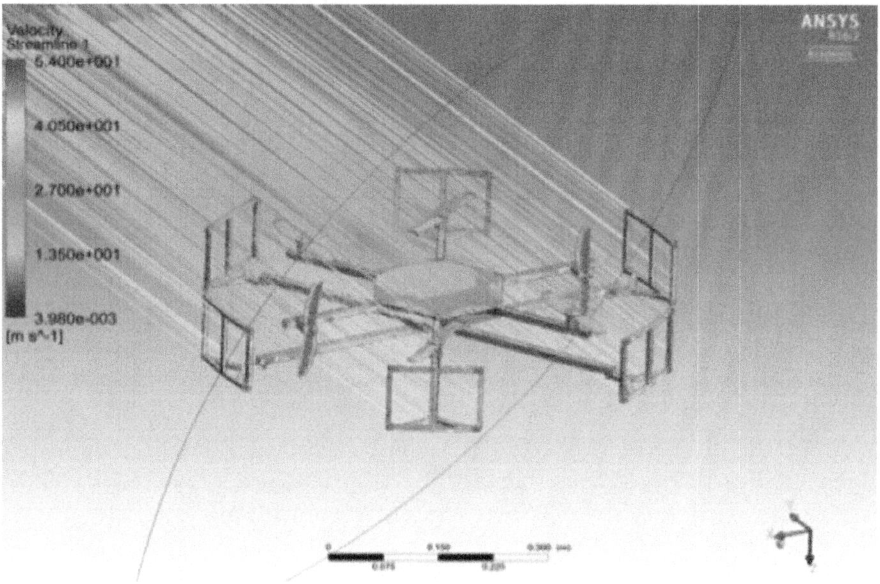

Fig. 22 Pressure and velocity streamline

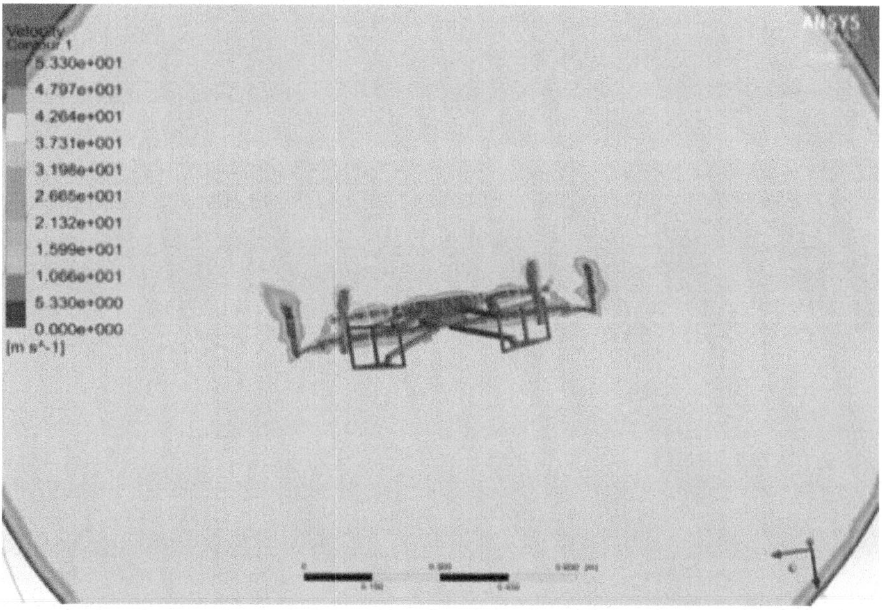

Fig. 23 Velocity distributions

Tilt-Hexacopter with Landing Gear—Forward Motion: Velocity 45 m/s

Figure 24 shows the variations of the primary fluid properties on the Tilt-Hexacopter, while in the pitch-up maneuvering. Figure 25 represents the planar view of velocity variation of the Tilt-Hexacopter with landing gear at the velocity of 45 m/s.

With Landing Gear: Optimized Wireframe Model—Forward Motion: Velocity 30 m/s

In CFX post processing, an advanced integrated fluid property variations of Tilt-Hexacopter is completed with the help of contour and streamline combined options. The combined plots are shown in Figs. 26, 28, 30 and 32 for the given input forward velocities of 30 m/s, 35 m/s, 40 m/s and 45 m/s respectively. Due to the implementations of no-slip boundary conditions, the velocities at the solid fluid interactions regions are zero so normal contour representations is not fit for velocity representations. Therefore with the help of planar view, the velocity variations around the vehicle are predicted and shown in Figs. 27, 29, 31 and 33 for the different forward velocities, which are finalized in the earlier stage.

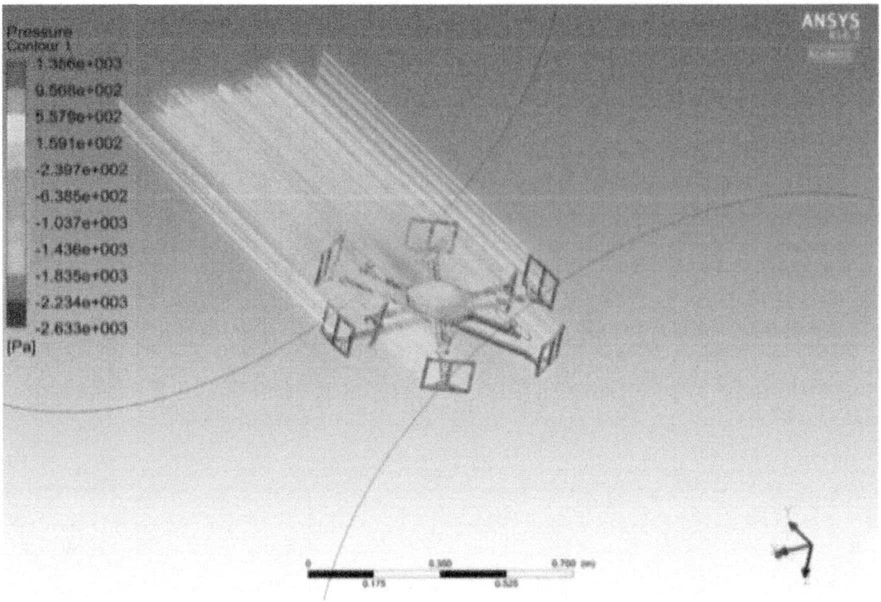

Fig. 24 Pressure and velocity streamline

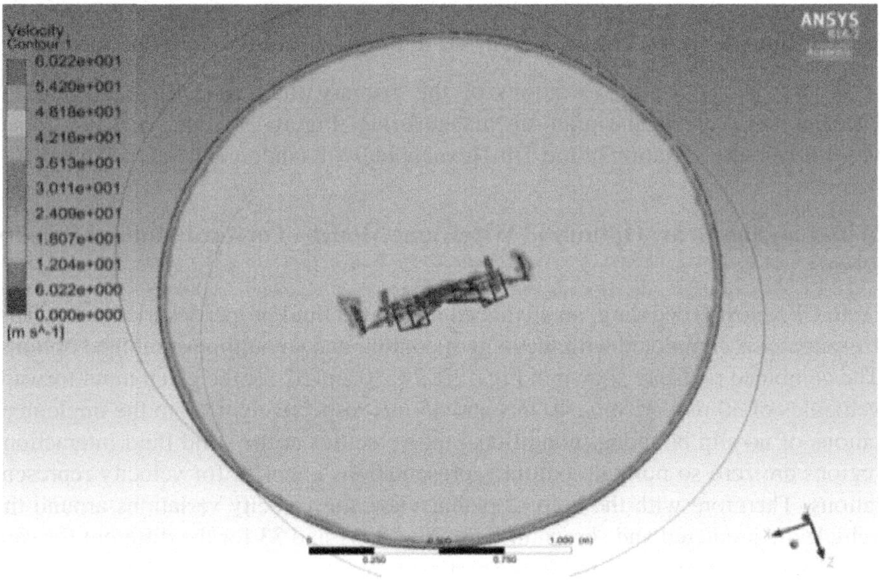

Fig. 25 Planar view of velocity variation

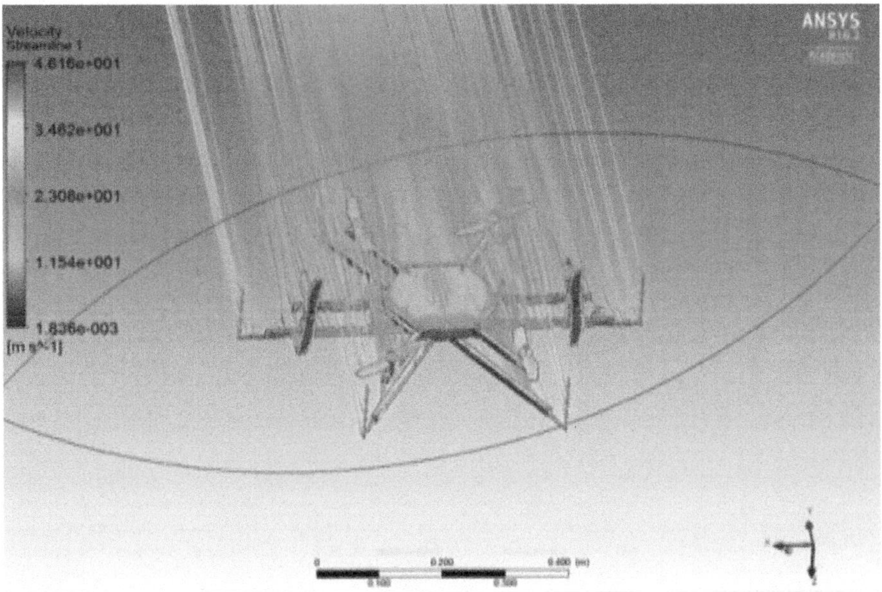

Fig. 26 Pressure and velocity Streamline

With Landing Gear: Optimized Wireframe Model—Forward Motion: Velocity 35 m/s

With Landing Gear: Optimized Wireframe Model—Forward Motion: Velocity 40 m/s

With Landing Gear: Optimized Wireframe Model—Forward Motion: Velocity 45 m/s

Drag force variations in the Y direction are plotted with respect to iterations, which are reveled in the Fig. 34.

3.4 Comparative Analysis of Forces

The comparative values of the forward motion case forces of different Hexacopter are noted and listed from Tables 1, 2, 3 and 4 for the different velocities of 30 m/s, 35 m/s, 40 m/s, and 45 m/s respectively.

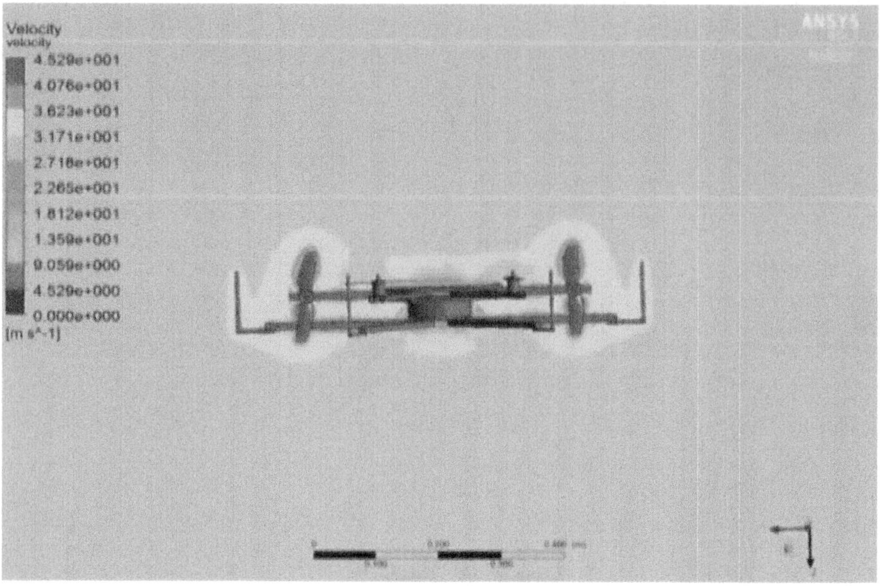

Fig. 27 Planar view of velocity variation

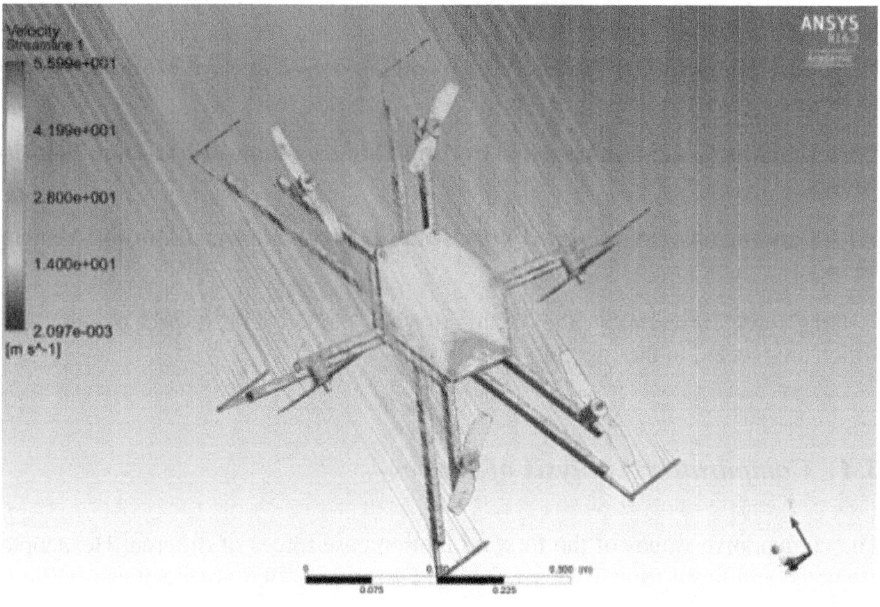

Fig. 28 Pressure and velocity streamline

Fig. 29 Planar view of velocity variation

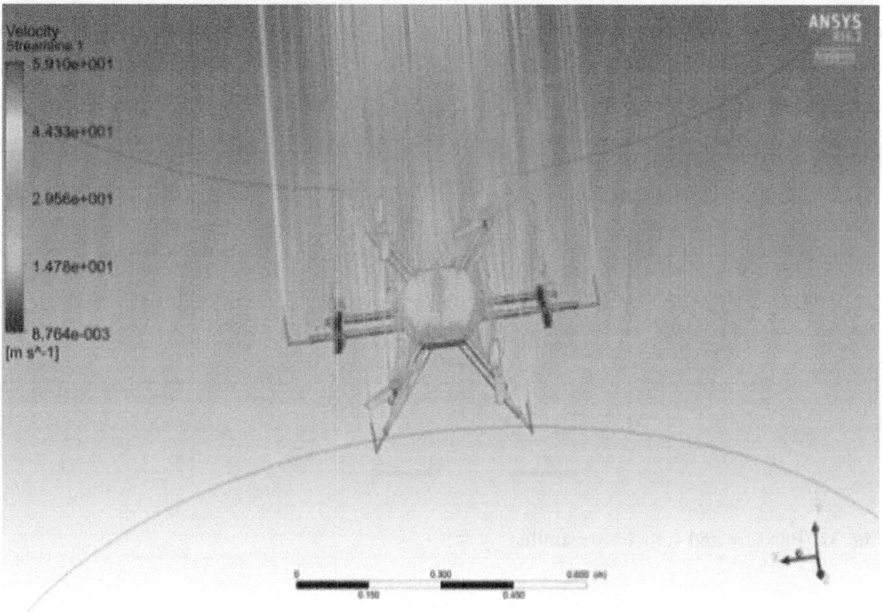

Fig. 30 Pressure and velocity streamline

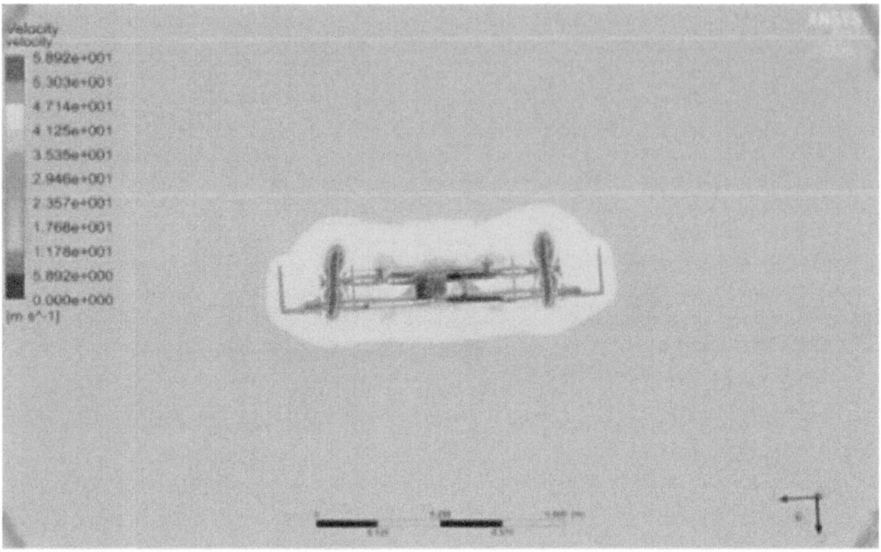

Fig. 31 Planar view of velocity variation

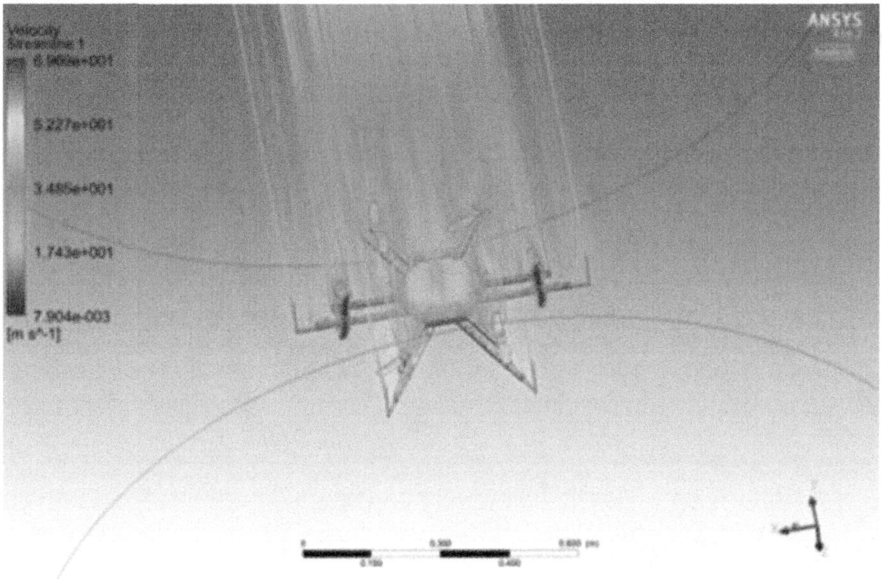

Fig. 32 Pressure and velocity streamline

Fig. 33 Planar view of velocity variation

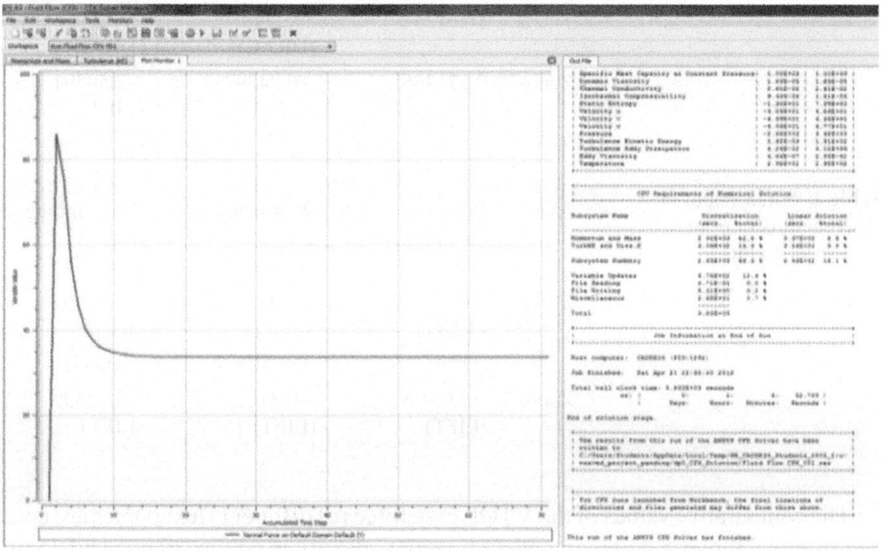

Fig. 34 Drag plot for 45 m/s

Table 1 Force values in different directions of velocity at 30 m/s

Axis	Without landing gear (N)	With landing gear (N) model-1	With landing gear (N) model-2
X	0.129794	0.010012	−0.0951157
Y-DRAG	9.31158	19.281	15.3641
Z-LIFT	0.1374944	−1.22648	−1.41748

Table 2 Force values in different directions of velocity at 35 m/s

Axis	Without landing gear (N)	With landing gear(N) model-1	With landing gear (N) model-2
X-axis	−0.152018	−0.0423309	−0.42279
Y-axis	12.7161	22.4304	20.9171
Z-axis	0.160803	−0.238343	0.305081

Table 3 Force values in different directions of velocity at 40 m/s

Axis	Without landing gear (N)	With landing gear (N) model-1	With landing gear (N) model-2
X-axis	−0.350575	−0.074354	−0.207123
Y-axis	16.6126	29.2348 (DRAG)	27.4234 (DRAG)
Z-axis	0.171466	−0.243979 (LIFT)	−2.50591 (LIFT)

Table 4 Force values in different directions of velocity at 45 m/s

Axis	Without landing gear (N)	With landing gear (N) model-1	With landing gear (N) model-2
X-axis	−0.278742	−0.0420509	−0.260072
Y-axis	20.9441 (DRAG)	37.0318 (DRAG)	34.5196 (DRAG)
Z-axis	0.336995 (LIFT)	−0.310252 (LIFT)	−3.22399 (LIFT)

The comparative values of the VTOL case forces of different Hexacopter are noted and listed in Tables 5 and 6, in which the input velocity is take as 30 m/s and 35 m/s respectively.

Table 5 Force values in different directions of velocity at 30 m/s

Axis	Without landing gear (N)	With landing gear (N) model-1	With landing gear (N) model-2
X-axis	0.289535	−0.3146	0.109387
Y-axis	0.0252483	0.370207	0.0795374
Z-axis (DRAG)	32.761	46.1335	35.8795

Table 6 Force values in different directions of velocity at 35 m/s

Axis	Without landing gear (N)	With landing gear (N) model-1	With Landing Gear (N) Model-2
X-axis	−0.31285	−0.482156	−0.45696
Y-axis	−0.026001	0.113019	0.00491986
Z-axis (DRAG)	46.0402	62.543	48.7616

4 Conclusion

The design of the Hexacopter with and without landing gear has been modeled by CATIA V5. The numerical simulations on the Hexacopter have been carried out with the help of Ansys Workbench 16.2, in which with and without propeller safeguard are playing a focal role. Comparative drag forces analyses have been analyzed with the help of Ansys CFX 16.2 for both the cases such as Hexacopter without landing gear and Hexacopter with the landing gear. From the numerical simulation, it is understood that the implementation of propeller safeguard cum landing gear has a chance to focal increment in the drag force on the hexacopter. The existing Hexacopter system has lack of capability of overcome these issues hence the propeller safeguard optimization additionally take place in numerical simulation. Finally, the design optimization of propeller safeguard cum landing gear in the hexacopter has been carried out and thereby the numerical simulation is carried out also the result are compared. Hence the optimized landing gear is the best method for damage production of for Hexacopter components, which have been derived from advanced numerical simulation techniques. This paper also suggested that instead of trial and error method production of multi-rotor UAV, production of multi-rotor UAV based on the numerical simulation results is best suitable by provides the high lifetime with a high probability of success.

References

1. Vijayanandh R et al (2018) Design, fabrication of Tilt-Hexacopter with image processing for critical applications. Int J Pure Appl Math. 118(9):935–945. ISSN: 1314-3395
2. Papachristos C et al (2011) Design and experimental attitude control of an unmanned Tilt-Rotor aerial vehicle. In: 15th international conference on advanced robotics (ICAR), electronic. ISBN: 978-1-4577-1159-6. https://doi.org/10.1109/icar.2011.6088631
3. Cetinsoy E et al (2011) Design and development of a tilt-wing UAV. Turk J Elec Eng Comp Sci 19(5):733–741. https://doi.org/10.3906/elk-1007-621
4. Sanchez A et al (2008) Autonomous hovering of a noncyclic tiltrotor UAV: modeling, control and implementation. In: Chung MP, Jin M (eds) 17th IFAC world congress. vol 17, no 1, COEX, South Korea
5. Shaikh S et al (2015) Automatic animal detection and warning system. Int J Adv Found Res Comput (IJAFRC) 2(Special Issue (NCRTIT 2015)) pp 405–410
6. Vijayanandh R et al (2017) Design, fabrication and simulation of hexacopter for forest surveillance. ARPN J Eng Appl Sci 12(12):3879–3884. ISSN 1819-6608
7. Hossain MS et al (2013) Development of an Autonomous y4 copter. Int J Inf Technol Control Autom (IJITCA) 3(2)
8. Lee J et al (2016) Fault tolerant control of hexacopter for actuator faults using time delay control method int'l. J Aeronaut Space Sci 17(1):54–63. Doi: http://dx.doi.org/10.5139/IJASS.2016.17.1.54
9. Mostafa Moussid et al (2015) Dynamic modeling and control of a hexarotor using linear and nonlinear methods. Int J Appl Inf Syst (IJAIS). Foundation of Computer Science FCS, New York, USA. 9(5):09–17. ISSN: 2249-0868
10. Artale V et al (2013) Mathematical modeling of hexacopter. Appl Math Sci 7(97)4805–4811. (Hikari Ltd). http://dx.doi.org/10.12988/ams.2013.37385
11. Vijayanandh R et al (2017) Numerical study on structural health monitoring for unmanned aerial vehicle. J Adv Res Dyn Control Syst 9(6):1937–1958

Review of Inpainting Techniques for UAV Images

Garima Kadian and Ganesh Khadanga

Abstract Occupying dead pixels, removing uninterested objects and shadows are often desired in the applications of an UAV to extract the natural and man-made feature boundaries. Image inpainting provides a mean to reconstruct the image. The basic idea behind inpainting methods is to naturally fill in absent or lacking portion of an image by using information from the surrounding area. Applications of this technique include the rebuilding of imperfect photographs and films, elimination of superimposed text, removal/replacement of unwanted objects, redeye correction, image coding. This paper reviews various image inpainting methods like PDE based image inpainting, wavelet-based inpainting, structural inpainting, exemplar-based image inpainting and textural inpainting with their variations. Image inpainting can also be used indirectly in squeezing image where some percentage of the original image is transmitted, and the whole image can be reconstructed on the other end using a pre-trained neural network. The critical reviews of each of these traditional methods along with the latest CNN based techniques are compared and suitability of these techniques for examining or repairing the UAV image is analyzed. In this paper, some of the existing quality assessment metrics like PSNR, MSE, ASVS, BorSal etc.related to image inpainting are also discussed.

Keywords UAV · Inpainting · CNN · Object removal · Shadow

1 Introduction

Unmanned Aerial Vehicles (UAV) is used across the world for civilan, commerical as well as military applications. The UAV images often encounter common problems such as stripe noise and bad pixels. Bad pixels are those pixels which are statistically distinct from neighboring pixels. The source of bad pixels includes calibration errors,

G. Kadian (✉)
CSED, Thapar Institute of Engineering and Technology, Patiala, India
e-mail: garima@thapar.edu

G. Khadanga
Civil Engineering Department, IIT Roorkee, Roorkee, India
e-mail: ganesh@nic.in

K. Jain et al. (eds.), *Proceedings of UASG 2019*, Lecture Notes in Civil Engineering 51,
https://doi.org/10.1007/978-3-030-37393-1_16

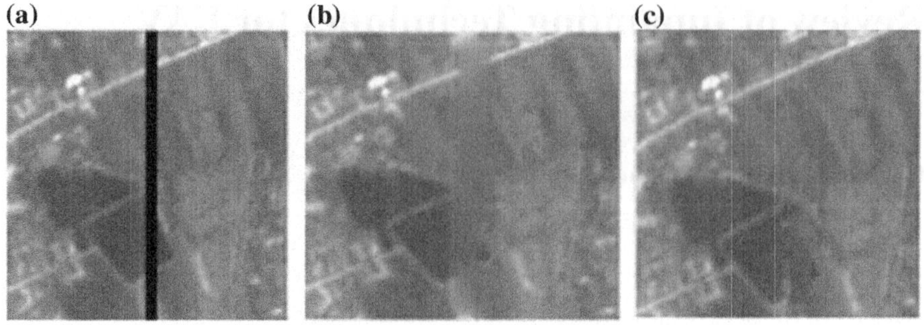

Fig. 1 Comparison of removing dead line from CBERS (China Brazil Earth Resource Satellite) image **a** 8-pixel dead line image **b** inpainting using PDE **c** inpainting using exemplar based technique

non-response of a detector, offset inequalities and relative gain of detectors. Bad Pixels are of two types named as warm and dead pixels. When measurement of a pixel has no correlation with the actual scene, then such pixel is termed Dead pixel. And warm pixels are those pixels which are brighter or darker than the healthy pixels [1]. In UAV images, destriping techniques are used to remove the stripe noise and dead pixel replacement methods to recover from dead pixels. But these techniques do not remove all stripes and lead to significant blurring within the image. So Image inpainting can be used for restoration from stripe noise and dead pixels in UAV images.

Image inpainting is a technique of reconstructing abscent or impaired region in an image in such a way that it is not easily detectable by an observer who does not know the original image. Image inpainting is also known as image retouching. Image inpainting has many applications such as eliminating object in a context of editing, restoring images from text overlays, and disillusion in image-based rendering (IBR) of viewpoints distinctive from those taken by the cameras and lost in secrecy in context of damaged image transmission [2]. All inpainting techniques assume that pixels in the familiar and unfamiliar parts of the image share the same geometrical structures and statistical features (Fig. 1).

1.1 Image Inpainting Problem

The goal of image in painting is to recover the region such that the inpainted area looks natural to human eye. An image A can be represented as:

$$\begin{aligned} \delta \subset Q^n &\rightarrow Q^m \\ A = k &\rightarrow A(k) \end{aligned} \tag{1}$$

Here k represent coordinates of pixel p_i such that $k = (i, j)$.

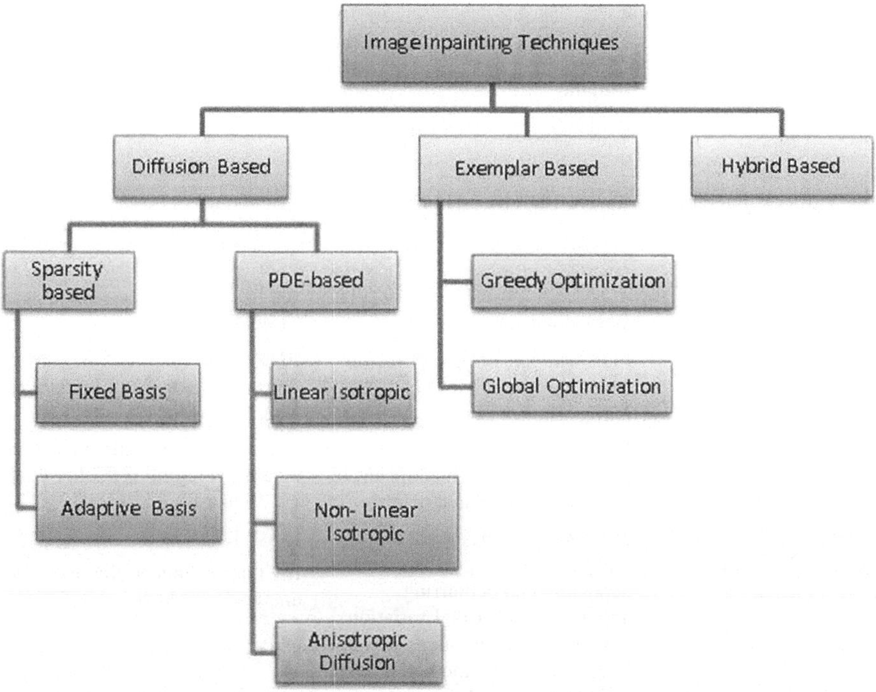

Fig. 2 Classification of image inpainting techniques

In image inpainting, the input image A is supposed to have gone through a deterioration, represented by N, which has eliminated samples from A. Due to which, δ is divided into two parts i.e $\delta = R\ V$, here R denotes known part of A and V is unknown part of A. The degradation can be denoted as $tt = NJ$. By applying inpainting techniques, the color components of pixel p_i located at position i in V is estimated (Fig. 2).

2 Image Inpainting Techniques

2.1 Diffusion Based Image Inpainting

In this technique, information from the known area is used to fill the unknown region. This technique works well when filling non-textured regions and mislaid regions as shown in Fig. 3. Partial Differential Equation (PDE) method and the variational method are two methods used by this technique. This algorithm first determines the local image geometry and later uses variational or PDEs technique to represent continuous change in the image and in its structures [2]. For instance, if the pixel is in

Fig. 3 Block diagram of diffusion-PDE based image inpainting technique

Table 1 A summary of papers based on diffusion based inpainting technique

Author name	Work done	Limitation
Bertalmino et al. [22]	PDE with anisotropic diffusion using laplacians is used. Navier-Stokes Equation is used	**PDE based display blur and does not work when texture area is large**
Telea et al. [23]	Estimates target pixels in one pass by using weighted means of previously calculated pixel	Blurring is produced when inpainting region is thicker than 10–15 pixel
Tschumperle et al. [24]	Trace based PDE model for multiple color images is used For denoising and deblurring application, TV (total variation) model is used Curvature Driven Diffusion (CDD) method is used. Both strength and geometry of isotopes determine the diffusion	Less smoothing across edges
Rudin et al. [25]		It violates connectivity principle and leads to blurring
Chan et al. [26]	Patch based approach for removing smaller non-linear objects and crack	Time taken by large image is long. And borken edges are also connected
Shen et al. [27]		Fails when object, patches and crack is large

a homogeneous area, the smoothing can be done in all directions if the pixel is placed on an image outline, the smoothing must be implemented along the outline direction and not beyond boundaries. This method is well suited for completing curves, lines and for inpainting small area. But, the weakness of this process is that it adds blur effect while filling large textured regions. Table 1 gives a summary of diffusion based inpainting technique.

2.2 *Texture Based Image Inpainting*

Also known as Sample based texture synthesis. This technique is used to construct a texture from a given sample see Fig. 4. The aim of this technique is to create a texture in such a way that the composed texture is larger than source sample with a similar visual characteristics [3]. All sample based techniques rely on Markov random fields

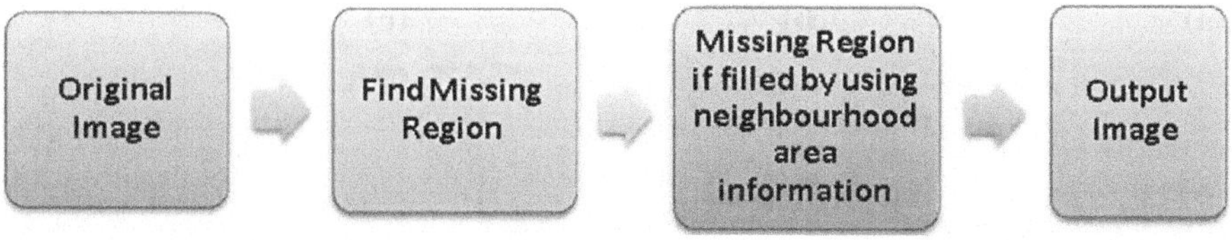

Fig. 4 Block diagram of texture based image inpainting technique

(MRF) modeling of texture. In this technique, entire patch is synthesized by learning from patches in the known part of the image. As this approach synthesizes whole patches at once, it is faster than pixel based approach [4].

Variants of Texture based image inpainting

Patch Stitching: Filling unknown part of the input patch leads to stitching together pieces of texture that are not consistent in term of color or contrast. The aim of patch stitching is to reduce boundary artifact and color bleeding. Stitching can be done by either using the quilting method (greedy method) or by blending method.

Distance Metric: Used to measure the correlation between images or between image patches. The distance metric is divided into two categories named as pixel-based and statistics based. In the former one, similarity is measured in term of cross-correlation or difference between pixel color values like SSD (sum of squared difference), normalized cross-correlation and Lp norm etc whereas in latter similarity is measured between probabilities of pixel color value in patches like Bhattacharyya distance [5], NMI (Normalized Mutual Information), Kullback- Leibler divergence etc.

PPO (Patch processing Order): In an image a missing region is composed of textures and structures. In PPO, patches of structure are filled first. PPO is the product of data term and confidence term [6]. Data term in PPO can be of several forms like Gradient based, Sparsity based and Tensor based etc.

Global Optimization: patch per patch progress in greedy method does not ensure global optimization. To improve the visual characteristic of inpainted image one can maximize analogy among the synthesized patch and original patches in the known area of the image [7].

Searching best match pixel fastly: Exemplar based inpainting approach uses *k*- NN (*k*-nearest neighbors) inside the known part of the image [8]. The Nearest Neighbors (NN) computes the gap from query patch to all feasible candidate patches (Fig. 5; Table 2).

2.3 *Exemplar Based Inpainting*

This technique is appropriate for reconstructing large target regions. It fills holes in the image by repeatedly synthesizing the target region by most identical patch in the

Fig. 5 Using GaoFen-2 RS imagery, comparison of clouds removal **a** Original image. **b** inpainting using SiLRTC [20] **c** inpainting using MRF [19]

Table 2 Difference between image inpainting techniques

Diffusion-based	Texture-based
Works well for small and sparsely scattered gaps	Works well in textured areas with similar patterns
Suitable only for piecewise smooth images but unable to restore texture	Not suitable for conserving edges or structure and for images with many small dispersed holes

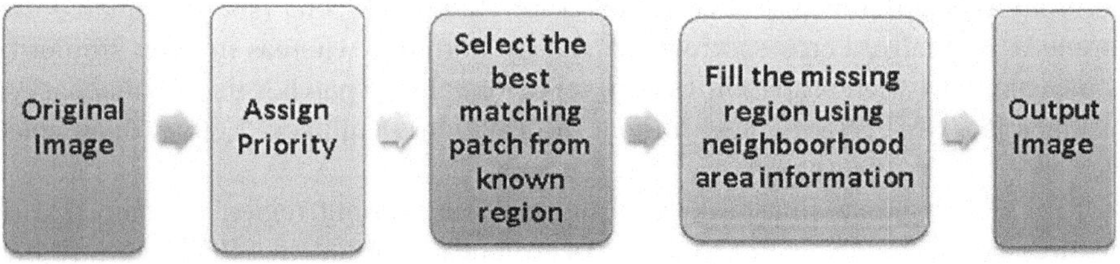

Fig. 6 Block diagram of exemplar based image inpainting technique

known area and copying the pixels from the most identical patch into the hole. This technique first assigns priority and then the selection of best matching patch is done (Fig. 6; Table 3).

2.4 Hybrid Based Inpainting

Natural Images comprises of both structure and texture. Area with homogeneous arrangement or is considered as texture and structures constitute primal outline of an images (like corners and edges). To deal with these images, two main methods have been considered. The first method is to combine different technique in one particular energy function using variation formulation [9], [10]. The Second strategy

Table 3 A summary of papers based on texture based and exemplar based inpainting technique

Author name	Work done	Limitation
Efros et al. [3]	Pixel per pixel propagation from source to the hole of the image. Quilting method is used for patch stitching	Complex and difficult when texture is not frontal and Cannot handle color inconsistency
Ashikhmin et al. [28]	Reduce the complexity by binding the search for best match patches among the candidates of surrounding pixel that have been previously inpainted	Unable to handle curved structure
Liang et al. [29]	By copying and sampling texture pattern from original image, entire patch is recovered	Not propagate accurate result for curved structure.
Bugeau et al. [5]	Use SSD method for measuring correlation between image patches	Fails when two patches have same distribution but are geometrically different.
Barnes et al. [30]	Use *k*-dimensional tree method to find the nearest patch by splitting the space along different coordinates Use nonparametric sampling and preserve small local structures	Less accuracy and require large amount of memory
Efros et al. [31]	Used the greedy approach to fill the target region	Very slow in speed
Criminisi et al. [32]		Less accurate when structure is complex

is to separate the texture and structure, and then inpainting them separately using convenient technique (i.e diffusion based or exemplar based) [5], [11] (Figs. 7, 8; Table 4).

Fig. 7 Block diagram of hybrid based image inpainting technique

Fig. 8 Generalized of skelton CNN [12]

Table 4 A summary of papers based on hybrid based inpainting technique

Author Name	Work Done	Limitation
Sun et al. [12]	Structure is identified using supervised learning Structure and texture is separated using variational method after which texture is inpainted using exemplar technique and structure is inpainted using PDE- diffusion based technique	Low accuracy
Bertalmio et al. [33]	Uses MCA (Morphological Component Analysis) to separate structure and texture	Not good output for nontextured with color variation image.
Elad et al. [34]	Integrate energy terms related to texture synthesis, coherence and geometry into one single energy function	Blurring image is produced if missing area is large
Aujol et al. [35]		Fails to rearrange texture pattern

2.5 CNN based inpainting

CNN (Convolutional neural network) algorithm detects and classifies objects in real time while being less expensive and performing better as compared to another machine learning methods. The problem in UAV images can be rectified by using CNN based inpainting. By using proper kernel, this technique inpaints image by convolving the neighbourhood of the target pixels. In [12], the value of a, b and c for convolving kernel are 0.0732, 0.1767 and 0.125 respectively. Here the central weight of kernel is zero because its related pixel in original image is unknown see Fig. 9 (Fig. 10; Table 5).

3 Quality Assessment Measures for Inpainted Image

The aim of image inpainting application is to reconstruct the original image such that the changes imported inside, outside or around the inpainted area are not detectable or distinguishable. The most accurate and reliable method is Subjective assessment

a	b	a
b	0	b
a	b	a

c	c	c
c	0	c
c	c	c

Fig. 9 Convolving kernel used by [21]

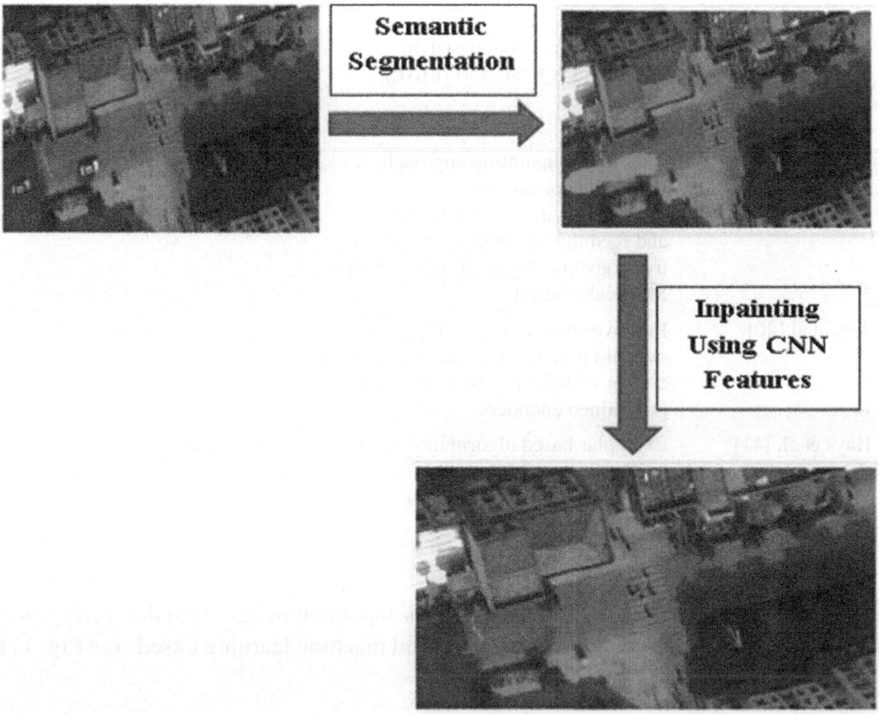

Fig. 10 Inpainting image using CNN [19]

methods [13], [14]. But these techniques are laborious, time-consuming and require a large number of viewer. Traditional metrics like MSE and PSNR were earlier used to classify the nature of inpainted images. But these metrics also are not well associated with perceptual quality evaluation [15]. To estimate the performance of various image inpainting approach, the metric of choice should be a qualitative analysis. Hence, we

Table 5 A summary of papers based on CNN based inpainting technique

Author name	Work done	Limitation
Richard et al [21]	The regions to be inpainted are convolved with a predefined diffusion mask repeatedly and the central weight of the diffusion mask is considered zero	Search similar neighborhoods only in the limited region and hence leads to high contrast damaged edges
Cheng et al. [36]	Proposed Multichannel Nonlocal Total Variation (MNLTV) model for textured images	This model can be used only for single band and some of the regularization parameters are fixed manually
Nalawade et al [37]	Used RBFNN (radical basis functional neural network) with DST (Discrete Shearlet Transform) to reconstruct the image	Due to limited computational resources the proposed algorithm is restricted to small inpainting regions and results often lack details and are blurry
Shen et al [38]	Proposed algorithm based on Maximum A Posteriori (MAP)	Fails when unsymmetrical structures are encountered Need separate network for each type of damaged pixel
Cai, et al [39]	Uses blind inpainting approach where the corrupted image is segmented into small sub-images and feed that through a network of five convolute layers followed by a MSE calculation	Fails when applied to image with large holes
Xie, et al [40]	Remove small damaged pixels and overlaid text by Combining sparse coding with deep neural nets using pre-trained encoders	
Hays et al. [41]	Exemplar based algorithm is used to inpaint the target regions and searches similar neighborhoods in all regions of image	Time consuming, and result in blurring of the inpainted image

can divide the quality assessment measure for inpainted images into three categories named as Saliency-based, Structural based and machine learning based (see Fig. 11).

3.1 *Structure Based*

Being Full Reference based, this metric requires information of both the original image as well as the inpainted image; to determine the quality of the inpainted image. Parameter Weight Image inpainting Quality (PWIIQ) [16] is one of the structure based metrics which uses luminance and gradient similarity to determine the quality of the inpainted image.

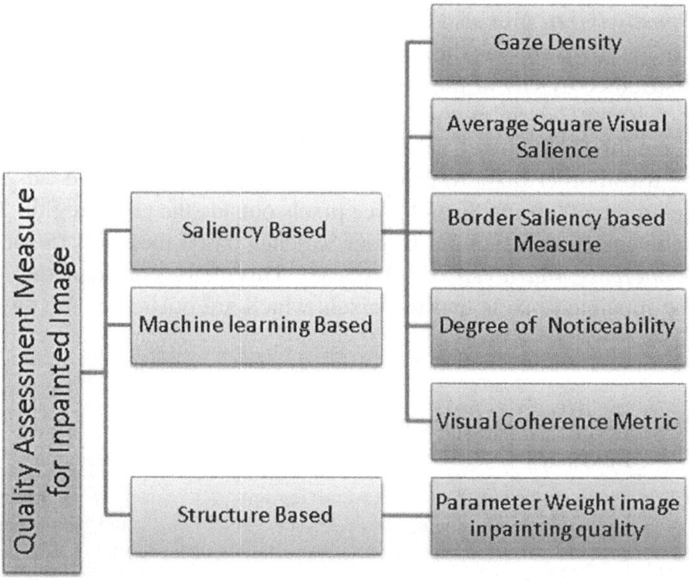

Fig. 11 Classification of quality assessment measure for inpainted image

3.2 *Saliency Based*

The saliency of the image highlights the area toward which the human vision is more responsive or interested. Hence, saliency can be used to estimate the visibility of various artifacts imported by inpainting techniques. In [17], inpainted image artifacts are categorized as in-region and out-region artifacts. In-region artifacts occur when different color and structures are introduced in the targeted region only. Due to which increased saliency in the inpainted image area is observed and hence disturbs the attention flow within the inpainted area. Outerregion artifacts appear when local colors and structures are not stretched to the target area by the inpainted technique. Due to which increase saliency in the neighbourhood of the inpainted region is observed. Some of Quality assessment metrics which uses the concept of saliency are:

Average Square Visual Salience (ASVS): Being Non-reference based, this metric does not require any information regarding the original image. This metrics is related to the in-region artifacts as it only acknowledge the inpainted pixels compared to the overall scene. As the value of this metric increase, the perceptual quality of the image decrease.

Degree of Noticeability (DN): Considering, in-region and out-region artifacts, [18] projected a metric named as DN. This metric identifies non-noticeable target regions and display any alteration in flow, in the surrounding of the inpainted region. As the value of the DN increases, perceptual quality decreases.

Gaze Density (GD): GD also consider both in-area and out-area artifact of the inpainted image. To overcome the deviations in textures and size GD of the inpainted image is distributed by GD of the original image.

Border Saliency based measures (BorSal): According to [19], by mapping saliency of neighborhood pixel, saliency change in the inpainted image is observed. This metric uses border pixel to calculate the normalized GD. One can extend the border pixel, three pixels inside and three pixels outside the target region. Enhanced version of this metric is Structural Border Saliency based measures (StructBorSal).

Visual Coherence MetricVisCoM (VisCoM): This metric considers the correlation between the inpainted pixels and the pixels which are outside of the target region (Table 6).

Table 6 Summary of quality assessment measure

Metrics	Description	Region used	Limitation
PWIIQ [16]	Statistical feature like luminance and gradient is used	Overall region	Require original image and fails when inpainted region is large
DN [18]	Original saliency is conserved and highlights shift in attention flow beyond target region	In-region and Out-region	Require original image and doesn't consider overall appearance of the inpainted image
BorSal [42]	Fast, uses single border area around the target image Inversely proportional to image quality. Used when fidelity is not important	Border–region	Require original Image
ASVS [19]	GD = 1 indicates no deviation of attention flow in the inpainted image. It also show shift in attention flow within and beyond target region	In-region	Doesn't consider overall appearance of the inpainted image
GD [16]	Visual coherence as well as structural information is considered	In-region and our-region	Require original image
VisCoM [43]		Overall region	Require original image

4 Conclusion

This paper examines various inpainting methods with a special focus on UAV images. The inpainting techniques are critically reviewed and gaps are indicated in the tables with features, limitations and sutability. Most of the methods works well for small area to be inpainted such as texture and PDE synthesis based inpainting techniques. They, cannot block the large disappearing area and also cannot recover the curvy sequence. Modified Oliveira algorithm packs the undesired objects in UAV images which are large without blur. Bilateral filter based approach protects the edges and eliminates the noise from UAV images. 8 neighborhood fast sweeping algorithm gives better results than Bertalmio's algorithm. Inpainting single and multiple regions in UAV images can be done by using the spatial contextual correlation algorithm. Poisson equation based approach gives good visual effects for large inpainting area. Using color distribution analysis, the consistency of texture and continuity at edges for a better visual quality can be obtained. Edges in the UAV images can be enhanced by using the extended wavelet transform. Non-linear diffusion tensor method repairs the corrupt zones and preserves discontinuities in UAV images. In future,3D image inpainting can be done using CNN algorithm and CNN based inpainting technique can be applied on UAV videos.

References

1. Ratliff BM, Tyo JS, Boger JK, Black WT, Bowers DL, Fetrow MP (2007) Dead pixel replacement in lwir microgrid polarimeters. Opt Express 15(12):7596–7609
2. Guillemot C, Le Meur O (2014) Image inpainting: overview and recent advances. IEEE Signal Process Mag 31(1):127–144
3. Efros AA, Leung TK (1999) Texture synthesis by non-parametric sampling. In: Proceedings of the 7th IEEE international conference on computer vision, vol 2. IEEE, pp 1033–1038
4. Wei LY, Levoy M (2000) Fast texture synthesis using tree-structured vector quantization. In: Proceedings of the 27th annual conference on computer graphics and interactive techniques. ACM Press/Addison-Wesley Publishing Co, pp 479–488
5. Bugeau A, Bertalmío M, Caselles V, Sapiro G (2010) A comprehensive framework for image inpainting. IEEE Trans Image Process 19(10):2634–2645
6. Criminisi A, Pérez P, Toyama K (2004) Region filling and object removal by exemplar based image inpainting. IEEE Trans Image Process 13(9):1200–1212
7. Drori I, Cohen-Or D, Yeshurun H (2003) Fragment-based image completion. In: ACM Transactions on graphics (TOG), vol 22. ACM, pp 303–312
8. Bentley JL (1975) Multidimensional binary search trees used for associative searching. Commun ACM 18(9):509–517
9. Bertalmio M, Vese L, Sapiro G, Osher S (2003) Simultaneous structure and texture image inpainting. IEEE Trans Image Process 12(8):882–889
10. Starck JL, Elad M, Donoho DL (2005) Image decomposition via the combination of sparse representations and a variational approach. IEEE Trans Image Process 14(10):1570–1582
11. Komodakis N (2006) Image completion using global optimization. In: 2006 IEEE computer society conference on computer vision and pattern recognition (CVPR'06), vol 1. IEEE, pp 442–452

12. Sun J, Yuan L, Jia J, Shum HY (2005) Image completion with structure propagation. In: ACM transactions on graphics (ToG), vol 24. ACM, pp 861–868
13. Fadili JM, Starck JL, Elad M, Donoho DL (2009) Mcalab: Reproducible research in signal and image decomposition and inpainting. Comput Sci Eng 1:44–63
14. Xu Z, Sun J (2010) Image inpainting by patch propagation using patch sparsity. IEEE Trans Image Process 19(5):1153–1165
15. Ardis PA, Brown CM, Singhal A (2010) Inpainting quality assessment. J Electron Imaging 19(1):011002
16. Gupta K, Kazi S, Kong T (2016) Deeppaint: a tool for image inpainting. Google Scholar
17. Oncu AI, Deger F, Hardeberg JY (2012) Evaluation of digital inpainting quality in the context of artwork restoration. In: European conference on computer vision. Springer, pp 561–570
18. Venkatesh MV, Sen-ching SC (2010) Eye tracking based perceptual image inpainting quality analysis. In: 2010 IEEE international conference on image processing. IEEE, pp 1109–1112
19. Schmidt U, Gao Q, Roth S (2010) A generative perspective on mrfs in low-level vision. In: 2010 IEEE computer society conference on computer vision and pattern recognition. IEEE, pp 1751–1758
20. Liu J, Musialski P, Wonka P, Ye J (2013) Tensor completion for estimating missing values in visual data. IEEE Trans Pattern Anal Mach Intell 35(1):208–220
21. Richard MMOBB, Chang MYS (2001) Fast digital image inpainting. In: Appeared in the proceedings of the international conference on visualization, imaging and image processing (VIIP 2001), Marbella, Spain, pp 106–107
22. Bertalmio M, Sapiro G, Caselles V, Ballester C (2000) Image inpainting. In: Proceedings of the 27th annual conference on computer graphics and interactive techniques. ACM Press/Addison-Wesley Publishing Co, pp 417–424
23. Telea A (2004) An image inpainting technique based on the fast marching method. J Graph Tools 9(1):23–34
24. Tschumperlé D (2006) Fast anisotropic smoothing of multi-valued images using curvature-preserving pde's. Int J Comput Vis 68(1):65–82
25. Rudin LI, Osher S, Fatemi E (1992) Nonlinear total variation based noise removal algorithms. Physica D 60(1–4):259–268
26. Chan TF, Shen J (2001) Nontexture inpainting by curvature-driven diffusions. J Vis Commun Image Represent 12(4):436–449
27. Shen J, Kang SH, Chan TF (2003) Euler's elastica and curvature-based inpainting. SIAM J Appl Math 63(2):564–592
28. Ashikhmin M (2001) Synthesizing natural textures. In: Proceedings of the 2001 symposium on interactive 3D graphics, Citeseer, pp 217–226
29. Liang L, Liu C, Xu YQ, Guo B, Shum HY (2001) Real-time texture synthesis by patch-based sampling. ACM Trans Graph (ToG) 20(3):127–150
30. Barnes C, Shechtman E, Goldman DB, Finkelstein A (2010) The generalized patchmatch correspondence algorithm. In: European conference on computer vision. Springer, pp 29–43
31. Efros AA, Freeman WT (2001) Image quilting for texture synthesis and transfer. In: Proceedings of the 28th annual conference on computer graphics and interactive techniques. ACM, pp 341–346
32. Barnes C, Shechtman E, Goldman DB, Finkelstein A (2010) Supplementary material for the generalized patchmatch correspondence algorithm. Retrieved from on Sep 9, 6
33. Bertalmio M, Bertozzi AL, Sapiro G (2001) Navier-stokes, fluid dynamics, and image and video inpainting. In: Proceedings of the 2001 IEEE computer society conference on computer vision and pattern recognition. CVPR 2001, vol 1. IEEE, pp I–I
34. Elad M, Starck JL, Querre P, Donoho DL (2005) Simultaneous cartoon and texture image inpainting using morphological component analysis (mca). Appl Comput Harmon Anal 19(3):340–358
35. Aujol JF, Ladjal S, Masnou S (2010) Exemplar-based inpainting from a variational point of view. SIAM J Math Anal 42(3):1246–1285

36. Cheng Q, Shen H, Zhang L, Li P (2014) Inpainting for remotely sensed images with a multichannel nonlocal total variation model. IEEE Trans Geosci Remote Sens 52(1):175–187
37. Nalawade VV, Ruikar SD Image inpainting using wavelet transform. Int J Adv Eng Technol E-ISSN, 0976–3945
38. Shen H, Zhang L (2009) A map-based algorithm for destriping and inpainting of remotely sensed images. IEEE Trans Geosci Remote Sens 47(5):1492–1502
39. Cai N, Su Z, Lin Z, Wang H, Yang Z, Ling BWK (2017) Blind inpainting using the fully convolutional neural network. Vis Comput 33(2):249–261
40. Xie J, Xu L, Chen E (2012) Image denoising and inpainting with deep neural networks. In: Advances in neural information processing systems. pp 341–349
41. Hays J, Efros AA (2008) Scene completion using millions of photographs. Commun ACM 51(10):87–94
42. Dang TT, Beghdadi A, Larabi MC (2013) Visual coherence metric for evaluation of color image restoration. In: 2013 colour and visual computing symposium (CVCS). IEEE, pp 1–6
43. Ardis PA, Singhal A (2009) Visual salience metrics for image inpainting. In: Visual communications and image processing 2009, vol 7257. W. International Society for Optics and Photonics, p 72571

A Fuzzy Sliding Mode Control Design for Quadcopter

Jagannath Samantaray and Sohom Chakrabarty

Abstract In this paper, an intelligent and robust control approach is used for the control of a highly nonlinear quadcopter. A sliding mode controller is designed for both altitude and attitude control of the quadcopter. The switching gain of the sliding mode controller is tuned based on the fuzzy control technique. Rule base of the fuzzy logic controller is designed based on vast simulation and stability of the designed algorithm is shown. Mathematical derivations for roll pitch, yaw and altitude control is shown step by step. Different uncertain conditions are simulated by taking random disturbances and robustness under uncertain condition are guaranteed with the fuzzy sliding mode control technique. Simulation results show that the designed control algorithm works satisfactorily for both altitude and attitude control. Qualitative analysis is done for the performance of the designed fuzzy sliding mode controller. Finally, trajectory tracking of the quadcopter with attitude control is done.

Keywords Quadcopter · Sliding mode control · Fuzzy logic control fuzzy sliding mode control · Altitude control · Attitude control trajectory tracking

1 Introduction

Quadcopters become popular in many fields like imaging, video recording, parcel delivery, defense sector, medical transportation etc. due to their agility, flexibility and programmable features [1, 2]. Application of quadcopter becomes feasible by controlling the altitude and attitude. The highly nonlinear dynamical equations related to its mechanical design makes it difficult to control. Furthermore, the control becomes more difficult when the quadcopter is subjected to an environment where it is disturbed by the aerodynamics of its design, environmental constraints like different wind profiles. Hence it is required to design a robust control algorithm which can control both altitude and attitude in spite of the presence of uncertainties.

J. Samantaray (✉) · S. Chakrabarty
Indian Institute of Technology Roorkee, Roorkee, Uttarakhand 247667, India
e-mail: samantaray.jagnnath2k9@gmail.com

S. Chakrabarty
e-mail: sohomfee@iitr.ac.in

K. Jain et al. (eds.), *Proceedings of UASG 2019*, Lecture Notes in Civil Engineering 51,
https://doi.org/10.1007/978-3-030-37393-1_17

Many control algorithms are already used for trajectory control i.e. altitude and attitude control of quadcopter. Controllers like sliding mode control (SMC) is used for trajectory tracking of quadcopter in [3–6]. Other robust techniques like feedback linearization in [7, 8], fuzzy logic controller in [9–11] and back-stepping controller in [6], and adaptive controller in [5] are reported.

In this paper, a fuzzy sliding mode controller (FSMC) is designed for a disturbance affected nonlinear mathematical model. Here the disturbance is assumed to be matched uncertainty in the sense of Gao [12]. Fuzzy logic controller (FLC) is designed based on the membership functions of error and change in error and used to tune the switching gain of the SMC. The switching gain of the SMC is tuned by FLC. Defuzzification is done with seven membership functions as more granulation leads to more accuracy. The existing works on FSMC [10, 13, 14] consider a simpler mathematical model and decide the fuzzy gains for a specific disturbance. The fuzzy rule base is designed with Lyapunov approach. But in this paper, it is done with input-output behaviour by doing multiple simulations for disturbance affected model. Both altitude and attitude control is done with FLC for time varying trajectories.

After a brief introductory section, the mathematical model of the quadcopter is given in Sect. 2. Then Design procedure is explained in Sect. 3. Sliding mode control design and Fuzzy sliding mode control design is explained in Sects. 3.1 and 3.2 respectively. Then, results are given in Sect. 4 and the results are briefly discussed. Finally, conclusions are highlighted in Sect. 5.

2 Mathematical Model of Quadcopter

The nonlinear dynamics of altitude (z), roll (Φ), pitch (θ) and yaw (Ψ) are given by the following equations [10].

$$\ddot{z} = \dot{g} - \frac{(u_1 + f_1)}{m}[\cos\Phi\cos\theta] \tag{1}$$

$$\ddot{\Phi} = \left(\frac{I_{yy} - I_{zz}}{I_{xx}}\right)\dot{\theta}\dot{\Psi} + \frac{(u_2 + f_2)}{I_{xx}} \tag{2}$$

$$\ddot{\theta} = \left(\frac{I_{zz} - I_{xx}}{I_{yy}}\right)\dot{\Phi}\dot{\Psi} + \frac{(u_3 + f_3)}{I_{yy}} \tag{3}$$

$$\ddot{\Psi} = \left(\frac{I_{xx} - I_{yy}}{I_{zz}}\right)\dot{\Phi}\dot{\theta} + \frac{(u_4 + f_4)}{I_{zz}} \tag{4}$$

where u_1, u_2, u_3, and u_4 are the control inputs to the quadcopter and f_1, f_2, f_3, and f_4 are the disturbances considered which may come due to aerodynamic constraints, different wind profiles etc. and affect the altitude, roll, pitch, and yaw dynamics respectively. It is assumed that $f_i < f_{im}$, where f_{im} is known for each i ∈ (1, 4). The parameters of the quadcopter is given in Table 1.

Table 1 Parameters of quadcopter

Variable	Value	Units
Mass of the quadcopter (m)	0.978	kg
Acceleration due to gravity (g)	9.81	m/s^2
Arm length (L)	0.302	m
Moment of Inertia in X-axis (I_{xx})	0.22	kg m^2
Moment of Inertia in Y-axis (I_{yy})	0.22	kg m^2
Moment of Inertia in Z-axis (I_{zz})	0.043	kg m^2

3 Design Procedure of Controller

Here our aim is to calculate the control inputs u_1, u_2, u_3, and u_4 so that the quadcopter will track the desired altitude, roll, pitch and pitch trajectories together.

Controller design is done in two steps. The first step is to design the sliding mode controller to track the desired profiles. Then in the next step, switching gain of the SMC is tuned by the fuzzy logic algorithm to get better robustness [12].

a. **Sliding Mode Controller Design for Trajectory Tracking**

Altitude Control

Let error be defined as

$$e_z = z_d - z \tag{5}$$

where z_d is the desired altitude trajectory. A sliding variable is chosen as

$$s_z = \dot{e}_z + c_z e_z \tag{6}$$

The constant plus proportional reaching law given in [12] for the above sliding mode is

$$\dot{s}_z = -k_2 s_2 - \eta_2 sign(s_2) \tag{7}$$

is utilized to calculate the control action. From (6) and (7), and using (1), we get

$$\Rightarrow \ddot{e}_z + c_z \dot{e}_z = -k_z s_z - \eta_z sign(s_z) \tag{8}$$

$$\Rightarrow \ddot{z}_d + \frac{u_1}{m}(\cos\theta\cos\Phi) - g - c_z e_z = -k_z s_z - \eta_z sign(s_z) \tag{9}$$

$$\Rightarrow u_1 = \frac{m}{\cos\theta\cos\Phi}\left[\ddot{z}_d + g + c_z e_z + k_z s_z + \eta_z sign(s_z)\right] \tag{10}$$

where $k_z > 0$ and $\eta_z > \frac{\cos\theta\cos\Phi}{m} f_{1m}$.

Roll Control

Let the roll error be defined as

$$e_r = \Phi_d - \Phi \tag{11}$$

where Φ_d is the desired roll trajectory to be followed. The sliding variable chosen here is given by

$$s_r = \dot{e}_r + c_r e_r \tag{12}$$

The constant and proportional reaching law is

$$\dot{s}_r = -k_r s_r - \eta_r sign(s_r) \tag{13}$$

Using (11), (12), and (13), the control input is derived as (15).

$$\Rightarrow \ddot{\Phi}_d - \ddot{\Phi} + c_r \dot{e}_r = -k_r s_r - \eta_r sign(s_r) \tag{14}$$

$$\Rightarrow u_2 = I_{xx}\left[\ddot{\Phi}_d + c_r \dot{e}_r - \left(\frac{I_{yy} - I_{zz}}{I_{xx}}\right)\dot{\theta}\dot{\psi} + k_r s_r + \eta_r sign(s_r)\right] \tag{15}$$

where $k_r > 0$ and $\eta_r > \frac{f_{2m}}{I_{xx}}$.

Pitch Control and Yaw Control

Similar Analysis is done like roll control and the control inputs are derived for pitch control and yaw control are given in (16) and (17), respectively.

Let the pitch error be e_p and sliding variable chosen for the pitch trajectory control be s_p, and given as $e_p = \theta_d - \theta$ and $s_p = e^{\cdot}{}_p + c_p e_p$ respectively. Then

$$\Rightarrow u_3 = I_{yy}\left[\ddot{\theta}_d + c_p \dot{e}_p - \left(\frac{I_{zz} - I_{xx}}{I_{yy}}\right)\dot{\Phi}\dot{\psi} + k_p s_p + \eta_p sign(s_p)\right] \tag{16}$$

where $k_p > 0$ and $\eta_p > \frac{f_{3m}}{I_{yy}}$.

Let the yaw error be e_y and sliding variable chosen for the yaw trajectory control be s_y, and given as $e_y = \Psi_d - \Psi$ and $s_y = e^{\cdot}{}_y + c_y e_y$ respectively. Then

$$\Rightarrow u_4 = I_{zz}\left[\ddot{\theta}_d + c_y \dot{e}_y - \left(\frac{I_{xx} - I_{yy}}{I_{zz}}\right)\dot{\theta}\dot{\Phi} + k_y s_y + \eta_y sign(s_y)\right] \tag{17}$$

where $k_y > 0$ and $\eta_y > \frac{f_{4m}}{I_{zz}}$.

b. **Fuzzy Sliding Mode Controller Design for Trajectory Tracking**

In this section, a fuzzy logic controller is designed to tune the switching gain of the sliding mode controller i.e. η for altitude, roll, pitch, and yaw control. Triangular

Table 2 Range of membership function

Membership function	Range
Negative Big (NB) Negative Moderate (NM) Negative Small (NS) Zero (Z) Positive Small (PS) Positive Moderate (PM) Positive Big (PB)	[−1 −0.6666] [−1 −0.6666 −0.3333] [−0.6666 −0.3333 0] [−0.3333 0 0.3333] [0 0.3333 0.6666] [0 0.3333 0.6666] [0.666 1]

Table 3 Fuzzy rule base

e/ce	NB	NM	NS	Z	PS	PM	PB
NB	NB	NB	NB	NB	NM	NS	Z
NM	NB	NB	NB	NM	NS	Z	PS
NS	NB	NB	NM	NS	Z	PS	PM
Z	NB	NM	NS	Z	PS	PM	PB
PS	NM	NS	Z	PS	PM	PB	PB
PM	NS	Z	PS	PM	PB	PB	PB
PB	Z	PS	PM	PB	PB	PB	PB

membership functions are chosen by doing multiple simulations of input and output model. Error (e) and change in error (ce) are taken as triangular membership functions. Crisp values of error and change in error are transformed to triangular membership functions as per MAMDANI type. The granulation and graduation are translated in the form of types of membership function and range of each membership. Scaling factor is also based on the input-output behaviour and is given in Table 2. Then Fuzzy rule base is framed after studying the behaviour of the output when excited by different kind of inputs with respect to different disturbance. This is given in Table 3. Output membership is chosen same as input membership function. De-fuzzified output is decided from the rule base by IF-THEN principle. For instance, error is Z and change in error is NS, then output is NS which is obvious from the rule base. The switching gains η_z, η_r, η_p and η_y are tuned as per the rule base designed in Table 3.

4 Results and Discussion

To check the effectiveness of the FSMC, simulation is done for a quadcopter by taking the parameter values as given in Table 1. A uniform random number having a minimum as −1 and maximum value as 1 is taken as disturbance in all the dynamics of the quadcopter. The value of k_z, k_r, k_p and k_y are taken 100 for faster reaching and η_z, η_r, η_p and η_y are decided from the FLC based on present values of error and change in error. A square wave is given as a desired altitude trajectory and the quadcopter is able to achieve the desired trajectory with the control input derived in (10) and it is shown in Fig. 1. Similarly, a sinusoidal trajectory is given as roll and pitch desired trajectory. Tracking of desired roll and pitch is shown in Figs. 2 and 3 respectively. Yaw trajectory tracking is given in Fig. 4. Sliding variables converge to zero in finite time, which is clear from Fig. 5. Control inputs required to achieve trajectory tracking are given in Figs. 6 and 7. Sudden jerks are due to the derivative terms included in the control input designed and in practice it can be be taken care

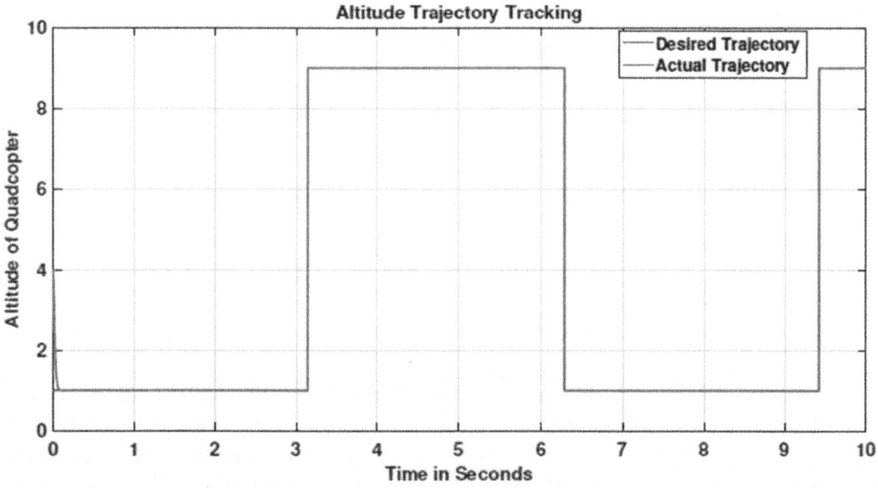

Fig. 1 Trajectory tracking of altitude

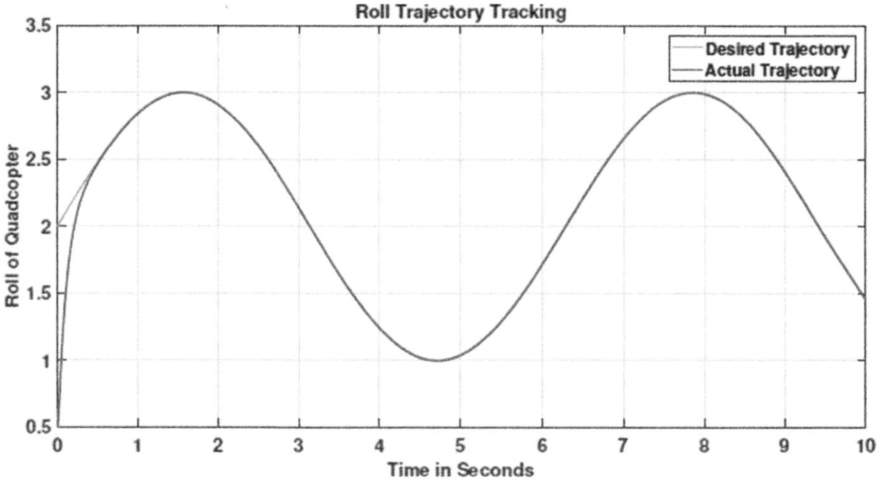

Fig. 2 Trajectory tracking of roll

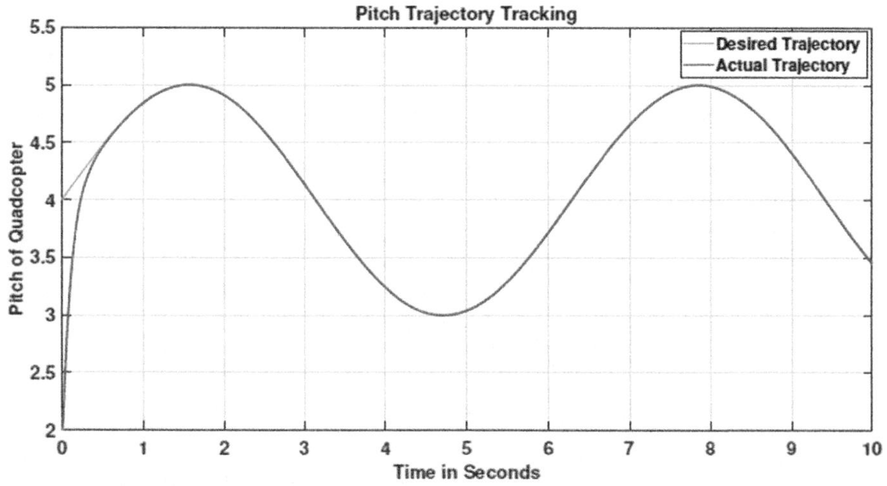

Fig. 3 Trajectory tracking of pitch

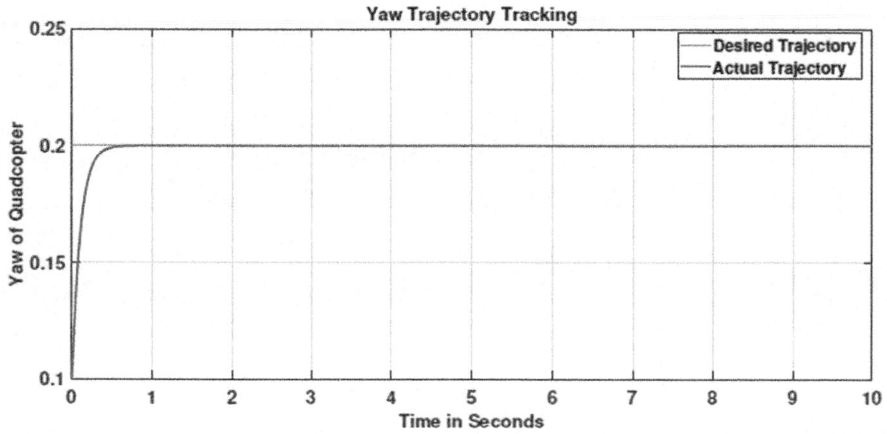

Fig. 4 Trajectory tracking of yaw

by replacing the *sign*(s) by $\frac{s}{}$ where s is a small number. This approach is called pseudo-sliding $|s| + s$ mode. With this method, chattering can be eliminated and can be implemented in hardware.

5 Conclusions

In this work, a fuzzy sliding mode controller is designed for both altitude and attitude control. FLC is designed based on the error and change in error MAM-DANI type.

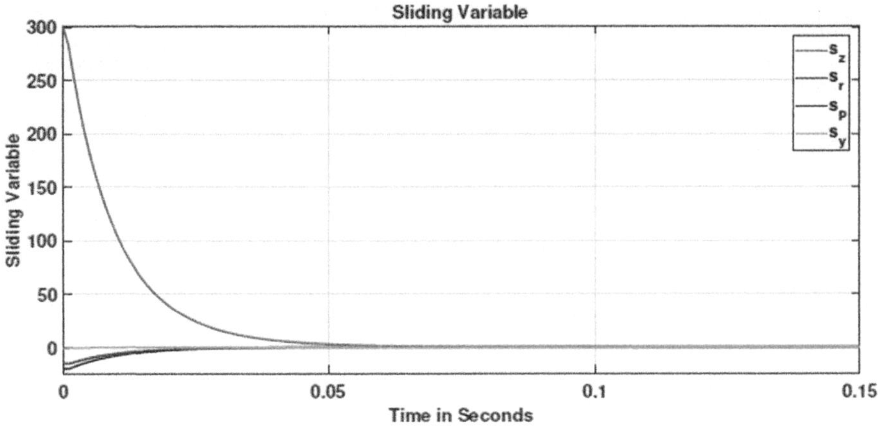

Fig. 5 Sliding variables of each dynamics

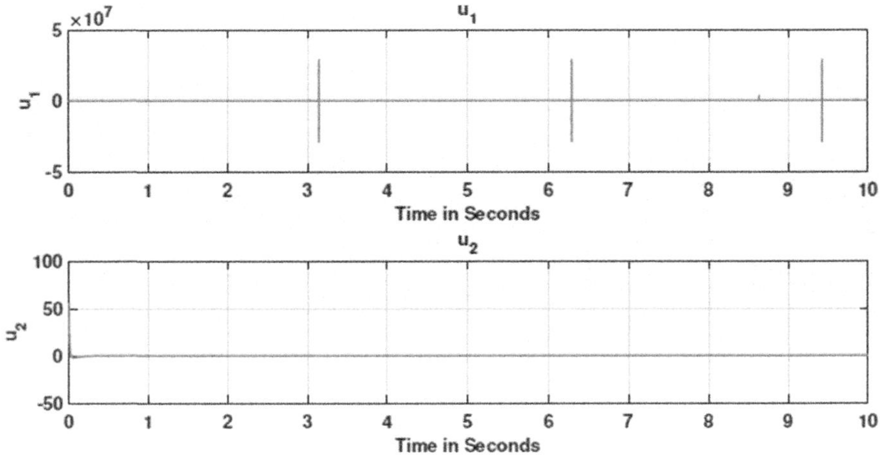

Fig. 6 Control inputs for altitude and roll control

FLC is used to regulate the switching gains of the SMC for achieving better robustness. A nonlinear model is considered with the matched disturbance and the designed FSMC is shown to work satisfactorily which is clear from the simulation plots. Future works can be done for trajectory tracking in discrete time and comparisons should be done with the continuous time.

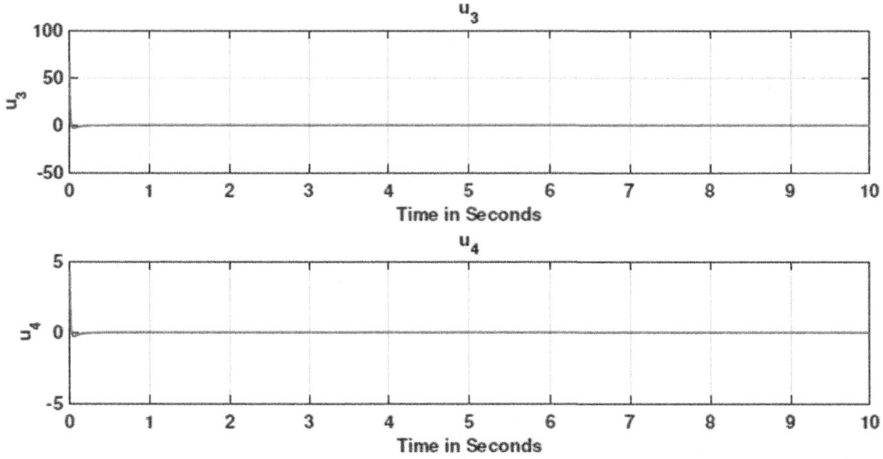

Fig. 7 Control inputs for pitch and yaw control

References

1. Wang N, Deng Q (2018) Finite-time disturbance observer based integral sliding mode control of a quadrotor. In: Proceedings—2018 33rd youth academic annual conference of Chinese association of automation, YAC 2018, pp 956–960
2. Zhao B, Tang Y, Wu C, Wu W (2018) Vision-based tracking control of quadrotor with backstepping sliding mode control. IEEE Access 6:72439–72448
3. Rios H, Falcon R, Gonzalez OA, Dzul A (2019) Continuous sliding-mode control strategies for quadrotor robust tracking: real-time application. IEEE Trans Industr Electron 66(2):1264–1272
4. Ruiz AGE, Alazki H, Rubio JEV, Salazar OG (2016) Embedded super twisting control for the attitude of a Quadrotor. IEEE Latin Am Trans 14(9):3974–3979
5. Huang Y, Zheng Z, Sun L, Zhu M (2018) Saturated adaptive sliding mode control for autonomous vessel landing of a quadrotor. IET Control Theory Appl 12(13):1830–1842
6. Fethalla N, Saad M, Michalska H, Ghommam J (2018) Robust observer-based dynamic sliding mode controller for a quadrotor UAV. IEEE access, 6(Dcm):45846–45859
7. Fang Z, Zhang Z, Jim L, Jian W (2008) Feedback linearization and continuous sliding mode control for a quarotor UAV. In: Proceedings of the 27th Chinese control conference, CCC, pp 349–353
8. Nguyen NT, Prodan I, Lefevre L (2017) Multi-layer optimization-based control design for quadcopter trajectory tracking. In: 2017 25th mediterranean conference on control and automation, MED 2017, pp 601–606
9. Sardar S, Kadri MB (2015) Autonomous control of a quadcopter via fuzzy gain scheduled PD control. In: Proceedings—12th international conference on frontiers of information technology, FIT 2014, pp 73–78
10. Tang Y, Zhang H, Gong J (2016) Adaptive-fuzzy sliding-mode control for the attitude system of a quadrotor. In: Proceedings—2015 Chinese automation congress, CAC 2015, pp 1075–1079
11. Santoso F, Garratt MA, Anavatti SG (2016) Fuzzy logic-based self-tuning autopilots for trajectory tracking of a low-cost quadcopter: a comparative study. In: ICAMIMIA 2015—international conference on advanced mechatronics, intelligent manufacture, and industrial automation, proceeding—in conjunction with industrial mechatronics and automation exhibition, IMAE, pp 64–69
12. Gao W, Wang Y, Homaifa A (1995) Discrete-time variable structure control systems. IEEE Trans Industr Electron 42(2):117–122

13. Deia Y, Kidouche M, Ahriche A (2016) Fully decentralized fuzzy sliding mode control with chattering elimination for a Quadrotor attitude. In: 2015 4th international conference on electrical engineering, ICEE 2015, pp 1–6
14. Huaman-Loayza AS (2018) Path-following of a quadrotor using fuzzy sliding mode control. In: 2018 IEEE XXV international conference on electronics, electrical engineering and computing (INTERCON), pp 1–4

Unmanned Aerial Vehicles: Vulnerability to Cyber Attacks

Susheela Dahiya and Manik Garg

Abstract With the increase of technological capabilities of automated systems, the use of unmanned aerial vehicles (UAVs) has also increased in many military and civilian applications. UAVs today play an important role in many other areas like wildlife surveys, weather monitoring, monitoring natural disasters affects etc. It is also expected that UAVs will be a major part of future smart cities. The amount and type of information present with UAVs makes it an extremely interesting target for cyber-attacks. However, the cybersecurity aspect of UAVs has not been fully considered while building UAVs. As a result, UAVs are more vulnerable to cyber-attacks. Potential security vulnerability may exist in the modules, which are responsible for their proper working or may exist during communication between UAVs and control station. Out of these two, communication security is critically important for the success of UAVs as they often carry sensitive information that adversaries might try to get hold of. Wi-Fi attacks such as Eavesdropping, Information Injection, Denial-of-Service, and Distributed DoS are the possible security threats to UAV communications. Recently, GPS spoofing attack, session hijacking, and compromised surveillance are also reported. The goal of this paper is to provide the different levels of vulnerabilities along with the prevention measures required at each level, some major attacks that can be performed on a UAV along with their cause, impact and the precautions required to avoid that attack. It has been observed that the most easily attackable vulnerability on the UAV system is flooding the UAV using the radio communication and the most harmful vulnerability is acquiring complete control through Man-in-the-Middle attack.

Keywords UAV · Vulnerability · Threat · Security · Network attacks

1 Introduction

Unmanned Ariel Vehicle is a drone without any direct human control. It is helpful in various tasks such as guided surveillance [1], weather monitoring [2], unmanned

---

S. Dahiya (✉) · M. Garg
School of Computer Science, University of Petroleum and Energy Studies, Dehradun, Uttarakhand, India

K. Jain et al. (eds.), *Proceedings of UASG 2019*, Lecture Notes in Civil Engineering 51,
https://doi.org/10.1007/978-3-030-37393-1_18

attacks, covert intrusions, enemy reconnaissance, aircraft maintenance and repair operations, military training [3], cargo transportation, disaster relief [4], rescue operations [5], search operations, tracking operations [6] etc. It is operated and guided from a remote location also known as the control center. A control center is typically a place with a transmitter and a controller (human) that gives commands to the UAV with the help of remote connection sending instructions through a communication channel using radio waves. Apart from radio waves, GPRS & EDGE technologies are also used for communication between the UAV transceiver and the control center transceiver.

The surveillance video data collected by UAVs is confidential and time sensitive that need to be shared on immediate basis. Any unauthorized access or delay in data transmission can result in mission failure [7]. Since there is a remote communication between the UAV transceiver and the control center transceiver through the atmospheric medium, it is susceptible to a huge number of cyber-attacks. The technologies used for communication with UAV such as radio, GPRS & EDGE works by means of packets contacting a small block of data. A packet traveling over a network needs to be secured from various threats that can in any way change the packet or result in sharing the data with unauthorized users.

The contents of this paper organized in eight sections. The first section aims at providing an introduction about the role of a UAV and the importance of its security. The second section deals with the various vulnerabilities associated with the UAV. Categorization of Attacks are discussed in the third section. The attacks that can be performed on a UAV are explained in the fourth section. The fifth section deals with the recent attacks on UAVs, their cause and cure. Prevention measures required on different levels of UAVs are explained in sixth section. The seventh section suggests some measure to protect UAVs against these attacks. Finally, section eight concludes the paper.

2 Vulnerabilities

In UAVs, the vulnerabilities can be at any of the following three levels: Transceiver Level, Control Center Level and Communication Channel Level. Out of these three levels, the threats at transceiver level and communication level are from outsiders but in case of control center, an insider can also be a threat [8]. The following subsections gives a brief overview of the vulnerabilities associated with each level.

2.1 Transceiver Level

The UAVs major component is its CPU along with its transceiver. The transceiver fulfills the function of transmitting and receiving packets to and from the control center through the communication channel. Since UAVs are also used for military surveillance thus, the traffic that is incoming as well as outgoing can be highly crucial and sensitive. The UAV should verify the incoming traffic for authenticity and integrity but if it fails to do so, there can be a flow of some illegitimate packets that can contain some wrong instructions for the UAV. The main vulnerability at this level is non-validation of the packets received. If any unwanted command is issued to the UAV and it acts upon it then, it may lead to unauthorized and unaccounted attacks. In addition to that, the whole surveillance system can be taken down.

2.2 Control Center Level

The control center is the brain of the UAV since all the control lies within the control center. The main threat to any asset is always the insiders i.e. any user who has rogue intentions. Other vulnerability lies on the network providing access to the control center where various attacks can be performed. If the control center is compromised, then the surveillance system can be hijacked. The access to control center network may contain vulnerabilities such as SQL injection which may lead to access to the network from a person who can be a threat. Also, if there are no Backup servers in the control center or load balancers are not set up then the whole control mechanism service can be taken down by flooding requests that will cause incomplete TCP Handshake [9].

2.3 Communication Channel Level

All the control instructions for the UAV travels through this medium. The data traveling needs to be encrypted else if intercepted can cause major harm to the UAV and the organization owning it. Apart from interception, the communication channel can be a medium to perform various types of active attacks. Various vulnerabilities from the OWASP top 10 list can be found over the communication medium. Since communication is mainly done using radio waves or GPRS/EDGE, the connection is very insecure and can be compromised easily. Also, while establishing a connection, using less secured protocols can also lead to compromised security [10].

3 Types of Attacks

The attacks that can be performed on a UAV depends on the vulnerability that the attacker is targeting. Since, there are three major targets in the whole UAV system i.e. UAV transceiver, communication channel, and the control center, there can be single or multiple vulnerabilities that can be targeted at a particular instance of time. The attacks that can be performed can be categorized into the following two categories.

3.1 Active Attacks

This type of attacks comes under the penetration testing part of the Ethical Hacking system. In such attacks, the main aim is to disrupt the services or perform a breach without caring about the interruption in the original transmission. These attacks are done in the real-time i.e. at $t = 0$ and are completed as soon as the required data or aim is achieved.

3.2 Passive Attacks

This type of attacks usually involves network monitoring, port listening or packet sniffing. In such attacks, the attacker usually sits on the network silently without the knowledge of the user and thus there is no interruption in the original transmission. The attacker captures the needed packets and afterward performs analysis on them to attain the required information such as secret keys or digital certificate algorithms.

4 Attacks and Their Risk Factors

Man-in-the-Middle attack, Denial of Service attack and Command Injection attack are the three main attacks that can performed on a UAV [11]. Each attack will cause a different kind of loss. That loss can be either a minor financial loss or a major industry collapse or a security threat. The cause and risk factor associated with the attacks that can be performed on a UAV system are explained in the following subsection.

4.1 Man-in-the-Middle Attack

This attack can be of both types either passive or active. In this attack, the attacker intercepts the traffic between the legitimate sender and the receiver and performs either reconnaissance or data tampering. This attack may lead to a data breach or a major loss in data integrity. Data Integrity is the most important part of the CIA triad and it needs to be preserved. This attack is possible by capturing the sharing of keys at the time of connection establishment. Other possible ways to attain this attack are IP spoofing, ARP poisoning, DNS poisoning, ARP spoofing, DNS spoofing, SSL hijacking, HTTPS spoofing and many more. The following image depicts a simple Man-in-the-Middle attack and demonstrates how the traffic is redirected from sender to attacker and then to the receiver [12] (Figs. 1 and 2).

The main concept on which this attack works is multiple TCP handshake establishments [13].

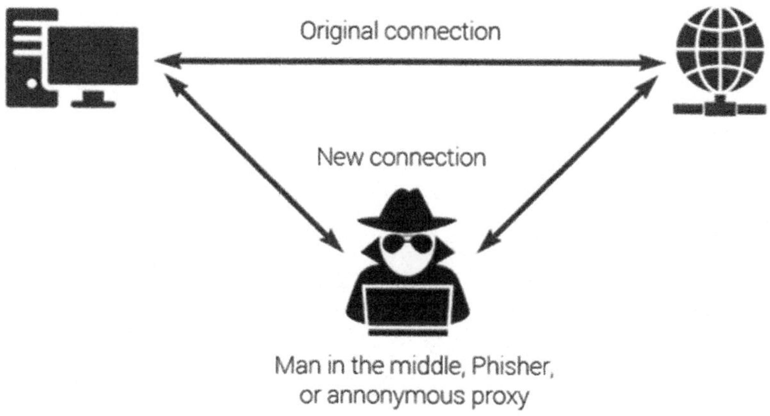

Fig. 1 Man-in-the-Middle attack [12]

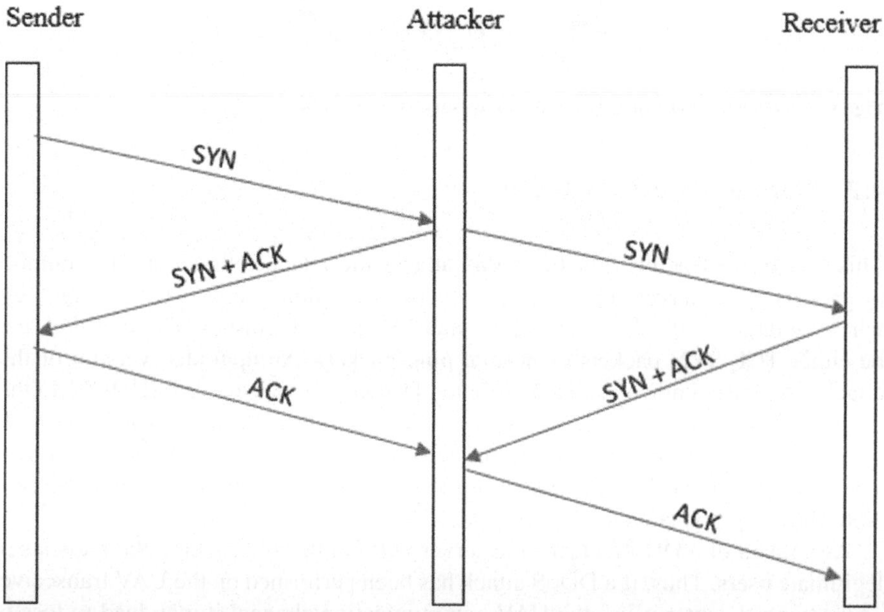

Fig. 2 Multiple TCP handshake in MITM attack

In the UAV communication channel if a MITM attack is performed it may lead to compromised surveillance, session hijacking, unauthorized activities, unauthorized attacks (in case of military UAVs), wrong data, change of projectile of UAV and much more.

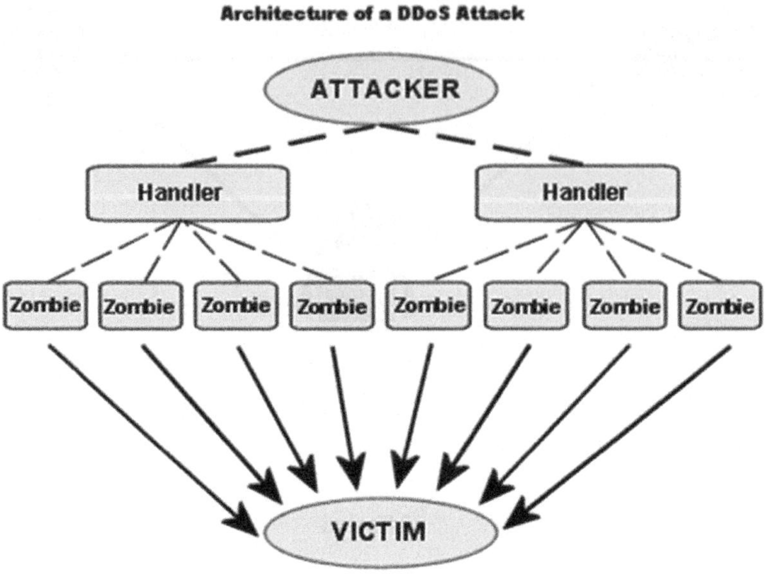

Fig. 3 Representation of distributed denial of service attack [14]

4.2 Denial of Service Attack

This is a type of active attack. In this attack, the attacker floods a huge number of packets to the target. These packets serve as multiple requests to the target and when the target is unable to serve these many requests, it crashes. These packets can be either TCP SYN packets or normal ping packets. An upgraded version of this attack also exists known as the Distributed Denial of Service attack (DDoS). In this attack instead of using a single source to flood packets, the attacker uses multiple sources controlled by him to flood packets. These sources are known as zombies. The complete network of zombies used to perform the attack is known as Zombie Net. The Fig. 3 explains the DDoS Attack.

As a result of a DDoS attacks the server will not be able to serve the requests of legitimate users. Thus, if a DDoS attack has been performed on the UAV transceiver or the control center using the UAV communication channel it may lead to loss of communication between the UAV and the control center. As a result, the UAV can be lost and some packets containing sensitive information may also be destroyed [14].

4.3 Command Injection Attack

This is also a type of active attack. In this attack, a piece of code is injected in the HTML based application. This injected code is malicious in nature and runs a script that can help in providing unauthorized access and data tampering. If a command injection vulnerability exists in the UAV control center or the UAV Drone, then the whole system can be compromised and taken in control by some external entity.

4.4 Privilege Escalation Attack

This is also a type of active attack. In this attack, the user performs operations that he/she is not authorized to perform by acquiring privileges of a higher authority. This happens due to weak access control systems or default passwords usage in admin systems.

4.5 IP Spoofing Attack

This is also a type of active attack. In this attack, the source of the requests is changed to a legitimate source by means of IP Spoofing. It refers to disguising the original IP address with some other fake IP address. Thus, if a firewall is configured in the UAV system to allow access from certain fixed IPs, this attack can be used to gain access.

5 Cyber Attacks on UAVs

UAV's could be hacked and turned into weapons. Thus, monitoring of UAVs is a big issue that needs to be addressed. Following are some attacks performed by using UAVs

I. In 2009, a terrorist group was found to have captured an unencrypted UAV video feed using SkyGrabber (a software for capturing free satellite videos) [15].
II. In 2011, Iran Cyber unit was able to acquire control over US army drones and received various sensitive information from these drones. This was also possible due to the weak security measures implemented on the drones [16].
III. In 2017, a Chinese drone was hacked. A Cyber Security Response Team from US did this and this team was able to find many vulnerabilities in this drone [17].
IV. On August 5, 2018, two drones were used by some terrorist organizations to carry out an attack on the President of Venezuela. These drones were packed with explosives. The main area of concern lies at how these drones were able to reach that near to the President [18].

6 Prevention of Vulnerabilities in UAV

There are various ways to prevent the UAV system from the attacks and threats that can exploit the vulnerabilities that exist in different modules of the UAV system. The prevention measures required at each level are explained in the following sub sections.

6.1 Communication Channel Level

The traffic that travels on the network, typically the sensitive information or the UAV control commands should be sent in an encrypted manner using some asymmetric cryptographic algorithms. Also, to provide safe integrity these packets should be associated with a hash value that can be checked at a later stage. Another measure that can be applied is checksum that also serves as a check to integrity. Also, while using wireless communication some secure protocols such as HTTPS, SSL, TLS, etc. should be used so as to serve requests in a better way using secure session management techniques.

6.2 Transceiver Level

Since this module is susceptible to various attacks such as DDoS & Session Hijacking so there should be some security and intelligent rules that should be implemented. The most secure way is to implement a firewall at the UAV that can filter the packets that are received. Rules should be configured in the firewall so that it only allows traffic from only one IP address i.e. from the control center server. This IP whitelisting will prevent DoS attacks as all other packets will be dropped immediately and also no malicious code will be able to reach the UAV. Since there are still chances of IP spoofing in this implementation thus an IDPS (Intrusion Detection & Prevention System) should be implemented to prevent any further intrusion.

6.3 Control Center Level

Since control center also has various vulnerabilities thus the first thing that shall be acted upon is access management. The access to the control center should be well maintained as it comes under a high-security zone. Measures such as Input Validation, Access log management, Concept of Least privilege should be implemented. In addition, since there is a huge possibility of DoS attack on the control center server, firewall and load balancers should be implemented with IP whitelisting rules. Also, the internet and the intranet should be bridged by a secure firewall or IDPS. To prevent data breaches or data destruction attacks, backup servers should be kept and updated at fixed intervals so that no crucial information is lost.

7 Results

It has been found that there are many vulnerabilities that exist on the UAV system like SQL injections, DoS attack, Man-in-the-Middle attack, Elevated Privileges and IP spoofing. Among these vulnerabilities, few are of high risk and few of low risk. The precautions required to avoid the vulnerabilities are as follows:

i. SQL/Command Injection vulnerability can be fixed by using input validation and strict type checking using sanitization of data received.
ii. To avoid DoS attack, the server and the UAV should be both able to handle huge number of packet flooding. This can be fixed using load balancers or IP whitelisting rules. Also, an IDPS or firewall should be setup that can detect and prevent from a possible DoS attack.
iii. Man-in-the-Middle vulnerability can be prevented by using secure transmission protocols, integrity checks, encrypted traffic and usage of VPNs (Virtual Private Networks).
iv. To prevent Elevated Privileges vulnerability strict access control mechanisms and access rights systems should be implemented. The network administrator should also regularly change default passwords and delete unused accounts.
v. IP spoofing vulnerability can be prevented by performing IP subnetting and masking. Also, there should be confidentiality involved while sharing these IP addresses as they are a crucial asset in this communication system.

8 Conclusion

The most easily attackable vulnerability on the UAV system is flooding the UAV using the radio communication. The most harmful vulnerability on the other hand is acquiring complete control through Man-in-the-Middle attack. These things need to be carefully addressed and worked upon. The amount of security measures to be deployed on a UAV depends upon the type of task it is being used for. Some crucial UAVs that are used for military or weather surveillance are more susceptible to attacks rather than normal event coverage drones. Also, the UAVs performing delivery operations can be targeted by some thieves. Specially the military drones that also have the capabilities to attack can be most targeted ones and need to have best security measures to be implemented.

Along with cybersecurity threats associated with UAV, we should always keep into consideration the other threats such as physical security threats, weather related issues, accidental collisions or intentional collisions and also most importantly drone capturing. Although UAVs bring a lot of automation and ease at many tasks, they also bring along a huge number of threats that needs to be addressed and worked upon based on their risk assessment depending on their impact and likelihood determination.

References

1. Kim A, Wampler B, Goppert J, Hwang I, Aldridge H (2012) Cyber attack vulnerabilities analysis for unmanned aerial vehicles. In: AIAA Infotech@Aerospace. https://doi.org/10.2514/6.2012-2438
2. Gupta SG, Ghonge MM, Jawandhiya PM (2013) Review of unmanned aircraft system (UAS). Int J Adv Res Comput Eng Technol (IJARCET) 2(4)
3. Udeanu G, Dobrescu A, Oltean M (2016) Unmanned aerial vehicle in military operations. Sci Res Educ Air Force 18(1):199–206. https://doi.org/10.19062/2247-3173.2016.18.1.26
4. Debusk W (2010) Unmanned aerial vehicle systems for disaster relief: Tornado Alley. In: AIAA Infotech@Aerospace 2010. https://doi.org/10.2514/6.2010-3506
5. Waharte S, Trigoni N (2010) Supporting search and rescue operations with UAVs. In: 2010 international conference on emerging security technologies. https://doi.org/10.1109/est.2010.31
6. Javaid AY, Sun W, Devabhaktuni VK, Alam M (2012) Cyber security threat analysis and modeling of an unmanned aerial vehicle system. In: 2012 IEEE conference on technologies for homeland security (HST), Waltham, MA, pp 585–590. https://doi.org/10.1109/ths.2012.6459914
7. Benkraouda H, Barka E, Shuaib K (2018) Cyber-attacks on the data communication of drones monitoring critical infrastructure. Comput Sci Inf Technol (CS & IT) 8:83–93. https://doi.org/10.5121/csit.2018.81708
8. Brauch H (2019) Security threats, challenges, vulnerability and risks international security, peace, development and environment, vol I. In: Security threats, challenges, vulnerability and risks
9. Aslanishvili I, Khvedelidze T (2015) Simple model for transmission control protocol (TCP). Int J Inf Models Anal 4(1)
10. Rafique S, Humayun M, Hamid B, Abbas A, Akhtar M, Iqbal K (2015) Web application security vulnerabilities detection approaches: A systematic mapping study. In: 2015 IEEE/ACIS 16th international conference on software engineering, artificial intelligence, networking and parallel/distributed computing (SNPD), Takamatsu, 2015, pp 1–6. https://doi.org/10.1109/snpd.2015.7176244
11. Gudla C, Rana MS, Sung AH (2018) Defense techniques against cyber attacks on unmanned aerial vehicles. In: International conference on embedded systems, cyber-physical systems, and applications (ESCS'18), pp 110–116
12. Secure BOX Page. https://securebox.comodo.com/ssl-sniffing/man-in-the-middle-attack/. Last accessed 2019/02/24
13. https://www.reuters.com/article/us-venezuela-politics-drones/apparent-attack-in-ve-nezuela-highlights-risk-of-drone-strikes-idUSKBN1KQ0MG. Last accessed 2019/03/01
14. DoS Attacks. https://www.thewindowsclub.com/ddos-distributed-denial-service-attacks. Last accessed 2019/02/24
15. He D, Chan S, Guizani M (2017) Communication security of unmanned aerial vehicles. IEEE Wirel Commun 24(4):134–139
16. https://doi.org/10.1109/mwc.2016.1600073wc

17. https://sputniknews.com/middleeast/201902221072659058-iran-hack-us-drones/. Last accessed 2019/03/02
18. https://www.forbes.com/sites/thomasbrewster/2017/04/25/vulnerable-quadcopter-d-rone-hacked-by-ut-dallas-cyber-researchers/#1b4103871037. Last accessed 2019/03/03

Perpetual Solar Potential of a Village by Machine Learning and Feature Extraction in UAV

A. Immanuel and K. Srinivasa Raju

Abstract The process of sustained rural development mandates uninterrupted supply of electricity and water without dependency of the urban infrastructure. Rural areas are often neglected causing power cuts for hours in villages particularly during summer. The study area, a village very near to coast, has high potential for harvesting solar energy. Scientific investigation of the feasibility and utilization of solar irradiance is to be evaluated, to identify the potential hotspots. UAV based Remote Sensing approach associated with socio-economic characteristics, was adopted for the study. A field survey was carried out for assessing the power consumption patterns of individual households. Feature extraction was performed using machine learning technique resulting in faster and efficient extraction of spatial information from UAV. A Regression analysis is carried out to correlate household power consumption with the physical and economic characteristic variables of individual buildings. The cumulative quantity of solar irradiance was used to quantify solar energy harvesting potential using UAV to promote sustained source of green energy for the village. A suitability analysis was performed for optimal location of the community solar power plant to make the village self-sufficient with respect to Electricity. The economic feasibility of the power plant was carried out to demonstrate the feasibility of the project.

Keywords Rural · Irradiance · UAV · Machine learning · Regression

1 Introduction

Our lives on earth depend on electricity as it is one of the most needed sources for our routine life. But how far will it go depending only on usage of coal. Solar Energy is a predominant substantial energy; the usage of solar energy with the photovoltaic cell is still limited and has some constraints in efficiency with the photovoltaic cell.

A. Immanuel (✉) · K. Srinivasa Raju
Institute of Remote Sensing, Anna University, Chennai 600025, India
e-mail: immiabraham@gmail.com

K. Srinivasa Raju
e-mail: ksrajuirs@gmail.com

K. Jain et al. (eds.), *Proceedings of UASG 2019*, Lecture Notes in Civil Engineering 51,
https://doi.org/10.1007/978-3-030-37393-1_19

Days are not far to utilize the solar energy as main source for maximum purposes, with less inevitable spills [1]. Generation of Solar energy for the next decade will be massive. The location of study area is along the coast. The amount of solar irradiance over the view shed [2] has been performed and found that the village has potentially high amount of view shed to the solar irradiance as there is no obstruction on any side [3]. A ground based survey conducted in order to collect the socio-economic data for individual houses, including the unit power consumption [4]. Cumulative amount of power consumed by the village helps in replacing the thermal electricity to self-depending solar electricity for the next decade. The effective locations were carried out and the automatic feature extraction technique [5] performed extracting the roof tops [6] from the UAV data which paved way for roof top solar panels that will be the primary location. Finding the suitable location for the array of panels will be the secondary location. Targeting the process through linear regression based feature extraction technique in turn benefits each household finding an effective location on the roof top, gaining enormous amount of solar energy for an uninterrupted supply of electricity.

1.1 Study Area

Implementing Solar energy through solar insolation was performed in Keerapakam village, Thiruvallur District, Tamil Nadu. The village possesses only one floor building, which benefits non-occurrence of huge shadow time in calculating the solar potential, shadows will reduce the potential of solar to fall directly on the surface. The site area has the spatial extent of 2.23 km^2 Fig. 1 also specifies point location of 13.43 latitude and 80.18 longitude. This is a northern part of Tamil Nadu. The consumption of electricity for the entire village should be calculated in order to know the actual usage of electricity distributed over the Grid from Devampattu.

Apart from the benefits in generating power by the solar panels, it also benefits Global Warming and Urban Heat [7], reduction of the emission of 579 kg of CO_2 per day and ashes too from the thermal power plant affects the health of living beings resulting in asthma, even affects plants as it covers the leaves imparts photosynthesis.

2 Methodology and Discussion

2.1 Data Required

For the initiation of the solar potential project, the foremost thing to have known is the total quantity of electricity that is required for the entire village. To attend this, the individual consumption of electricity from household has to be found out by primary and secondary survey. Primary survey involves direct collection of socioeconomic

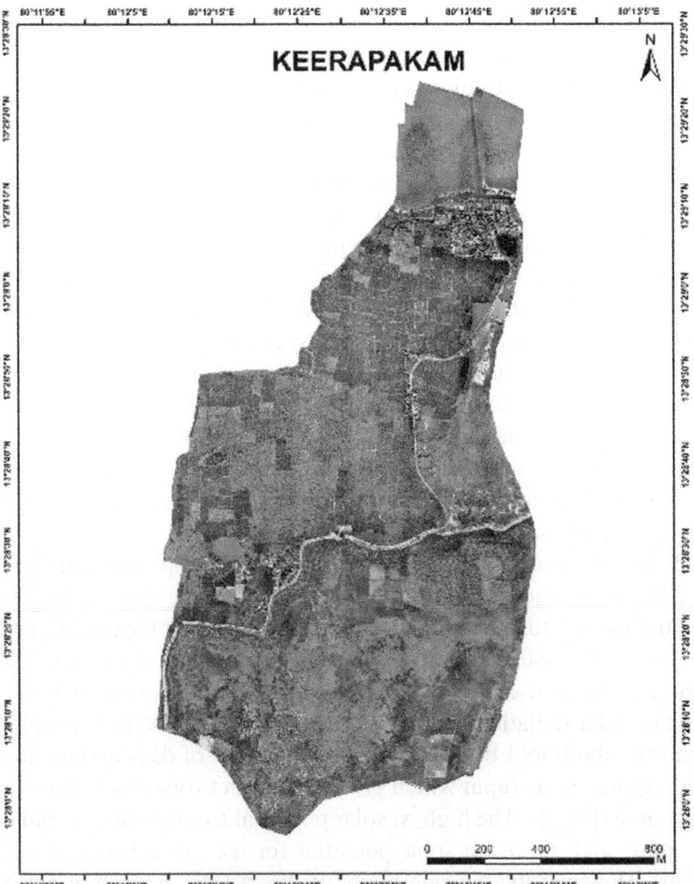

Fig. 1 Study Area Map (Keerapakam)

data, while secondary collection involves collection of consumption data for a couple of years. Also, collection of UAV raster data 4 cm high resolution imagery for extracting the rooftop, helps to perform feature based extraction process for every house, includes the optimal location of solar panels enabling the consumption of electricity.

2.2 Solar Radiation and Potential

The emission of solar radiation from the sun which is a natural star as it comes through, gets distorted further and further by the atmospheric features (aerosols, fog, rain, cloud, haze, dust) and finally intercept the surface of the earth in three possible ways such as direct, diffuse and reflected radiation, the one which meets the surface directly without any scattering through the atmospheric windows are direct solar radiation. All these radiations come together called as total solar radiation. The amount of solar radiation and potential is calculated when the view shed comes under sunlight [2] during day time. View shed is said to be the sun's position as well as sky direction to take the radiation for the respective location.

Estimation of the solar potential gives the accurate solar insolation over the area [8], as the obtained data is so finer, giving an absolute potential of solar energy that falls on the surface without disturbance. Interceptions are shadows from tree, buildings, etc. which obstructs the falling radiation of the sun, which imparts solar energy on the surface. Coal might get exhaust at a period; demand will rise as this goes on. Government of India still pursues efficiency on the usage of solar energy. Unlike Thermal, it is a predominant, dependable energy in which the energy depends on the efficiency of the solar and it has been improving over the years. Suits best for the rural areas as the thermal energy might not be an effective energy in reaching the village, having several hours of power cuts in crisis times. Limitations of the solar power are dependable on the efficiency of the solar panel.

The subsequent calculation of the potential from rooftop-mounted array of panels relied on GIS data including building footprint and zoning characterization [9]. The solar potential for the study area is calculated by overlaying the digital surface model data which is a high resolution data of the study area. A 4 cm resolution DSM (Digital Surface Model) data is used for the solar radiation, which is very finer in resolution. Calculation of solar radiation is done by importing the DSM [10] in solar radiation tool which is a inbuilt tool from ArcGIS. The number of days and the total number of hours is required as an input which gives the perfect solar insolation map over the entire study area (Fig. 2). The highest solar potential from the above said calculation is 14,781 Wm^2 and the mean solar potential for the given keerapakam village is 5101.8 Wm^2. However the average hours of direct solar insolation per day for the 13.478 latitude is 5.5 ½ h and the final value taken for the solar potential is 5100 Wm^2.

3 Estimation

Initiating a design for the solar panel involves: (1) Load Quantization, (2) Estimation of number of panels and battery bank, (3) Cost estimation of the system. The analysis has been performed to differentiate the consumption of units [11]. Such details are collected and the usage (consumption) of electricity for the particular houses Fig. 3 are picked up through the TANGEDCO website and summed up.

Three distinct colors indicating three categories namely orange (up to 100), pink (100–200), Green (200–500) units of thermal power used by each household Fig. 4 and are analyzed cluster wise from the collected socioeconomic data. Thus by doing so, the total amount of electricity consumed for the given two years is found to be 37 MW. The value represents the daily consumption of electricity which is 616.66 kWh/day. The maximum power generated by the utilization of a solar panel

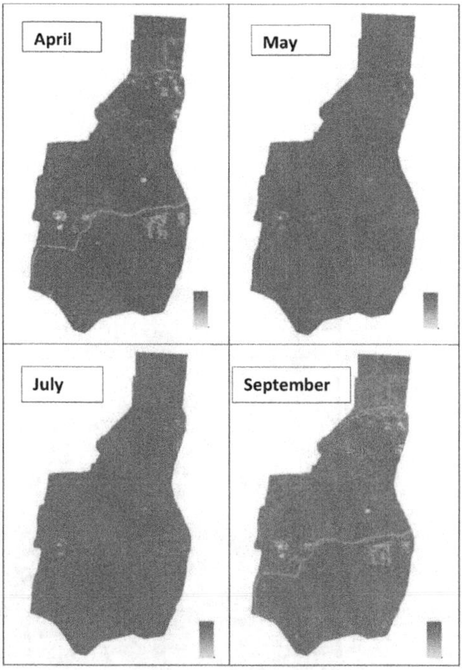

Fig. 2 Solar Potential over Keerapakam on April (top left), May (top right), July (bottom left) and September (bottom right)

in full incidence to sunlight for five and half hours per day is 52 kWh. The total number of solar panels required for the village is 356 solar panels. Incurring the losses such as System loss and Storage loss are taken into consideration which is also called as operating factor. Those losses are included; system loss 0.9%, Storage loss 0.75% calculating the power by adding losses for one panel is not 35 kWh. By this calculation, now 528 numbers of solar panel is needed to fulfill the requirement of electrical supply for the overall village.

Since the required amount is 617 kWh/day, the array of solar panels thus produces the power is to be stored by the big inverters, the choice of the inverter should be 3000 Ws. It is to be noted that 206 numbers of 3000 W inverter is essential and are wired parallel by the technique called solar stacking to store the DC power from the solar to convert into AC. A panel covers 2 m^2 of the area and the total area required for the erection and installation of the solar panel is 1056 m^2 area. The foremost part in the installation of solar power plant is the cost estimation that actually depends on the materials required. However quality, high efficient and cost effective materials focusing the next decade should be chosen. The efficiency [12] of the solar panels has increased over the years.

Fig. 3 Part of a major settlement at Top of Keerapakam

3.1 *Automatic Feature Extraction Technique*

The feature extraction technique being conducted [5] to extract the buildup area for choosing the optimal location for the solar panels. It has found that the roof top of each house suits best for the panel adoption. The extracted roof top data in Fig. 5 consists of all types of roof layers such as asbestos, tiled, concrete, thatched, etc. In these, some might be unsuitable for the fixation of solar panel.

The evaluation of the adequate available roof surfaces is the most crucial stage in implementation of roof-integrated solar panel [13]. From the rooftops, only concrete roof tops in Fig. 6 are evitable for adopting the solar panel. The User should provide some samples of the rooftops; here the samples were given as concrete and asbestos. Though this technique gives certain results, the use of such power supply systems is limited by the structural features of the roofs [14]. When each surface differs, the

Fig. 4 Electricity Consumption of each household in units; Orange(up to 100), Pink (100–200) and Green (200–500)

samples which the user adds increases. By this feature extraction the concrete roof tops are found suitably best for this study which paved way in running the samples to have the full extraction of concrete roof tops alone.

4 Suitable Location of Solar Panels

Usually the location of the solar panel nowadays will be on roof top, the length, breadth and area of the roof top for the entire village has been calculated by digitizing each house's roof top using ArcGIS software with the high resolution imagery from Drone. There are about 335 houses in the village and the area of each has been calculated and displayed. As per the houses' consumption mentioned, the panel

Fig. 5 Roof Top visibility of various types

which suits best will be placed on roof top, thus producing the electricity based on their consumption.

Overall 43% of the land surface is cultivable agriculture land, 22% are built up and 35% are water body. Nevertheless there is no such called waste land or barren to find the optimal location. Although roof top gives best results for the location of panels, an alternate optimal location such as placing on tank is to be found out since half of the area covered with water body and does suits the location for the erection of 528 solar panels as an array covering 1056 m^2. It acts as back water cooling system for the panels neglecting the operating factor of losses. The water body covering the top of the village which is Pulicat Lake suits the suitables location to hold 426 solar panels about 852 m^2 is perfectly suitable for the solar panels (Figs. 7 and 8).

5 Conclusion

The source of energy thus produced by the renewable resources will eliminate the demand of electricity for the next decade. In spite of having renewable energy by other ways, still the solar insolation over the years drastically increased in efficiency. The

Fig. 6 Automatic extraction of concrete roof top

Fig. 7 Optimal location of Solar panel in Pulicat lake

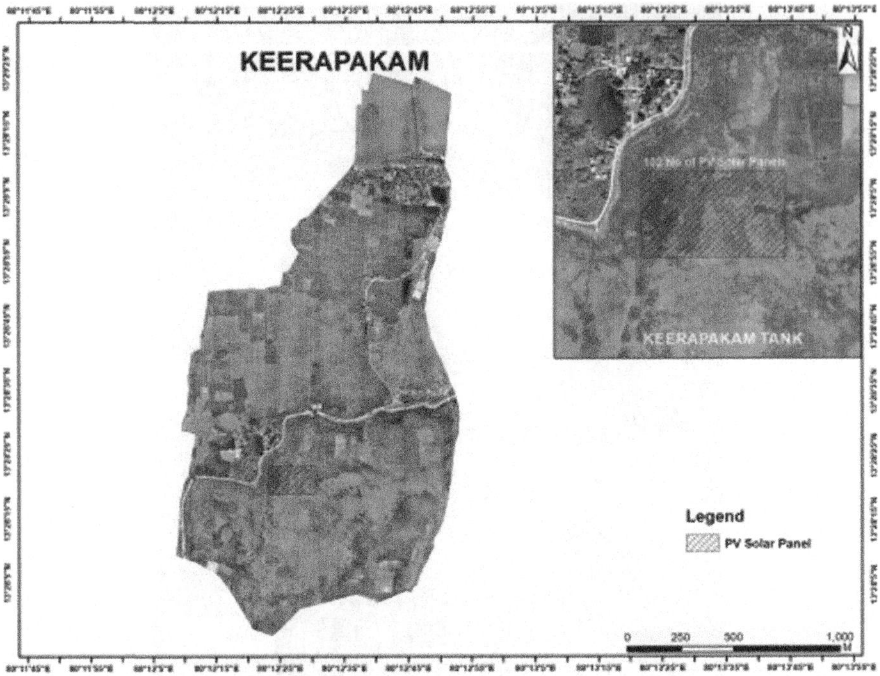

Fig. 8 Suitable sites for Solar panel

project specified here is a future determined and at times needs some maintenance. Roof top location plays a vital role in self-dependency project as there are no outside disturbances in fixing the panels over their own house. If a neighbor dwells under a thatched or asbestos roof then arises the problem of locating a possible location. Since the area required for the panel will not be sufficient. In that case a possible way of locating a panel has to be some other place rather than placing it on the roof top. Land acquisition is the perplexing role without depleting the available agricultural land; it is possible to find out the optimal location for the installation of the enormous grid of solar panels on the tank as a floating solar panel in village level.

Government of India supports subsidizing the initiation of renewable energy [12] like solar power projects by Jawaharlal Nehru National solar mission scheme [15] up to 50% bearable helps people depend more likely on renewable energy. The other half expense shared among the houses on the basis of consumption. The work for the project seems little yet it stands mightily as an 18.5 MW power producing for a year. As it welcome Government, NGO, researches and students to have an interest rather than depending on thermal energy but also to depend on the renewable energy.

Apart from the benefits in generating power by the solar panels, it also benefits Global Warming and Urban Heat, reducing emission of 579 kg of CO_2 per day and ashes too from the thermal power plant affects the health of living beings resulting in asthma, even affects plants as it covers the leaves imparts photosynthesis.

References

1. Stauffer NW (2015) The future of solar energy: a summary and recommendations for policymakers. MIT energy initiatives
2. Rich PM, Dubayah R, Hetrick WA, Saving SC (1994) Using view shed models to calculate intercepted solar radiation applications in ecology. In: American society for photogrammetry and remote sensing technical papers, pp 524–529
3. Datta A, Karakoti I (2010) Solar resource assessment using GIS & remote sensing techniques
4. Gerland P (2019) Socio-economic data and geographic information systems: sources, scope, indicators and issues
5. Omitaomu O, Kodysh JB, Bhaduri BL (2012) Modeling and analysis of solar radiation potentials on building rooftops. In: ASME 2012 international mechanical engineering congress and exposition, p 6. https://doi.org/10.1115/imece2012-86805
6. Dayal K, Chauhan I (2018) UAV photogrammetry for feature extraction and mapping of corrugated industrial rooftops. In: ISPRS—international archives of the photogrammetry, remote sensing and spatial information sciences, vol XLII–129–133. https://doi.org/10.5194/isprsarchivesXLII-4-129-2018
7. Masson V, Bonhomme M, Salagnac JL, Briottet X, Lemonsu A (2014) Solar panels reduce both global warming and urban heat island. Front Environ Sci 2
8. Yousuf MU, Siddiquia M, ur Rehman N (2018) Solar energy potential estimation by calculating sun illumination hours and sky view factor on building rooftops using digital elevation model. J Renew Sustain Energy 10:013703
9. McIntyre JH (2012) Community-scale assessment of rooftop-mounted solar energy potential with meteorological, atlas, and GIS data: a case study of Guelph, Ontario Canada. Energy Sustain Soc 2:23. https://doi.org/10.1186/2192-0567-2-23
10. Agugiaro G, Nex F, Remondino F, De Filippi R, Droghetti S, Furlanello C (2012) Solar radiation estimation on building roofs and web-based solar cadastre. ISPRS Ann Photogram Rem Sens Spat Inf Sci I–2. 2012 XXII ISPRS congress, 25 Aug–01 Sept 2012, Melbourne, Australia
11. Sundaray S, Mann L, Bhattacharjee U, Garud S, Arun K, Tripathi TERI (2014) Reaching the sun with rooftop solar New Delhi. The Energy and Resources Institute, 62 pp
12. Abhishek Jain (2017) Subsidy on solar PV systems through NABARD in India. https://www.bijlibachao.com/solar/proceduretogetsubsidyonsolar-pv-systemsthroughnabardinIndia.html, 5 Sept 2017
13. Mansouri Kouhestani F, Byrne J, Johnson D, Spencer L, Hazendonk P, Brown B (2019) Evaluating solar energy technical and economic potential on rooftops in an urban setting: the city of Lethbridge, Canada. Int J Energy Environ Eng 10(1):13–32. https://doi.org/10.1007/s40095-018-0289-1
14. Dausa YU, Yudaeva IV, Stepanchuka GV (2017) Reducing the costs of paying for consumed electric energy by utilizing solar energy. Econ Ecol Solar Eng. https://doi.org/10.3103/s0003701x18020056
15. Jawaharlal Nehru National Solar Mission (2010) (JNNSM) https://mnre.gov.in/resolution, 11th Jan 2010

References

Comparison of Performance of Artificial Neural Network (ANN) and Random Forest (RF) in the Classification of Land Cover Zones of Urban Slum Region

Deepak Tyagi, Mohd. Anul Haq, Gazi Rahaman, Prashant Baral and Joydip Datta

Abstract India is one of the world's largest economies and economic growth has remained continuous. This has led to accelerating urbanization which requires proper planning and monitoring. As the urban areas are expanding, urban slum areas are also increasing along with it. These growing urban slum areas require proper observation so that existing resources can be employed to provide these regions with the best possible livelihood conditions. For this purpose, urban slum areas as well as surrounding land resources should be well identified and classified so that the existing land resources can be appropriately utilized for future implementation of development activities. Machine learning classification algorithms are found to be very suitable for the identification and classification of remotely sensed images. Their efficiency in feature identification and extraction has established these algorithms as important tools in decision making. In this study, our major objective is to identify and classify different land cover zones in the urban slums areas of Chingrajpara area of Chhattisgarh using remotely-sensed images. For this purpose, high-resolution images, collected using unmanned aerial vehicles (UAVs), are used and these images are classified into different land cover features using two different machine learning algorithms Artificial Neural Network (ANN) and Random Forest (RF). The results obtained show that the overall accuracy achieved by ANN and RF are 72.6% and 84.35% respectively. The study highlights the role and importance of landcover classification for future planning and management.

Keywords Unmanned aerial vehicles (UAVs) · Artificial neural network (ANN) · Random forest (RF)

1 Introduction

Urbanization is growing at an alarming rate around the globe. As per the reports, at present, more than half of the world's population i.e. 54.827% [1] is residing in the urban regions. Urban areas are often considered as the epicenters of almost all

D. Tyagi (✉) · Mohd. A. Haq · G. Rahaman · P. Baral · J. Datta
NIIT University, Neemrana, India
e-mail: deepak.tyagi18@st.niituniversity.in

K. Jain et al. (eds.), *Proceedings of UASG 2019*, Lecture Notes in Civil Engineering 51,
https://doi.org/10.1007/978-3-030-37393-1_20

the world's growth potential areas. As India is one of the world's largest economies, the trend of urbanization is also increasing here at a fast pace. As per the reports, the urbanization is increasing at a constant rate from the last decade and currently, 33.6% [2] of India's population is residing in urban areas and this is estimated to increase by 41% by 2030 [3]. This increasing high growth rate is often accompanied by inadequate development of infrastructure, be it in the terms of transport, housing, utilities or other basic needs of the people. This together with the large share of people with informal low-paid employment have made the conditions more difficult. This leads to the growth of informal settlements, also known as slums.

Urban and urban slum settlements are usually differentiated on the basis of administrative boundaries and income level. Further, the census-based method is used which provides the demographic data based on typical headcount. However, this data does not provide detailed spatial information about the concentration or location of slum-dwellers. Thus, systematic quantification of urban slums areas requires a proper method to identify them and define them in terms of the spatial domain [4]. This needs to be done in the most consistent manner so as to support programs which are geographically defined and especially targeted towards slum areas. The identification of such zones is crucial for defining proper urban slums zones. Once such areas are defined, it becomes easier for the government or other planning bodies to deploy any kind of development activities. Different geographical, demographical and other attributes can also be stored by the planners and government bodies to record current situation and store as evidence for future references.

The image analysis using UAV images has been recognized as a better way to identify different geographical features [5]. The images captured by UAV helps to easily detect many more features compared to conventional remote sensing approach as it provides an abundance of details. Moreover, UAVs can be deployed easily and at any need of time to fulfill the need of mapping and rapid monitoring. The user can define a particular spatial-temporal scale as per the need of the study [6]. UAVs are also beneficial for mapping informal settlements which are distributed over large areas. Additionally, continuous monitoring at high resolution can provide an accurate base map and information related to spatial and temporal changes based on urban scenes which can be used to identify newer buildings, different variations in terrains and other information [7]. Different areas can be classified on the basis of their texture, and high compactness of one feature with the other as compared to dwelling sizes. The high compactness in the urban area can point towards the presence of slum region [4].

In the present study, we have used very high-resolution UAV images to properly classify the different land cover zones [8] in the urban slum region of Bilaspur district, Chhattisgarh. In order to classify the land-cover features different techniques have been used in the past but the most promising results are generated using machine learning techniques. Therefore, the current study deals with the classification of high-resolution images using supervised image classification algorithms Artificial Neural-Network (ANN) [9] and Random Forest [10]. These high-resolution images were obtained using UAV data have been to classify the terrain. Performance of the two machine learning algorithms has been assessed based on Producer Accuracy, User Accuracy and Overall Accuracy of classification.

2 Study Area

The present study is concentrated on the Chingrajpara urban slum area of Bilaspur district, Chhattisgarh (22° 5′ 34.46″ N–22° 5′ 24.84″ N; 82° 10′ 5.48″ E–82° 10′ 9.02″ E) covering a total area of 0.0592 km^2 (see Fig. 1). This region is famous for its aromatic rice variety, handloom, and Soft Kosa silk sarees. The region enjoys a rich climatic condition which varies from hot summer from March to May with maximum temperatures reaching up to 45 °C. This is followed by Monsoon season from the mid of June to the end of September. The average rainfall recorded during these months is 280 mm. After the end of monsoon, mild winters remain till the month of March with the minimum temperature falling up to 10 °C. According to the earlier census reports of 1991, Chingrajpara is the biggest slum out of 49 slums of Bilaspur district. This region is unable to fulfill the basic needs of livelihood of people and the development of the area. For this purpose, all the present features of the area need to be properly identified and to be planned judicially for the development of people.

3 Methodology

3.1 Workflow

The overall comparative study workflow between ANN and Random Forest can be differentiated into four major steps: (1) UAV data acquisition and preprocessing (2) Creating orthomosaic and mosaic correction (3) Image classification (4) Accuracy assessment (see Fig. 2).

3.2 Data Collection and Preprocessing

A set of 85 images captured ground sampling distance of 2.19 cm from a height of 150 m with DJI Phantom 4pro and DJI Inspire 2 was provided [11]. These images were then processed in Pix4Dmapper Pro version 4.3.31 and orthomosaic was generated. This orthomosaic generated was further corrected using Ortho Projection correction in Pix4Dmapper. The next step was the creation of Region of Interest for the training of supervised classification algorithms.

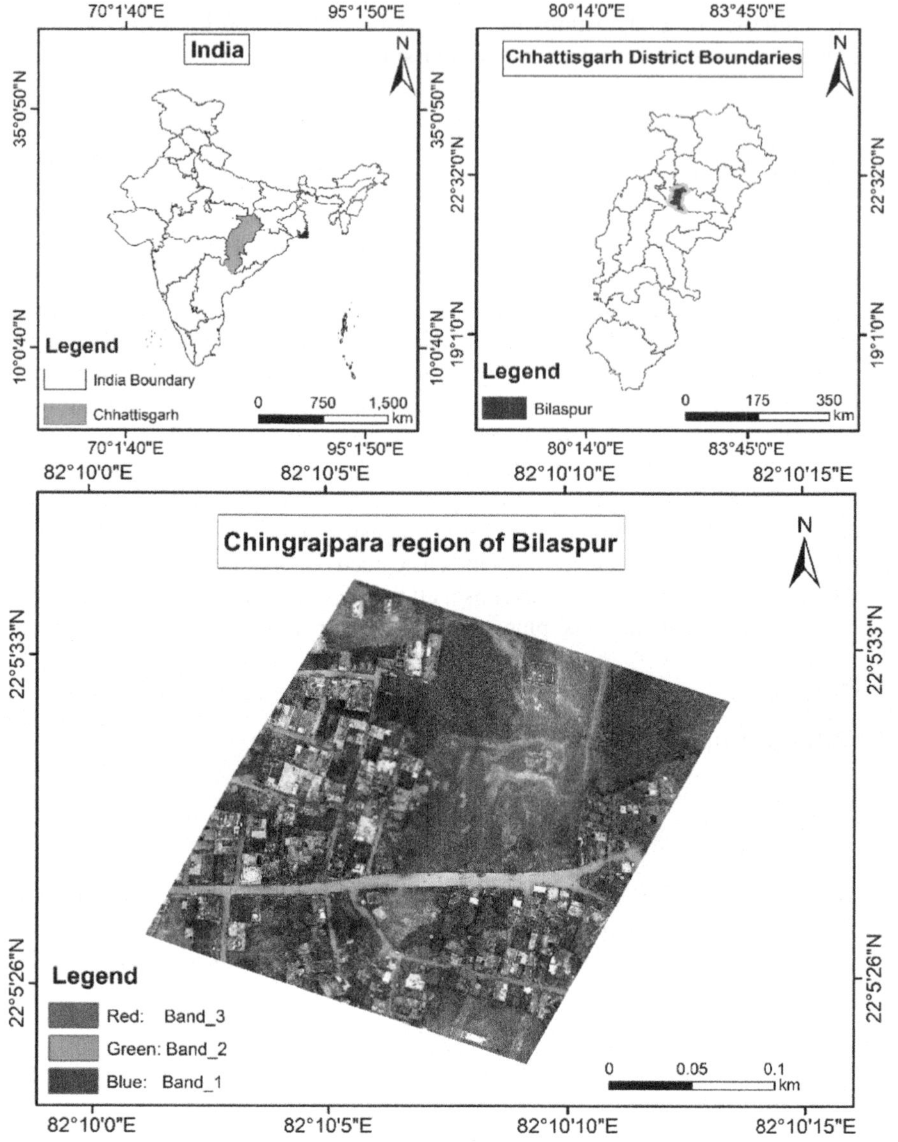

Fig. 1 Study area map

3.3 Artificial Neural Network (ANN)

Artificial Neural Network is computing systems which are simple, interconnected and are able to recognize the patterns, just like neurons of the human brain. In order to duplicate the human brain the scientist had first developed the neural network into

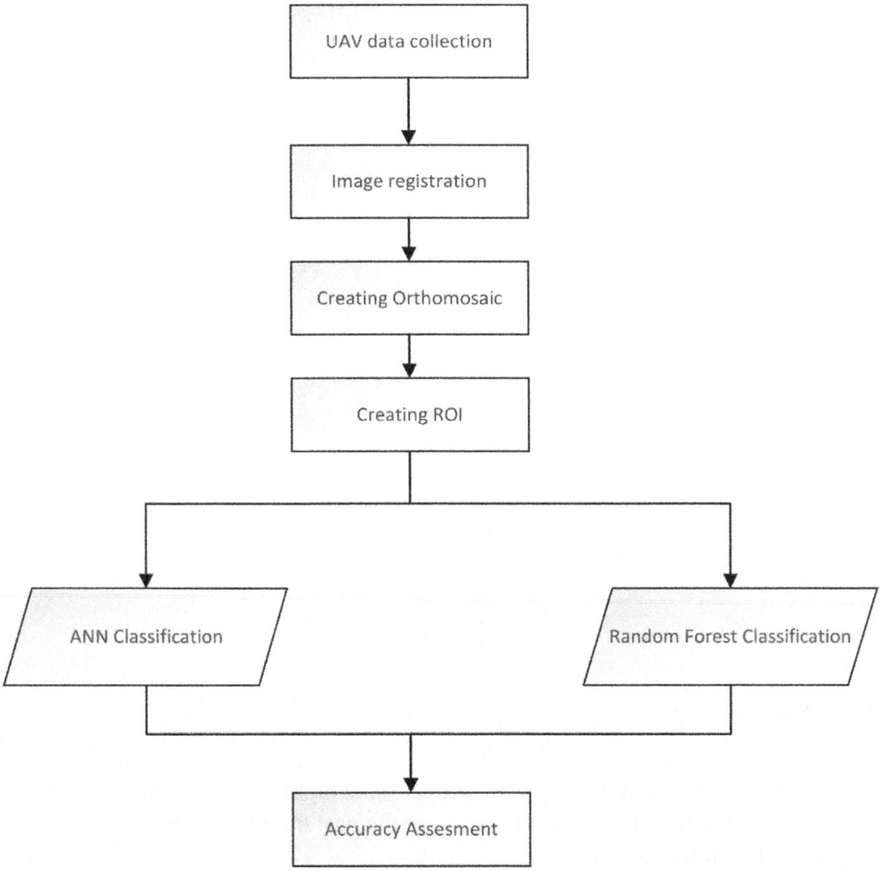

Fig. 2 Flow diagram of study

the computing world. The initial processing unit or a node was developed which is known as an artificial neuron. In the computing world these neurons are connected to each other forming an artificial neural network.

The ANN has the ability to develop an internal representation of the signal that it receives as input and convert them into a pattern in order to enhance its self-learning from the dataset. This self-learning capability or programming is accomplished through the dynamic adjustment of interconnection strengths associated with each neuron [12]. This process is also known as back propagation (see Fig. 3), as this uses desired outcome pattern and predefined input i.e. training the dataset to initiate feedback to the neural network. This network of backpropagation runs again and again through the training dataset until the most accurate weights are achieved that satisfy the defined input values to the desired output from the dataset.

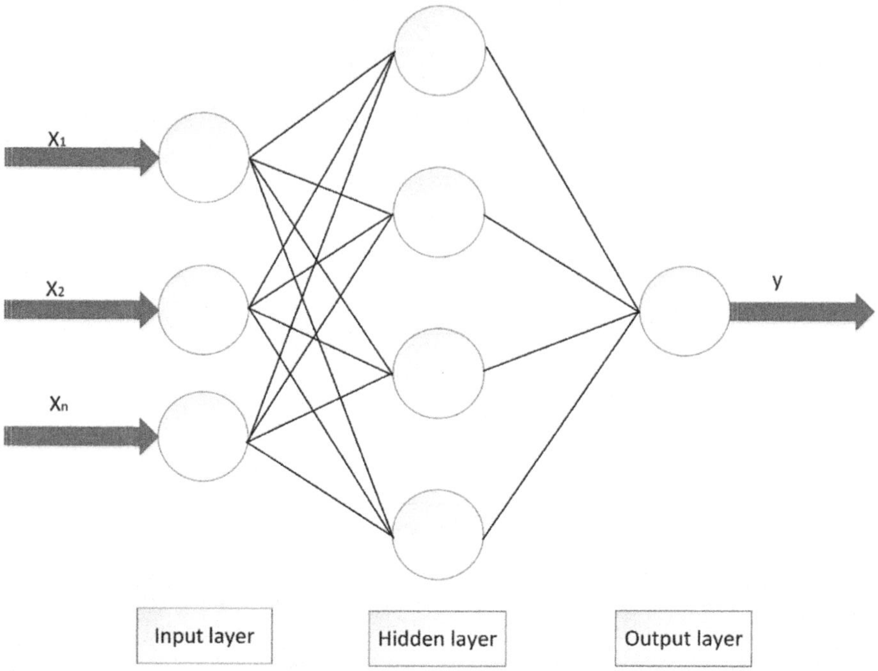

Fig. 3 ANN architecture

This pattern generated by neurons segregates the input signals to one of the output class generated. ANN stores knowledge from the past trained experience and makes it available for future datasets.

In the present study, ANN has been applied using ENVI software (ENVI 5.3, Research System Inc., USA) for which training data set was provided. ANN was executed by using the hyperbolic function [13], which is expressed as Eq. (1):

$$\sigma(x) = \frac{e^x - e^{-x}}{e^x + e^{-x}} \tag{1}$$

For the purpose of current classification, training rate: 0.2, training threshold contribution: 0.9, training RMS exit criteria: 0.1, training momentum: 09, hidden layers: 2 and number of training iterations: 1000 were used generate the current result.

For the purpose of current classification, 0.2, 0.9, 0.1, 09, 2 and 1000 were used for the parameters training rate, training threshold contribution, training RMS exit criteria, training momentum, hidden layers and number of training iterations respectively to generate the current result.

3.4 Random Forest

Random Forest is a decision tree based ensemble classifier [14] that performs exceptionally well as compared to other machine learning classifications. This is non-parametric and does not require assumptions on the distribution of the data [15, 16]. It can be used to solve the multi-class problems. The major advantages of applying Random Forest in remote sensing are:

- It can run efficiently on large datasets.
- Can handle thousands of variable inputs with deletion of the variable.
- Provides estimation of the variables that are important in the classification.
- Computes proximities between pairs of cases that can be used in locating the outliers.
- Generates an internal unbiased estimate of the generalization error.
- Relatively robust to the outliers and the noise.
- Gives good predictive performance even if the predictive variables are noisy.
- This reduces the need for variable preselection.
- Interpretability, ease of use and simplicity.

These extraordinary features make it suitable for the classification process of multisource data or multispectral data. Ensemble learning algorithms like Random forest, Boosting and Bagging are gaining more interest because they are more accurate and robust to the noise than single classifiers [17]. Among these Random Forest is the most popular one due to the above reasons. Moreover, Random Forest is a collection of classification and regression tree (CART) classifiers it can be defined as Eq. 2:

$$\{h(x, \theta_k), \quad k = 1, 2 \ldots i \ldots\} \tag{2}$$

where h can be defined as Random Forest classifier, x as an input variable and $\{\Theta_k\}$ as the independently and distributed random vectors variables used for generation of each Cart tree (see Fig. 4). Each of the classifiers contributes with a single vote for the assignation of the most frequent class of the input variable. Tree design needs identification of suitable attribute selection to measure which attribute increases dissimilarity between classes. There are many approximations for proper selection of attributes which are to be used for induction in decision trees. Some of the commonly used are Gini index, Chi-Square and gain-ratio. The prominently used is Gini Index to measure for the best split selection [18], it measures the impurity of a given element with respect to the rest of the classes. For this purpose, we have included Gini Index as shown in Eq. (3).

$$\text{Gini Coefficient} = 1 - 1 - \sum_{c} p^2(c \mid t) \tag{3}$$

Here; c = number of classes, t = node of a tree, p = relative frequency of c.

Moreover, Random Forest increases the diversity [15] of the tree by taking the inputs from different training datasets created by Bagging or Bootstrap aggregating.

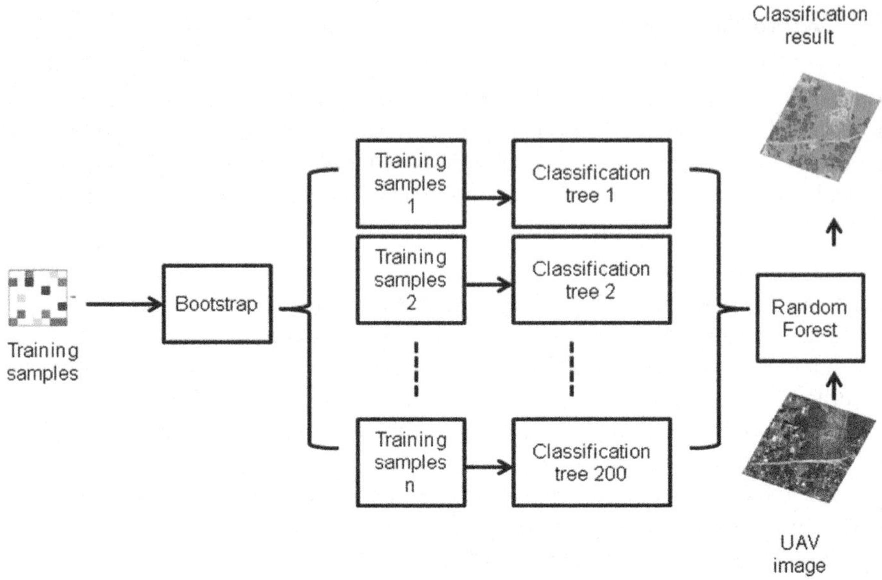

Fig. 4 Random forest working

For the current study Python 3.6 software was utilized in order to classify the image with the help of training dataset. For the purpose of classification, 200 decision trees, minimum value 1 for number of samples in node, and 0.0 for minimum impurity were used to generate the current output.

3.5 Accuracy Assessment

The accuracy assessment of the classified images was done by generating the confusion matrix based on the random validation samples selected from the classified image and verified using the high-resolution image. The main reason for accounting on visual inspection is that the dataset is of very high resolution i.e. 2.19 cm which makes objects identification very easy. This error matrix generated was used to estimate the user accuracy, producer accuracy, and the overall accuracy. The user accuracy is defined as the probability of classified pixel of the feature class to the actual feature on the ground. The producer accuracy tells us about the ratio of the total number of correctly classified pixel to total ground truth pixel. The overall kappa index was also derived which helps to quantify the performance of classified images.

4 Result and Discussion

In the present study, two machine learning algorithms ANN and Random Forest were used to determine the different land cover zones in the urban slum region. This region was classified into built up, open land, road, vegetation and water body zones (see Figs. 5 and 6). In order to quantify the output generated accuracy assessment of both the machine learning algorithms was performed by creating the confusion matrix. The overall accuracy achieved in ANN and Random Forest are 72.86 and 84.35% with a kappa coefficient value of 0.6247 and 0.7793 respectively as represented in Table 1.

From our results, we can ascertain that random forest proves to be a better technique to identify the land cover zones for the present regions using high-resolution UAV images. Random Forest classification works better in classifying the water bodies, urban features like roads and buildings. The output of ANN classification shows that the results generated are not that accurate as compared to Random Forest, the mixing of classes like roads and buildings are frequent. The accuracy achieved by Random Forest is more for the current scenario but for different terrain we might get better results through ANN. ANN generally performs well when the dataset is

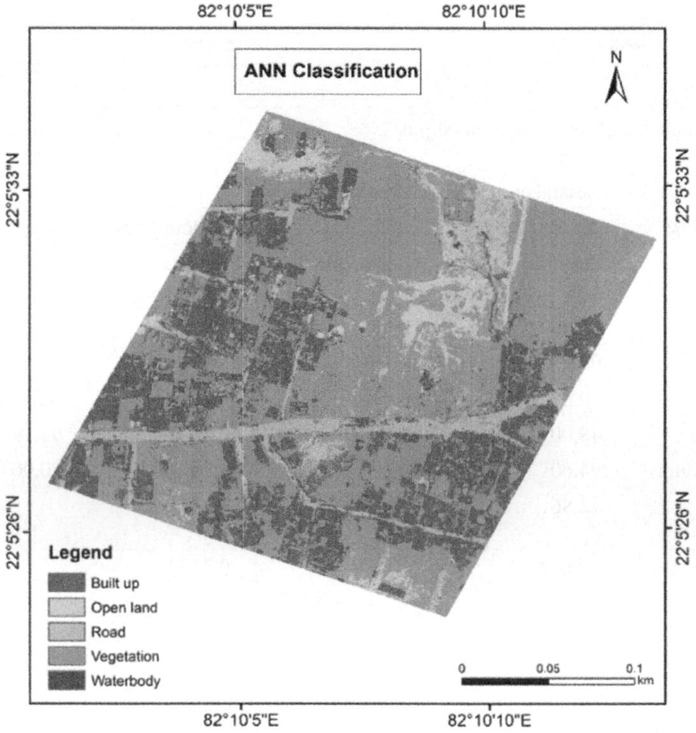

Fig. 5 ANN classification output

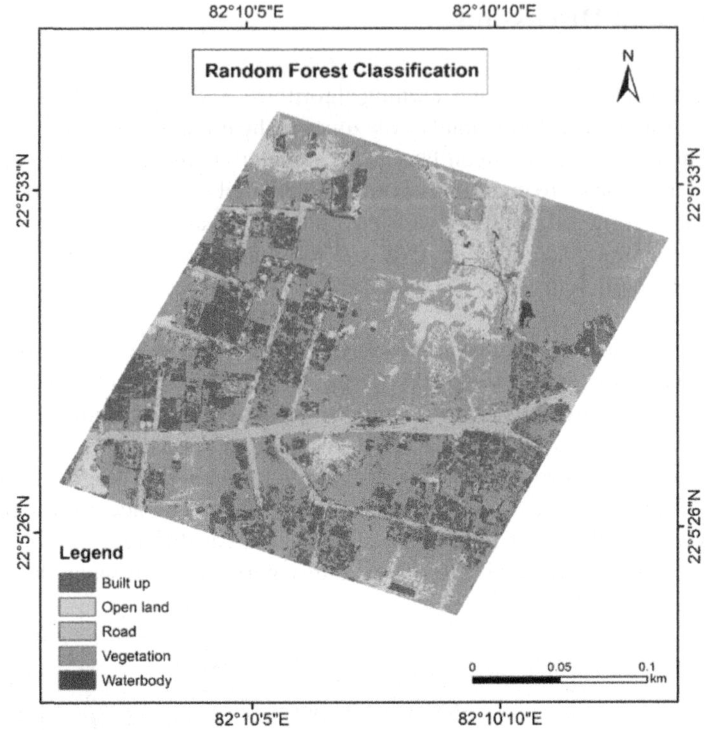

Fig. 6 Random forest classification output

Table 1 Accuracy assessment of ANN and random forest

Feature class	ANN		Random forest	
	Producer accuracy (%)	User accuracy (%)	Producer accuracy (%)	User accuracy (%)
Built up	62.86	70.97	80.95	77.27
Vegetation	94.55	88.14	85.71	85.71
Roads	60.87	66.67	64.71	68.75
Open land	48.00	50.00	90.38	92.16
Surface water	94.60	40.00	98.28	80.00
Overall accuracy (%)	72.86		84.35	
Kappa coefficient	0.6247		0.7793	

large and more training points are provided to it. Moreover, the light is unavailable to reach the areas which are under the trees or in the shadow of a building or other features. This factor changes the original spectral signature of the area and this in turn accounts in the decrease of classification accuracy.

5 Conclusion

In the present paper, a comparative study between ANN and RF was conducted to classify different land cover zones using high-resolution UAV imagery of Chingrajpara, Chhattisgarh area. The obtained images were processed using Pix4Dmapper and further machine learning algorithms ANN and Random Forest were applied using Envi 5.3 and Python respectively. The result generated by RF consisting of 200 decision trees gave the best result for urban land cover classification with an overall accuracy of 84.35% and Kappa coefficient index of 0.7793. The comparative study between the two classifiers depicts that Random Forest performs better when compared to ANN.

Acknowledgements We are highly thankful to Dr. Kamal Jain, Professor, Indian Institute of Technology-Roorkee for providing the processed UAV images as well as Digital Surface Model (DSM) and Digital Terrain Model (DTM) of the study area.

References

1. Jiang L, O'Neill BC (2017) Global urbanization projections for the shared socioeconomic pathways. Glob Environ Change 42:193–199
2. Chandrasekhar S, Sharma A (2015) Urbanization and spatial patterns of internal migration in India. Spat Demogr 3(2):63–89
3. Ghosh S, Kanjilal K (2014) Long-term equilibrium relationship between urbanization, energy consumption and economic activity: empirical evidence from India. Energy 66:324–331
4. Kit O, Lüdeke M, Reckien D (2012) Texture-based identification of urban slums in Hyderabad, India using remote sensing data. Appl Geogr 32(2):660–667
5. Poblete-Echeverría C, Olmedo GF, Ingram B, Bardeen M (2017) Detection and segmentation of vine canopy in ultra-high spatial resolution RGB imagery obtained from unmanned aerial vehicle (UAV): a case study in a commercial vineyard. Remote Sens 9(3)
6. Feng Q, Liu J, Gong J (2015) Urban flood mapping based on unmanned aerial vehicle remote sensing and random forest classifier-a case of yuyao, China. Water (Switzerland) 7(4):1437–1455
7. Gevaert CM, Persello C, Sliuzas R, Vosselman G (2017) Informal settlement classification using point-cloud and image-based features from UAV data. ISPRS J Photogram Remote Sens 125:225–236
8. Tiwari A, Dixit A (2015) Unmanned aerial vehicle and geospatial technology pushing the limits of development. Am J Eng Res 4(01):16–21
9. Hepner GF, Logan T, Pitter N, Bryant N (1989) Artificial neural network classification using a minimal training set: comparison to conventional supervised classification. Photogram Eng Remote Sens 56(4):469–473

10. Ahmad MW, Mourshed M, Rezgui Y (2017) Trees vs neurons: comparison between random forest and ANN for high-resolution prediction of building energy consumption. Energy Build 147:77–89
11. Jain K (2019) Urban slum dataset for Chingrajpara slum area (Chhattisgarh)
12. Satir O, Berberoglu S, Donmez C (2016) Mapping regional forest fire probability using artificial neural network model in a Mediterranean forest ecosystem. Geomat Nat Hazards Risk 7(5):1645–1658
13. Karlik B (2015) Performance analysis of various activation functions in generalized MLP architectures of neural networks. Int J Artif Intell Expert Syst 1(4):111–122
14. Cutler DR et al (2007) Random forests for classification in ecology published by: ecological society of America. Ecology 88(11):2783–2792
15. Rodriguez-Galiano VF, Ghimire B, Rogan J, Chica-Olmo M, Rigol-Sanchez JP (2012) An assessment of the effectiveness of a random forest classifier for land-cover classification. ISPRS J Photogram Remote Sens 67(1):93–104
16. Rahmati O, Pourghasemi HR, Melesse AM (2016) Application of GIS-based data driven random forest and maximum entropy models for groundwater potential mapping: a case study at Mehran Region, Iran. CATENA 137:360–372
17. Dietterich TG (1990) Ensemble methods in machine learning
18. Breiman L (2004) Technical note: some properties of splitting criteria. Mach Learn 24(1):41–47

Identification of Urban Slums Using Classification Algorithms—A Geospatial Approach

K. Nivedita Priyadarshini, V. Sivashankari and Sulochana Shekhar

Abstract Urbanization is a dynamic phenomenon which plays a vital role in the global scenario and it is estimated to increase in the near future. Rapid urban sprawl attracts rural communities in search of employment and basic amenities. Improved transport network and communication have rendered the rural population to upgrade job prospectus by migrating from outskirts to the city center. This in turn, has raised the density of informal settlements into the urban landscape which is termed as 'urban slums'. In this study, an attempt has been made to discriminate formal and informal settlements for Chingrajpara, Chhattisgarh by employing various classification algorithms using Unmanned Aerial System (UAS) dataset. Incorporating pixel-based approaches like Maximum Likelihood and Mahalanobis distance classifiers, ensemble decision tree namely Random Forest classifier, back propagation algorithm such as Neural Net classifier and object-based image analysis using feature extraction to geometrically rectified datasets yields classified results with diverse accuracies. Selection of representative training samples favors for acquiring reliable accuracies. This study also addresses the suitable classifier that outperforms for Very High Resolution (VHR) datasets depending on the accuracy assessment. Since UAV data produces excellent resolution images, the land cover feature appears distinct. Among the array of advancements, point clouds provide 3D information that exhibits true ground features. Thus the resultant classified images are validated using elevation information estimated from point cloud datasets. Methodical results serve the urban planners and spatial analysts for systematic designing, thus alleviating random growth of informal settlements as VHR UAV datasets are a boon to the field of geospatial technology.

Keywords Informal settlements · Ensemble · Random forest · Neural net · Point clouds

K. N. Priyadarshini (✉) · V. Sivashankari · S. Shekhar
Department of Geography, Central University of Tamil Nadu, Thiruvarur, Tamil Nadu, India
e-mail: nivi.darshini@yahoo.com

K. Jain et al. (eds.), *Proceedings of UASG 2019*, Lecture Notes in Civil Engineering 51,
https://doi.org/10.1007/978-3-030-37393-1_21

1 Introduction

Urban landscapes with the persistence of shanty dwellings are termed as "urban slums". Rapid urbanization attracts the rural communities into urban frontiers paving way for construction of informal settlements in a haphazard manner and in a piecemeal fashion. The unplanned slums lack procurement of basic amenities, inefficient infrastructure coupled with the shortage of living space [1, 2]. It is estimated that half of the global population resides in urban areas with associated slum growth. Most of the slum households prevailing across Asia shares about 10–30% of urban area [3, 4]. It is estimated by the UN that about one billion people live in a slum and it is expected to increase triple the time by 2050 [5, 6]. Slum proliferation and poverty could be eradicated by providing access to land and facilities for slum dwellers [7]. Addressing the growth of slum in order to propose holistic solutions could be solved by detecting factors related to these settlements [8]. Advancement in remote sensing has brought the significance of urban slums to limelight in the temporal variation of urban morphological studies. Multispectral satellite imageries render spectral bands which provide ample details about the topographical features [9]. Recent techniques using Unmanned Aerial Systems (UAS) have replaced these traditional multispectral datasets due to its level of accuracy. Identification of slums from the urban landscape encounters many challenges. Conventional multispectral data provides coarse images of meter level accuracy whereas UAS datasets are of centimeter level accuracy. Interpretations of land use/land cover features are apparent using UAS datasets [10]. By utilizing the precise datasets, policy makers and planners can execute concerted actions for formulating measures like slum rehabilitation, promoting green spaces, management of waste disposal sites etc. [11]. In this study, classification algorithms are applied to the urban slum structures to infer the land cover composition based on its distribution pattern, type of housing/roofing, the texture of building materials, compactness and surrounding infrastructures. Segregation of features depends upon the ontology and is extracted using various classifiers. Accuracies associated with the classifier results are assessed and are further discussed. The study also poses a major research question for the choice of classification technique to be applied for acquiring better discrimination between formal and informal settlements that are clearly explained under Sect. 3.2.

1.1 Identification of Urban Slums

Slums can be stratified into groups from core urban using roofing textures and building footprints. Spatial metrics of local slum ontology can be derived from built-up the environment through textural feature contrast [12]. In today's scenario, researchers show a keen interest in extracting the physical characteristics of global slums to improvise slum habitats. Monitoring temporal changes in urban slums using multispectral and Very High Resolution (VHR) commercial sensors have increased due to the availability of robust datasets [13]. High Resolution (HR) UAS datasets has driven the attention of urban planners to map the formal/informal settlements to regularize dynamic slum growth [14].

This paper attempts to employ diverse classification algorithms to the VHR UAS datasets for mapping and monitoring urban slum for Chingrajpara slum area, Chhattisgarh [15].

The objectives of the study include (i) to classify formal/informal settlements from UAS mosaic dataset for interpretation of land cover features (ii) to elucidate point cloud information with respect to elevation for distinguishing formal/informal settlements (iii) to resolve suitable measures for proper utilization of open areas, forbid dumping of solid waste near households and foster growth of greenery.

1.2 Analysis of Point Cloud Information

Among the array of technological advancements, photogrammetric point cloud datasets, yield accurate three-dimensional information about earth's surface. Point cloud information generated from airborne captures the existing land cover at nominal point densities. For customizing the specific application, this study focuses on to extract the elevation information from point cloud data provided for Chingrajpara slum area, Chhattisgarh [15] to discriminate formal and informal settlement distribution. As the visual interpretation of photogrammetric point cloud data is precise, differences in elevation could be seen clearly and verified using associated DSM [16].

1.3 Classification Algorithms—An Overview

Unmanned Aerial Vehicles (UAV) employs a sensor that acquires imageries of the ground scene with higher accuracies. Unlike other multispectral satellite products, UAV datasets provide clear-cut details about the topographic features. The enhanced dataset portrays RGB image exhibiting the characteristics of urban slum settlements thus easing the process of homogeneous training samples collection. Various algorithms of classification are applied to UAV imageries to check if the datasets respond the same as like other multispectral sensor imageries. Selection of representative training samples is an integral part of performing supervised classification. Choice of training samples for classification is done at its best to obtain good accuracy. Homogeneous samples are extracted with ease in UAV datasets due to its better Ground Sampling Distance (GSD). Minimal GSD of datasets provides maximum accuracy for classified results [17]. In this paper, an attempt has been made to distinguish formal and informal settlements using pixel-based and ensemble classification algorithms for the UAS dataset of Chingrajpara slum area, Chhattisgarh.

2 Study Area and Datasets

Urban slum landscape used in this study is Chingrajpara which is one among the wards of Bilaspur City, Chhattisgarh, India. The study area extends between 22° 5′ 27.16″ N latitude and 82° 10′ 6.17″ E longitude. Bilaspur city has the largest and ancillary coal field of India having the South Eastern Coalfields Ltd. It is also ranked as the second largest jurisdiction for supplementary of electricity to the state of Chhattisgarh. Many people residing in slum regions are under the low poverty line and hence they migrate towards the city-side. Growth of coal fields attracts rural communities in search of better employment thus resulting in an equivalent increase of informal settlements. The reason behind this activity is primarily classified based upon work, business, education, marriage or moved after birth and some other causes. The study area collected with UAS covers around 1500 houses over the Chingrajpara region. The houses here are divided into four types based on the building materials. Nearly 500 Hut houses, 600 Kutcha houses, 390 Semi-Pucca houses and 10 Pucca houses are present in this whole ward. A portion of this area is considered for study which includes formal settlements with haphazard arrangement of the informal settlements. Since informal settlements lack proper arrangement patterns, solid wastes are disposed of inappropriately in open areas and sewage water intrudes into living areas thus affecting households. The UAV dataset provided for this study shown in Fig. 1, covers a total area of 0.13 km^2 which includes apparent ground scene suitable

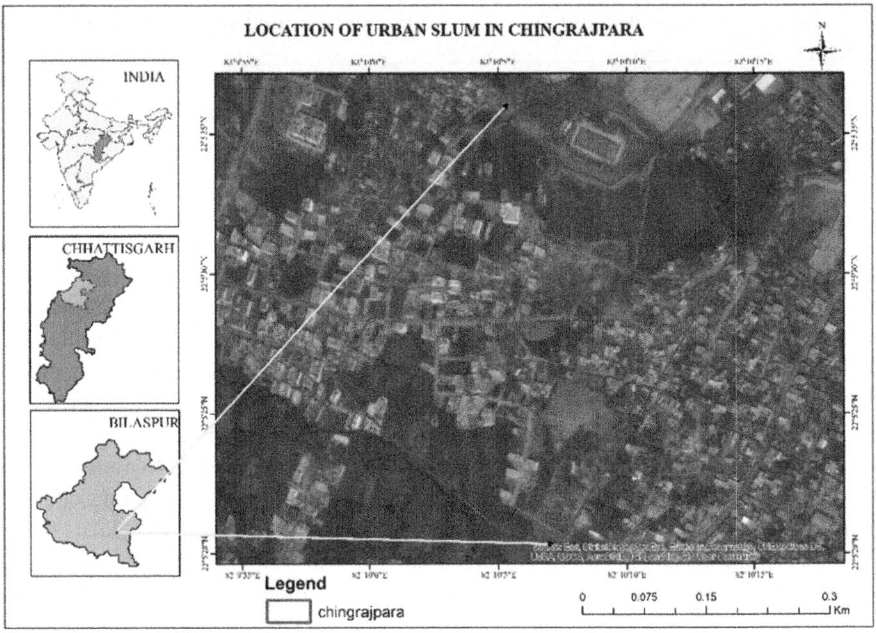

Fig. 1 Study area map

for processing. The datasets were taken at the flying height of 100 m associated with the Ground Sample Distance of 2.19 cm. Around 85 number of images covered over the places that were processed for orthomosaic. The dataset also contains point cloud source file that can be manipulated for classification and future processing [15]. Diverse software platforms were used for classification and result interpretation of orthomosaic and point cloud datasets. Some of the softwares utilized for acquiring classified results were: ENVI 5.5, SNAP, Arc GIS, Trimble Real works (point cloud data processing) and eCognition.

3 Methods

3.1 Selection of Training Samples

Reference vector samples collected need to be pure pixels exhibiting homogeneous characteristics for effective accuracy of the resultant image. Collection of representative samples does not have any direct impact on classification accuracy but affects the overall producer and user accuracies [18]. Hence, an orderly scheme for selection of samples with respect to image interpretation keys is necessary and it is shown (see Fig. 2) [14].

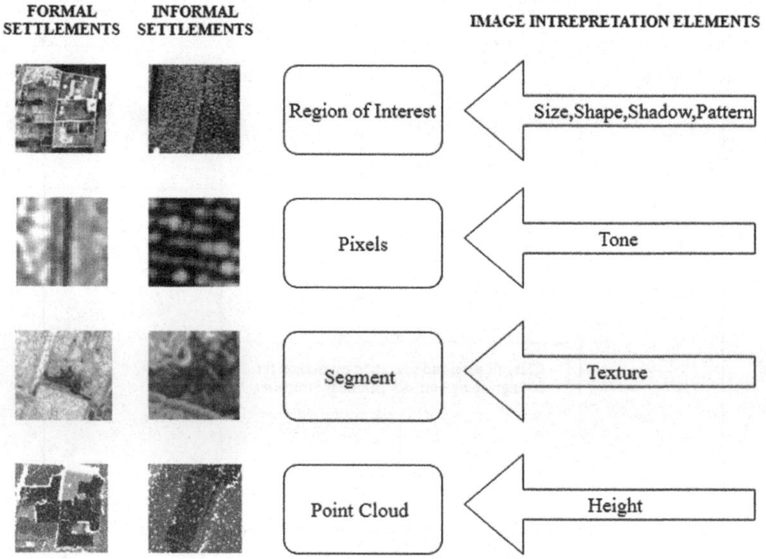

Fig. 2 Selection of representative training samples with respect to image interpretation keys

3.2 Orthomosaic Dataset Classification

UAV datasets are collected at varying altitudes, atmospheric conditions and camera calibrations hence need to be organized for applying classification. Orthomosaic datasets are geometrically corrected representing the true ground features that are the prime source for classification. The objective of this paper is to incorporate traditional pixel-based classification approach and decision tree algorithms to VHR UAV datasets to analyze the capability of how well it differentiates urban slum textures is described below using a formulated workflow in Fig. 3. VHR datasets are efficient for pixel level classification because of its accuracies. Quantifying vegetation, formal

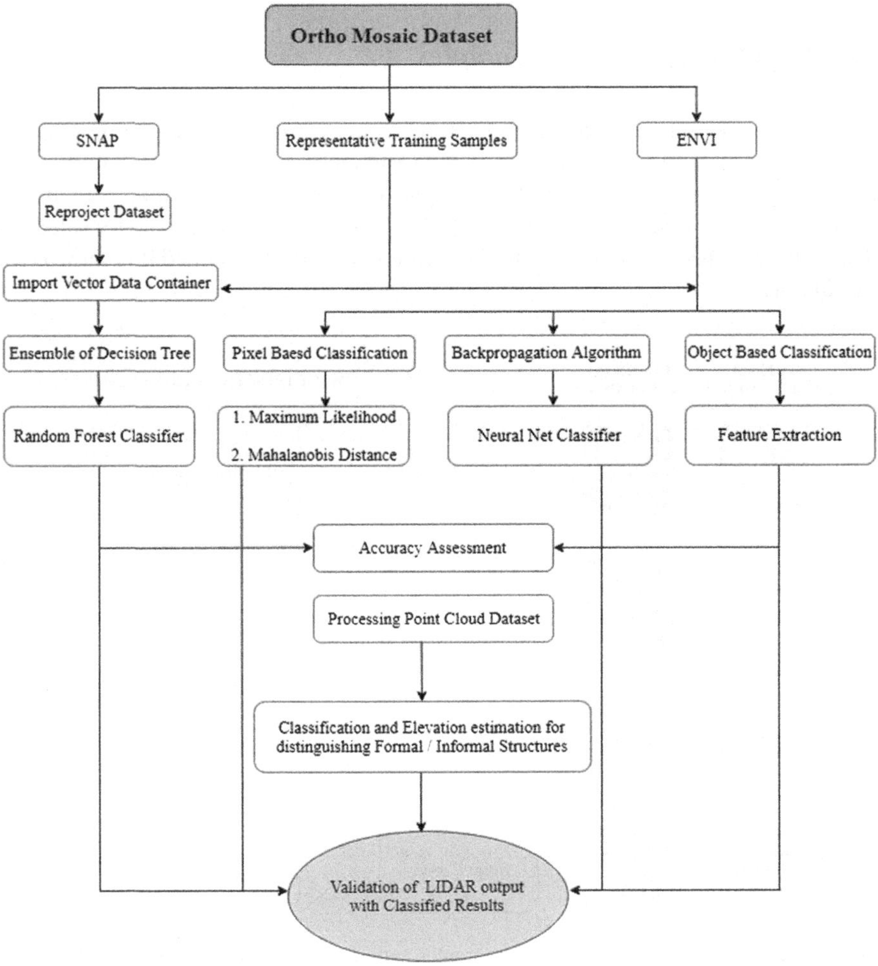

Fig. 3 Methodology

concrete urban structures, randomly distributed slums waste dumping sites can be extracted from the study area due to finer spatial resolution. Spatial resolution of the dataset for Chingrajpara slum area, Chhattisgarh is 2.19 * 2.19 cm thus providing enhanced land cover patterns.

Object-based image classification (OBIA), pixel-based classification and back propagation algorithms are major classification approaches used widely for HR data classification. Characteristics of HR and VHR imageries shows pure pixels that can be uniquely identified in pixel-based classification whereas in OBIA neighboring pixels influences single pixels thus resulting in misclassification [19]. As the dataset used in the study is highly accurate pixel based approach is more preferable than OBIA approach. Some object-based analysis techniques have also been explained by implementing segmentation toolsets of eCognition for better comparison.

An ensemble decision tree classifier approach is used in this paper to check which classifier technique outperforms well for the provided representative samples. The applied classifier is Random Forest classifier, a machine learning algorithm that bags training samples randomly to train decision trees [20].

Classification algorithms used for analyzing land cover features of the study area are:

Pixel-based classification approach Some of the algorithms include:

Maximum Likelihood Classifier This is a pixel-based classifier, in which the highest probability is assigned to the vector of a class among the probabilities of vectors assigned to numerous other classes. It is a parametric classifier, and hence performance depends on how well data matches with samples. Assigning of pixels is based upon user-defined threshold values. In order to reduce unclassified pixels, the threshold value is to be given higher than the class probability value. Chosen training samples are at a distance with respect to Jeffries-Matusita (JM) distance. Finally, every pixel in the dataset is associated with a desired land cover type [17, 21, 22]. The mathematical expression for calculating discriminated functions of pixels is given in the equation:

$$g_i(x) = \ln \rho(\omega_i) - 1/2 \ln|\Sigma_i| - 1/2(x - m_i)T\Sigma - 1_i(x - m_i) \tag{1}$$

where; i represents class, x is band dimensional data, $\rho(\omega_i)$ is the probability of classes occurring in the image, m_i is the mean, $|\Sigma_i|$ and $\Sigma - 1$ are the determinant of the covariance matrix and its inverse in class ω_i, respectively [21].

Mahalanobis Distance Classifier This classifier is direction-sensitive and uses the covariance matrix by taking into account the variability of classes. Pixels closer to the representative training samples are classified. Mahalanobis distance classifier is similar to maximum likelihood classifier but considered as a distance squared measure [23]. Parametric pixel based classifiers do not fix regular sample sizes due to higher dimensionality of the datasets (Hughes phenomena).

$$d(x - m_i)^2 = \ln|\Sigma_i| + (x - m_i)T\Sigma_i^{-1}(x - m_i) \tag{2}$$

Here the class covariance are reverse mean for all i, $\Sigma_i = \Sigma$.

Back Propagation algorithm This algorithm uses the following classifier:

Neural Net Classifier Back Propagation Neural Network (BPNN) are complex computations used for class separation. Artificial neural networks are non-parametric classifiers that are aggregated based upon assumptions of data distribution. It constitutes a series of layers connected to the primitive processing part, the neurons [24]. For classification of very high-resolution imageries, logistic regression and deep convolution neural networks can be used for segmentation and feature extraction [25].

BPN algorithm comprises of 3 layers namely input layer, hidden layer and output layer. Reference calculation formula for determining hidden neuron is given as;

$$L = (m + n)/2 + c \tag{3}$$

where, L is the number of hidden neurons, m is input neurons, n is the number of output neurons and c is a constant.

Ensemble decision tree It is explained by using the following algorithm:

Random Forest Classifier This classifier is a machine learning algorithm and it is being used for both Classification and Regression Tree (CART). Random Forest (RF) builds ensemble decision trees and merges them together using the bagging method to attain maximum accuracy and prediction as shown in Fig. 4. The classifier adds extra randomness to the data by selecting the best feature for splitting each node. Random forest is an ensemble of decision trees which means a collection of decision trees to formulate set of rules for predictions. Random forest classifier shows lower variance thus creating many CART trees with respect to bootstrapping [19]. Classification using RF yields good accuracy when used with VHR imageries.

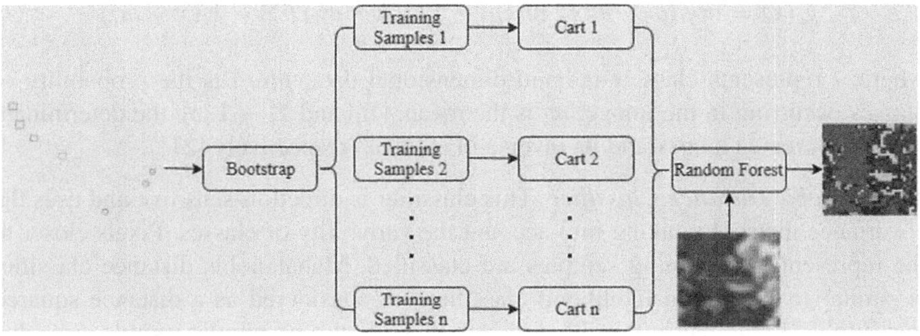

Fig. 4 Random forest classifier

Object Based Image Analysis The Object Based Image Analysis (OBIA) is explained using:

Feature extraction Land cover features are extracted through OBIA by understanding the spatial, spectral and textural characteristics of segments. Urban slum UAV datasets are segmented using multiresolution image segmentation algorithm using a scale parameter of 50 and shape/compactness factor of 0.5 using eCognition. For the provided dataset bands 3 and 4 representing multispectral sensor like properties are used for greenery detection thus resembling the process of NDVI. Arithmetic feature for these two bands produces results that fairly delineates vegetation present in the study.

3.3 Processing LAS Dataset

Photogrammetric point cloud datasets provide dense point cloud information representing real-world features in x, y and z coordinates. Point clouds carry elevation details that are three dimensional in nature and enhance the classification of terrain landscapes. Various point cloud processing toolsets are introduced in software like Arc GIS and ENVI from which .las files could be visualized. Classification of UAV mosaiced dataset using conventional classifiers may not provide subtle results as like point clouds classification because of the density of point arrangement. Photogrammetric points are collected dense based upon the number of returns. The point cloud information provided for this study contains five returns from varied elevations. Hence, processing the data for the classification of land cover produces precise results and can be compared with general classification algorithms.

Classification and elevation estimation of photogrammetric point cloud data Each point in a point cloud defines the returned laser pulses from the target. Numerous points fall into one class and are differentiated using class codes for better understanding. As the quality of the point cloud is higher, minute feature is apparently visible, thus favoring interpretation. In this study, the area comprises of both high raised buildings and low roof structures. Point clouds associated with formal concrete structured buildings are grouped at a higher elevation around 338.6 ft from the ground. Informal settlements appear condensed in between built-up areas with metal sheet rooftops or improper tiled roofing at a minimum elevation of 317.3 ft above the ground. Employing LAS toolsets, point clouds can also be generated from photogrammetrically orthorectified data. Generated las data could be visualized for slope, aspect, elevation, return and classification. Huge file ground classify is applied for bare earth extraction providing step size (in meters), projection information and last return as a default, as it is applied for bare earth. For extraction of buildings, elevation wise classified results are manipulated.

4 Results and Discussion

4.1 *Analysis of Classified Results*

Results from applied classification algorithms such as pixel based, back propagation, ensemble and OBIA are interpreted to examine if the classes are aggregated properly. Relevant classes for the VHR UAV datasets were; formal settlements, informal settlements, road, greenery, sewage leaks, solid waste disposal site and open spaces. Each algorithm includes a systematic selection of training samples as per image interpretation elements. Number of training sites was collected for pixel-based classification approach since accuracy depends on maximum samples collected in a distributed manner and the results are shown in Fig. 5. It is observed that all the classification algorithms have given appropriate results of which random forest outperforms well. OBIA using feature extraction used segmentation that shows formal concrete rooftops, open spaces clearly since the segments are well defined by multiresolution image algorithm are shown in Fig. 6. Major limitation of OBIA is that it was not capable of classifying informal settlements due to its coarse arrangements and low lying roof structures present inbetween raised buildings, but greeneries and high raised buildings appeared clear using feature extraction. Hence OBIA classification depends upon the distribution of materials to be classified. Accuracies of every algorithm are discussed further in Sect. 4.2. These results are validated with photogrammetric point cloud data elevations for verification purpose.

4.2 *Accuracy Assessment*

A confusion matrix displays the reference class and the classified data using which the overall accuracy and kappa coefficient is estimated and compared. The overall accuracy is calculated as a ratio by adding the total number of correctly classified sites to that of the total number of reference sites. This provides a percentage value which is the overall accuracy. The kappa coefficient is a value to evaluate the obtained overall accuracy [17]. Accuracies of each classification results are shown below in Figs. 5 and 7. Overall accuracies obtained for class results were, MLC—84.4%, MD—74.8%, NN—82.7% and RF—92.6%.

4.3 *Analysis of Point Cloud Classified Results*

Formal/informal structures are well classified with respect to point cloud input information as shown in Fig. 8. Three-dimensional path profile is generated in Fig. 9 to check exacted ground-based classification results. Classified point clouds represent

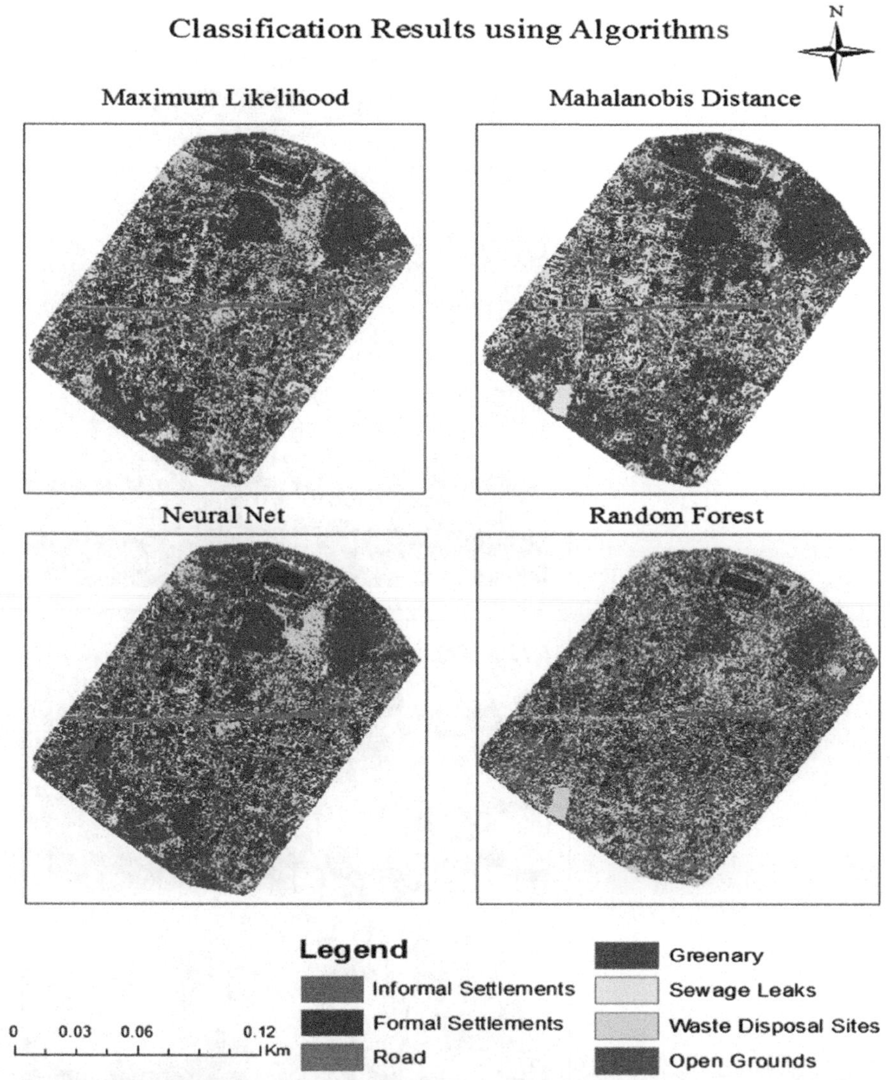

Fig. 5 Classification results

details of intensity, elevation, laser returns that discriminate urban and slum structures. Results obtained using .las datasets are highly accurate yielding ideas for urban planners to ensure unnecessary occupation of open spaces and greenery.

Fig. 6 OBIA—segmentation

4.4 *Validating Classified Results from Orthomosaic and Point Cloud Datasets*

Classified results from UAV and photogrammetric point clouds exhibit similar results when compared. The densification of points are helpful for identifying exact materials from the clumsy terrain. The classification algorithms produced satisfactory results

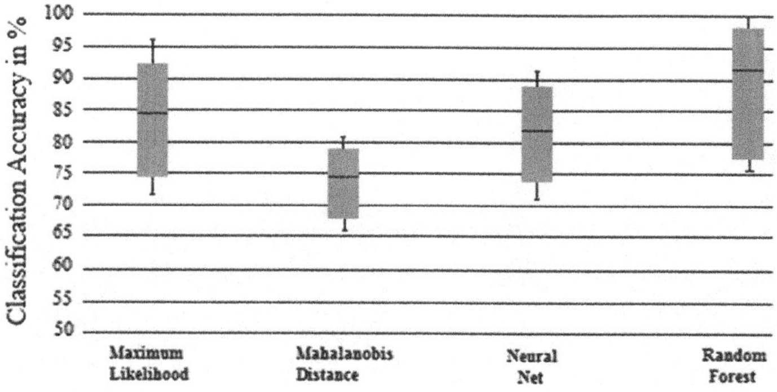

Fig. 7 Accuracy results

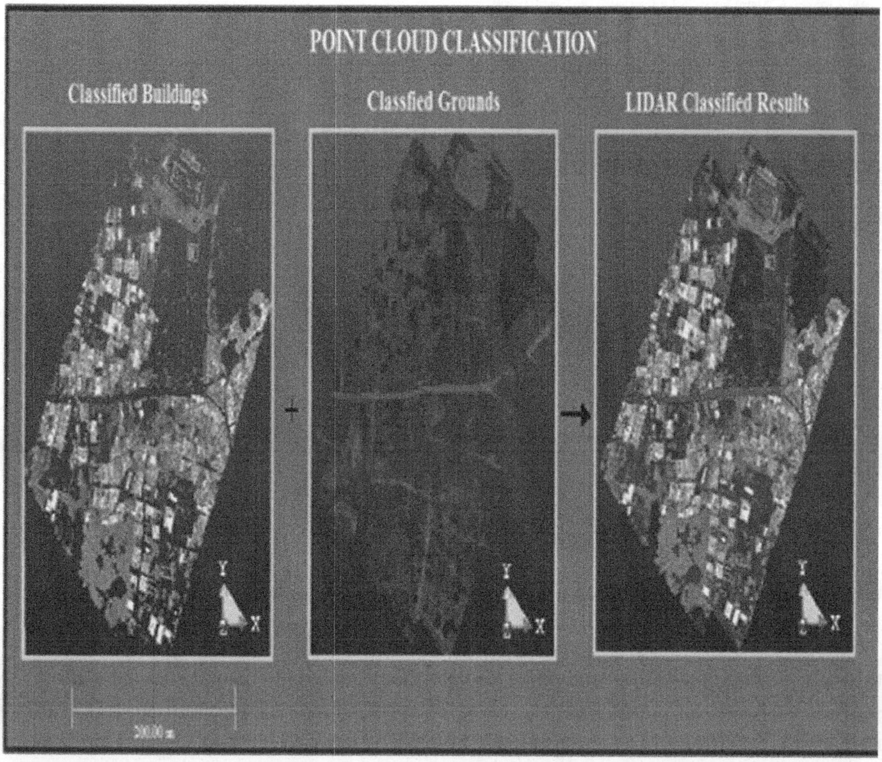

Fig. 8 Point cloud classification

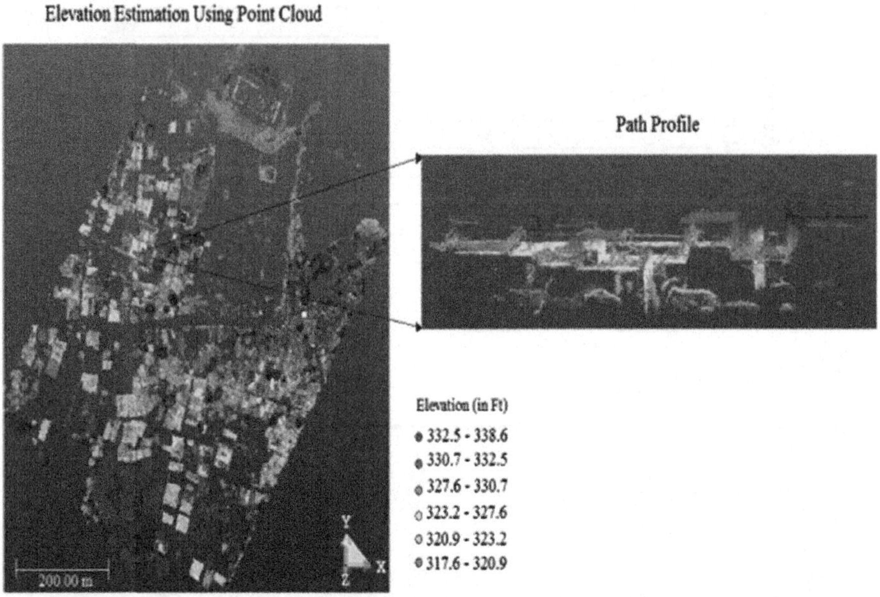

Fig. 9 Elevation based classification and path profile

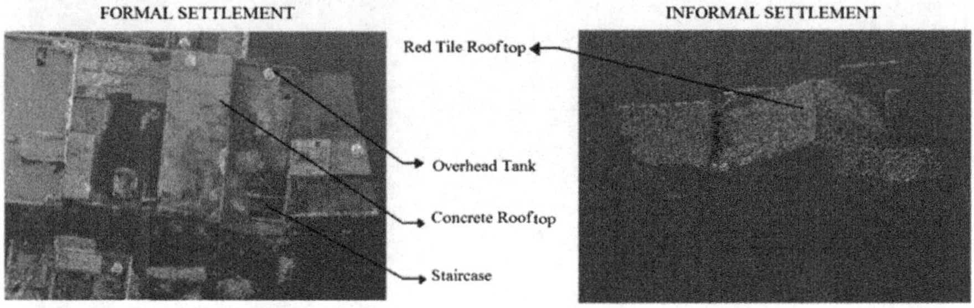

Fig. 10 Verification of formal/informal settlements using point cloud

that are cross-checked using point clouds because of its feasibility. Validation of surface features using path profile retrieves robust results that can be used for future research (Fig. 10).

4.5 Discussion

Growth of urban slum landscape is dynamic with subsequent requisite for living spaces. Reducing the social inequalities between urban and slums strengthens the socio-economic status of developing countries like India. Planners and decision-makers need to focus on upgradation of informal settlements inside urban areas as it triggers the local economical development. Physical upgrading of slums by providing access to basic facilities and proper network of drainage, forbid waste disposal near households and fostering green spaces improvises physical health and well being of communities. Though it is non-viable for absolute eradication of informal settlement growth, measures can be taken to improvise its spatial distribution in a well-planned manner.

4.6 Future Works

Using the high-resolution UAV datasets, classification algorithms can be manipulated in depth for urban slum studies. Object-based classification approached of segmentation can be focused in particular to bring out a precise grouping of pixels with respect to spatial, spectral and texture elements. The segmentation process can be improvised by defining appropriate thresholds for defining objects. Airborne retrieval of photogrammetric point clouds is a mature remote sensing technique that serves various applications. These point clouds are to be well optimized for urban planning to achieve cognitive results in order to slacken the growth of deprived informal settlements.

5 Conclusion

Visualizing the growth of slums in through image processing describes the spatial distribution of slums apparently inside the urban sphere. Detection of informal settlements from satellite imageries is a challenging task because of its variability and morphology. In that case, UAV imageries are ideal for the extraction of land cover types present in the real world scene. Unlike other remote sensor data, UAV provides accurate imageries that are utilized for producing explicit results. Spatial metrics for quantifying the growth of informal settlements can be studied in detail using UAV and its corresponding point cloud data. Results can be manipulated for future research works in enhancing segmentation techniques that can be used for planning purposes. Effective solutions practiced for regularization of urban slum landscapes influence the economic welfare which outcomes a vision of a slum-free India in the near future.

Acknowledgements This study was supported by ESRI technical team by providing essential software package and corresponding trial version licenses.

References

1. Saxena A (2008) Monitoring of urban fringe using remote sensing and GIS techniques
2. Rahman G, Alam D, Islam S (2008) City growth with urban sprawl and problems of management for sustainable urbanization. ISOCARP Congress

3. Shekhar S (2012) Detecting slums from Quick Bird data in Pune using an object oriented approach. In: International archives of the photogrammetry, remote sensing and spatial information sciences, vol XXXIX, pp 519–524. XXII ISPRS Congress, Melbourne
4. https://ourworldindata.org/urbanization
5. Taubenböck H, Kraff NJ (2015) The global face of urban poverty? Settlement structures in slums. In: Taubenböck H, Wurm M, Esch T, Dech S (eds) Global urbanization. Springer Spectrum, Berlin
6. Friesen J, Rausch L, Pelz PF, Fürnkranz J (2018) Determining factors for slum growth with predictive data mining methods. MDPI, Urban Science
7. UN-HABITAT (2011) Annual report 2010. United Nations Human Settlements Programme, Kenya
8. Rausch L, Friesen J, Altherr L, Meck M, Pelz P (2018) A holistic concept to design optimal water supply infrastructures for informal settlements using remote sensing data. MDPI Remote Sens 10(2)
9. Ben-Dor E, Levin N, Saaroni H (2001) A spectral based recognition of the urban environment using the visible and near infrared spectral region (0.4-1-1 mm). A case study over Tel-Ariv, Israel. Int J Remote Sens 22:2139–2218
10. Jain S (2007) Use of IKONOS satellite data to identify informal settlements in Dehradun, India. Int J Remote Sens 28(15):3227–3233
11. Shekhar S (2012) Modeling the probable growth of slums by using geoinformatics. Indian Soc Educ Environ 1(8):588–598
12. Kohli D, Sliuzas R, Stein A (2016) Urban slum detection using texture and spatial metrics derived from satellite imagery. J Spat Sci 61(2):405–426
13. Kuffer M, Pfeffer K, Sliuzas R (2016) Slums from space—15 years of slum mapping using remote sensing. MDPI Remote Sens 8(6)
14. Sliuzas R, Kuffer M, Gevaert C, Persello C, Pfeffer K (2017) Slum mapping: from space to unmanned aerial vehicle based approaches. In: Joint urban remote sensing event. IEEE, Dubai, pp 1–4
15. Jain K (2019) Urban slum dataset for Chingrajpara slum area (Chhattisgarh)
16. Kim AM, Olsen RC, Kruse FA (2013) Methods for LiDAR point cloud classification using local neighborhood statistics. In: Turner MD, Kamerman GW (eds) Laser radar technology and applications XVIII, vol 8731. SPIE
17. Priyadarshini NK, Kumar M, Rahaman AS, Nitheshnirmal S (2018) A comparative study of advanced land use/land cover classification algorithms using Sentinel-2 data. In: The international archives of the photogrammetry, remote sensing and spatial information sciences, vol XLII(5), pp 665–670. ISPRS TC V mid-term symposium, Dehradun
18. Zhen Z, Quackenbush LJ, Stehman SV, Zhang L (2013) Impact of training and validation sample selection on classification accuracy and accuracy assessment when using reference polygons in object-based classification. Int J Remote Sens 34(19):6914–6930
19. Feng Q, Liu J, Gong J (2015) UAV remote sensing for urban vegetation mapping using random forest and texture analysis. MDPI Remote Sens 7(1):1074–1094
20. https://towardsdatascience.com/decision-tree-ensembles-bagging-and-boosting266a8ba60fd9
21. https://www.harrisgeospatial.com/docs/MaximumLikelihood.html
22. Ahmad A, Quegan S (2012) Analysis of maximum likelihood classification on multispectral data. Appl Math Sci 6(129):6425–6436
23. Gao J (2008) Digital analysis of remotely sensed imagery, 1st edn. McGraw-Hill Professional
24. Caetano M (2007) Image classification. Advanced training course on land remote sensing. ESA
25. Priyadarshini NK, Kumar M, Kumaraswamy K (2018) Identification of food insecure zones using remote sensing and artificial intelligence techniques. In: The international archives of the photogrammetry, remote sensing and spatial information sciences, vol XLII(5), pp 659–664, ISPRS TC V mid-term symposium, Dehradun

Estimation of Forest Tree Heights and Crown Diameter Using High Resolution Images from UAV: A Case Study of Kalesar, Haryana

Nitheshnirmal Sadhasivam, C. Dineshkumar, S. Abdul Rahaman and Ashutosh Bhardwaj

Abstract The Very High Resolution (VHR) imagery obtained from Unmanned Aerial Vehicle (UAV) has the potential to generate huge and comparably wealthy information on forest trees than the traditional forest inventories. Due to its effectiveness in providing faster and practically more reliable measurements on individual trees, UAV data has now been highly utilized in forestry applications. In the present study, VHR images from a UAV covering an area of about 0.35 km^2 were used to estimate the tree heights and their crown diameter from a moist deciduous forest. UAV images with an average ground sample distance of 11.86 cm are processed for generating point cloud, Digital Elevation Model (DEM), Digital Terrain Model (DTM), ortho-images, and Digital Surface Model (DSM). From the generated DTM and DSM, Canopy Height Model (CHM) is obtained after which local maxima filter has been applied for raster smoothening to detect the crown top from which the individual tree heights were estimated. The diameter of the tree crown is assessed using the Inverse Watershed Segmentation (IWS) method. The result of this study shows that the trees present in the Kalesar forest area have a maximum height of 46 m. The result of this research is highly useful for the planners and decision-makers in biomass estimation and forest management.

Keywords UAV · DTM · DSM · CHM · Inverse watershed segmentation

1 Introduction

Estimation of forest tree heights and crown diameter plays a vital role in managing and sustaining the most valuable forests ecosystem. Mostly the assessment of forest

N. Sadhasivam (✉) · S. Abdul Rahaman
Department of Geography, Bharathidasan University, Tiruchirappalli, India
e-mail: nirmalgsarath@gmail.com

C. Dineshkumar
Department of Civil Engineering, SRM Institute of Science and Technology, Chennai, India

A. Bhardwaj
Photogrammetry and Remote Sensing Department, Indian Institute of Remote Sensing, Dehradun, India

K. Jain et al. (eds.), *Proceedings of UASG 2019*, Lecture Notes in Civil Engineering 51,
https://doi.org/10.1007/978-3-030-37393-1_22

attributes was carried out using the light detection and ranging (LiDAR) technique [1–4]. However, the cost incurred during the flight of the airborne LiDAR system foils the wide usage of the application in forest metrics. Advancement in the field of remote sensing with the help of Unmanned Aerial Vehicle(UAV) techniques have extended the limits in forestry by producing Very High Resolution (VHR) images and 3D data using point cloud processed from the ortho-mosaic image which is easily available at a reasonable price. Using this application, we can compute the structure and composition of the forest at high accuracy using different spatial resolution with the help of statistical algorithms [5]. The dense images produced through UAV flight is utilized in the generation of Digital Surface Model (DSM), Digital Terrain Model (DTM), Digital Elevation Model (DEM) and photographic point clouds which can be used in the estimation of Canopy Height Model (CHM). A DSM represents the earth's surface along with its features such as buildings and trees while a DTM represents the bare earth. A CHM is the depiction of the highest point in a tree from the base ground landscape and is derived by the subtraction of DSM and DTM [5]. Determining CHM of a forest using DSM and DTM derived from the UAV ortho-mosaic images largely aids in the estimation of forest tree heights and crown diameter using the local maxima and inverse watershed segmentation respectively. However, the point clouds generated from orthomosaic image can be used in the 3D visualization of the individual tree structure [6]. Local maxima filter helps in the determination of tree height by identifying the highest CHM pixel value using window size defined by the user [7]. Watershed segmentation is the widely used boundary detection segmentation [8]. However, it is called as an inverse watershed segmentation (IWS) due to its advantage in the detection of concave-shaped tree crowns in the inverted CHM in which the crowns are bifurcated as individual watersheds or basins. In this way, IWS is very helpful in the detection of crown diameter. Using ground truth data, an accuracy of 94, 85% was achieved in studies estimating tree heights and crown diameter using the local maxima filter and IWS method [1, 9]. Due to these advantages in using VHR UAV images, the present study is aimed at detection and delineation of tree height and their respective crown diameter in Kalesar forest of Haryana using an automated workflow. The aim of this study is achieved by utilizing (a) local maxima method for tree height detection and (b) inverse watershed segmentation for tree crown delineation. However, in this study validation of acquired results with ground data is not carried out due to certain limitations of the dense forest.

2 Study Area and Dataset

The present study utilizes forest and mountain area located in the Kalesar area of Yamuna Nagar district of Haryana. It extends from 77° 32′ 03.57″ E longitude to 77° 32′ 25.60″ E longitude and from 30° 21′ 22.63″ N latitude to 30° 21′ 03.33″ N latitude, covering a total area of about 0.35 km^2. The study area is present 278 m above the mean sea level (MSL). The Kalesar area of Yamuna Nagar district of

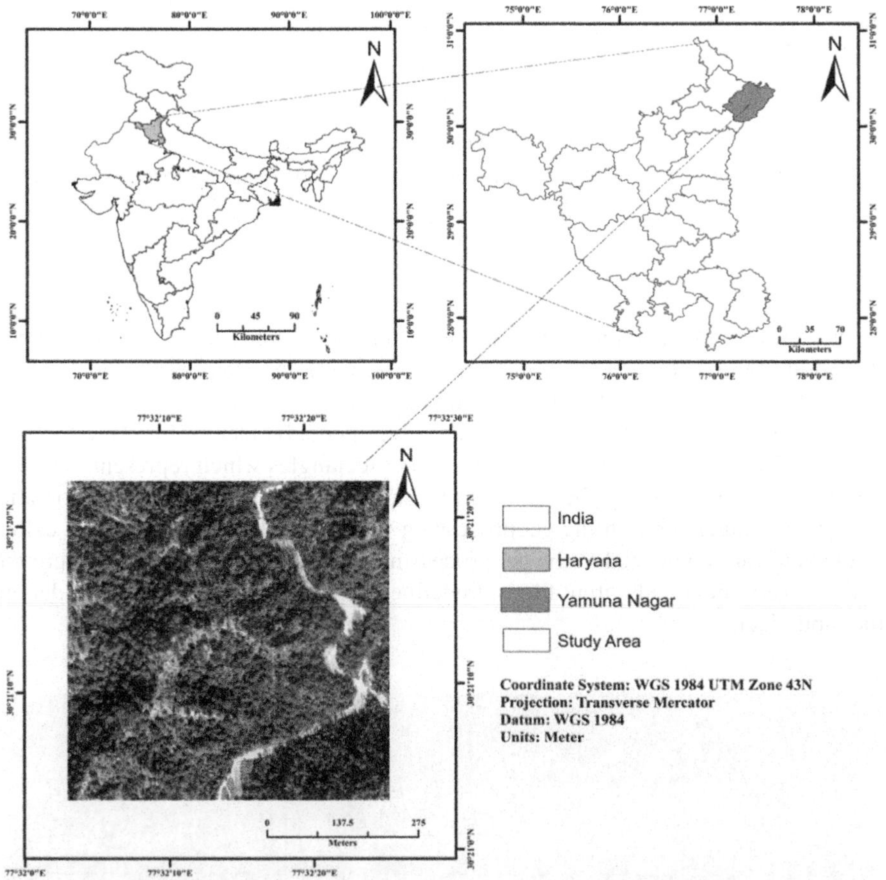

Fig. 1 Location of Kalesar forest area in Haryana

Haryana consists of Sal trees which is a sub-deciduous tree. The Sal tree grows up to a height of 45 m (Fig. 1).

In this study, VHR images were acquired from DJI Phantom 4pro UAV [10]. The UAV has inbuilt internal and external routing parameters with a high-resolution RGB camera. To generate good quality images, windless dry days were selected for performing UAV air flights. The UAV camera and flight parameters are provided in Table 1.

3 Methodology

In this present study, the VHR ortho-mosaic images of the study area obtained from UAV are used in generating DSM and DTM by the Pix4D mapper using Structure

Table 1 UAV camera and flight parameters

Parameters	Values
Spatial resolution	11.86 cm (RGB, DSM, and DTM)
Drone utilized	DJI Phantom 4pro
Camera model	FC6310_8.8_4864 × 3648 (RGB)
Sensor dimension	11.407 (mm) × 8.556 (mm)
Information	Highly dense urban
Area covered	0.35 km^2

from Motion (SfM) and photogrammetric techniques which are further used for the detection of tree height and crown diameter. Figures 2 and 3 represents the 3D view of DSM and 3D view of classified ground points respectively.

In this study, a model has been created using model builder tool in ArcGIS 10.5, shown in Fig. 4. The model consists of yellow rectangles which represent the tools utilized for geoprocessing the variable present in the green circles, which contains the altered values through the geoprocessing tools [11]. The input data also called as elements are connected to the geoprocessing tools through connectors where the path of every input and output file is predefined. The blue circles are used to denote the input data.

Fig. 2 3D view of DSM

Fig. 3 3D view of classified trees using point cloud

3.1 Creation of Canopy Height Model

The uppermost crown part of a forest is known as the canopy. The height of the forest above the ground topography is called as CHM. It consists of various attributes of trees present in a forest. In this study, CHM serves as the main component in extracting tree heights using image smoothening and local maxima filters along with crown diameter detection using IWS method.

The input data is used to derive the Canopy Height Model (CHM) through the raster calculator tool using Eq. 1.

$$\text{CHM} = \text{DSM} - \text{DTM} \quad (1)$$

Following the derivation of CHM, the model splits into two processes for detection of (a) tree height and (b) crown diameter.

3.2 Calculation of Tree Heights

The morphological filter along with focal statistics tool of ArcGIS 10.5 was used for the identification of the local maxima parameter setting. In this study for the detection of tree height, derived CHM is smoothened using low pass filter for reducing the noise

Fig. 4 Model for extraction of tree height and crown diameter

effect and for regulating the heights of CHM [12, 13]. The usage of low pass filter helps in identifying the kernel size which helps in the next step, eliminating multiple local maxima identified in a single crown. Following, low pass filtering, local maxima determined using focal statistics tool. Focal statistics is a morphological filter which identifies the pixel having the highest value in the top of a tree crown. In this study to eliminate multiple local maxima in a single crown, various kernel size where experimented among which a kernel size of five cell units is suited for the tree height detection in this study. From focal statistics, the following Eq. (2) is used in the raster calculator to identify the highest pixel value.

$$\text{Con}(\text{“CHM”} == \text{“focal statistics”}, 1) \tag{2}$$

Through this process, the identification of tree height is completed.

3.3 Inverse Watershed Segmentation

In this study, for the detection of individual crown diameter, inverse watershed segmentation (IWS) method was utilized [14]. The CHM is inverted using the raster calculator using the following Eq. (3).

$$((\text{“CHM”} - \text{Z\_Max}) * -1) + \text{Z\_Min} \tag{3}$$

Whereas Z_Max and Z_Min are the maximum and minimum of CHM.

The resultant inverted CHM is used to determine the flow direction using hydrology tool present in spatial analyst of ArcGIS 10.5. After computing flow direction, the resultant raster is given as an input in the hydrological drainage basin tool [15]. As there are large gaps seen between the tree crowns, a Boolean logic was used, where the CHM is converted into a raster having two-pixel values of 0 and 1. Later, this layer is multiplied with the basin raster using the raster calculator and finally, the resultant segments are converted into polygons, which represent the individual crown diameter.

4 Results and Discussion

4.1 Canopy Height Model

In the present study, DSM is subtracted from DTM to derive the CHM. Figure 5 represents the DSM and DTM which is used to obtain the CHM of the study area. The value of DSM ranges between 432.7 and 374.2 m, whereas the DTM ranges from 409.3 to 329.5 m respectively. The CHM of the study area ranges between 46.07 and

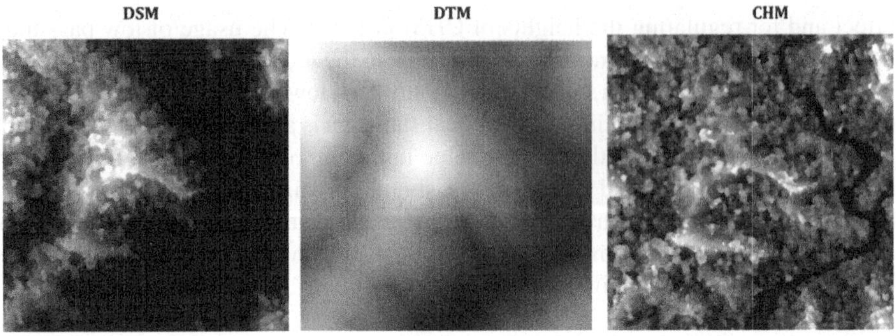

Fig. 5 DSM, DTM and CHM of the study area

3.1, where the maximum value of CHM represents the highest tree pixel present in the study area. From this, it can be seen that the highest treetop has a height of about 46 m. The lowest value of CHM is observed in the river channel, which flows through the study area.

4.2 Estimated Height of Forest Trees

The derived CHM is smoothened using a low pass filter, followed by applying a morphological local maxima filter using focal statistics which is shown in Fig. 6. Through the focal statistics, tool kernel window size is given as five cell units after several experimental tests.

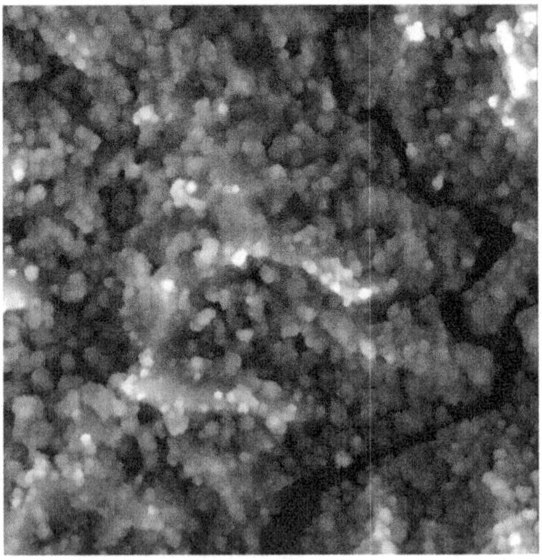

Fig. 6 Local maxima filter using focal statistics

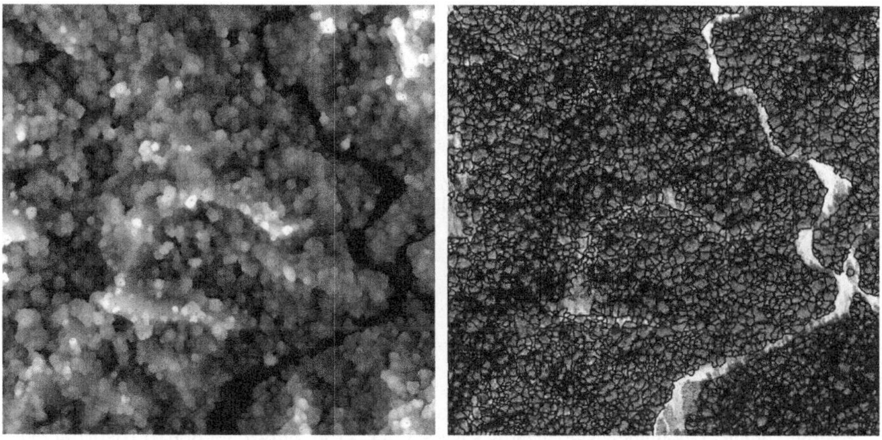

Fig. 7 Local maxima seeds identified as tree tops (left) and tree crowns derived using IWS

The local maxima filtering identifies the pixel having the highest pixel value using pixel reflectance which could be the possible treetop from which, tree height is determined. In the present study, after applying local maxima filter, based on the kernel window sizes the highest pixel values having high reflectance were determined from which height of the trees present in the study were obtained (Fig. 7). Nearly 2140 treetops were identified using local maxima filtering where the maximum height of treetop ranges from 12.3 to 46 m. This value corresponds to the average Sal tree height present in the study area.

4.3 Estimated Crown Diameter

Segmentation of tree crown using IWS can be seen in Fig. 7. The diameter of the tree crown polygons was estimated ranging from 5.2 to 22. 5 m. It was observed that there lies a high correlation between the tree height and crown diameter, as the crown diameter increases with tree height and vice versa.

5 Conclusion

In this study, DSM and DTM derived from VHR UAV ortho-images are used to obtain the CHM, which serves as a significant component in detecting the tree height and crown diameter in the study area. Individual tree height can be estimated using CHM as an input, with the help of local maxima method where the pixel present in each treetop having the highest pixel value is detected as a treetop. The crown diameter is also detected by giving CHM as an input in IWS method where each tree crown is segmented and is converted into polygon to calculate the diameter of the crown. This paper suggests that the methodology used in this study can be utilize ed for semi-automatic detection of tree height and crown diameter which helps in the estimation of biomass and site productivity of forest ecosystems.

Acknowledgements The authors of this study acknowledge Dr. Kamal Jain, Professor, Indian Institute of Technology-Roorkee for providing datasets in order to carry out this study.

References

1. Birdal AC, Avdan U, Türk T (2017) Estimating tree heights with images from an unmanned aerial vehicle. Geomatics Nat Hazards Risk 8:1144–1156. https://doi.org/10.1080/19475705.2017.1300608
2. Breidenbach J, Næsset E, Lien V, Gobakken T, Solberg S (2010) Prediction of species specific forest inventory attributes using a nonparametric semi-individual tree crown approach based on fused airborne laser scanning and multispectral data. Remote Sens Environ 114:911–924. https://doi.org/10.1016/J.RSE.2009.12.004
3. Wallace L, Lucieer A, Watson CS (2014) Evaluating tree detection and segmentation routines on very high resolution UAV LiDAR Data. IEEE Trans Geosci Remote Sens 52:7619–7628. https://doi.org/10.1109/TGRS.2014.2315649
4. Selkowitz DJ, Green G, Peterson B, Wylie B (2012) A multi-sensor lidar, multi-spectral and multi-angular approach for mapping canopy height in boreal forest regions. Remote Sens Environ 121:458–471. https://doi.org/10.1016/J.RSE.2012.02.020
5. Mohan M, Silva C, Klauberg C, Jat P, Catts G, Cardil A, Hudak A, Dia M, Mohan M, Silva CA, Klauberg C, Jat P, Catts G, Cardil A, Hudak AT, Dia M (2017) Individual tree detection from unmanned aerial vehicle (UAV) derived canopy height model in an open canopy mixed conifer forest. Forests 8:340. https://doi.org/10.3390/f8090340
6. Kattenborn T, Koch B, Sperlich M, Kattenborn G (2014) Potential of unmanned aerial vehicle based photogrammetric point clouds for automatic single tree detection. In: Gemeinsame Tagung 2014 der DGfK, der DGPF, der GfGI und des GiN (DGPF Tagungsband 23/2014), pp 1–6
7. Popescu SC, Wynne RH, Nelson RF (2002) Estimating plot-level tree heights with lidar: local filtering with a canopy-height based variable window size. Comput Electron Agric 37:71–95. https://doi.org/10.1016/S0168-1699(02)00121-7
8. Carleer, AP, Debeir O, Wolff E (2005) Assessment of very high spatial resolution satellite image segmentations. Photogramm Eng Remote Sens 71:1285–1294. https://doi.org/10.14358/PERS.71.11.1285
9. Panagiotidis D, Abdollahnejad A, Surový P, Chiteculo V (2017) Determining tree height and crown diameter from high-resolution UAV imagery. Int J Remote Sens 38:2392–2410. https://doi.org/10.1080/01431161.2016.1264028
10. Jain K (2019) Forests and mountains dataset for Nagli area (Haryana)
11. Nitheshnirmal S, Thilagaraj P, Rahaman SA, Jegankumar R (2019) Erosion risk assessment through morphometric indices for prioritisation of Arjuna watershed using ALOS-PALSAR DEM. Model Earth Syst Environ 1–18
12. Dralle K, Rudemo M (1996) Stem number estimation by kernel smoothing of aerial photos. Can J For Res Revue 26(7):1228–1236
13. Falkowski MJ, Smith AMS, Hudak AT, Gessler PE, Vierling LA, Crookston NL (2006) Automated estimation of individual conifer tree height and crown diameter via two-dimensional spatial wavelet analysis of lidar data. Can J Remote Sens 32:153–161

14. Edson C, Wing MG (2011) Airborne light detection and ranging (lidar) for individual tree stem location, height, and biomass measurements. Remote Sens 3:2494–2528
15. Wannasiri W, Nagai M, Honda K, Santitamont P, Miphokasap P (2013) Extraction of mangrove biophysical parameters using airborne LIDAR. Remote Sens 5(4):1787–1808

Object Based Automatic Detection of Urban Buildings Using UAV Images

Nitheshnirmal Sadhasivam, C. Dineshkumar, S. Abdul Rahaman and Ashutosh Bhardwaj

Abstract Detection of urban buildings is a pre-requisite in urban planning and management. Very High Resolution (VHR) images acquired from Unmanned Aerial Vehicle (UAV) can play a dominant role in extraction of urban features effectively. In recent years, Geographic Object Based Image Analysis (GEOBIA) has been highly utilized for classification of VHR images, than the traditional pixel based classification owing to its novel paradigm and very high accuracy. The present study aims at detecting the dense urban buildings more precisely and reliably through GEOBIA using the orthomosaic image, Digital Surface Model (DSM) and Digital Terrain Model (DTM) processed from VHR UAV images. Dense urban buildings in Khanjarpur area of Roorkee, covering an area of about 1.63 acres was selected for this experimental study. As the object-based classification involves both segmentation and classification, multi-resolution segmentation algorithm is utilized for segmentation and to select suitable values of parameters such as scale, compactness and shape for building detection and extraction. Classification has been executed after segmentation with a formulated set of rules. Further, the classification accuracy is verified through reference data obtained through heads-up digitization of buildings from the VHR UAV orthomosaic image. The extracted buildings achieved a overall accuracy of 88.1% and 76.3% as cross verified with reference buildings using object based and area based accuracy measures respectively.

Keywords UAV · DSM · Object-based building detection · Segmentation · Urban planning

N. Sadhasivam (✉) · S. Abdul Rahaman
Department of Geography, Bharathidasan University, Tiruchirappalli, India
e-mail: nirmalgsarath@gmail.com

C. Dineshkumar
Department of Civil Engineering, SRM Institute of Science and Technology, Chennai, India

A. Bhardwaj
Photogrammetry and Remote Sensing Department, Indian Institute of Remote Sensing, Dehradun, India

K. Jain et al. (eds.), *Proceedings of UASG 2019*, Lecture Notes in Civil Engineering 51,
https://doi.org/10.1007/978-3-030-37393-1_23

1 Introduction

With advancements in the field of remote sensing, usage of Very High Resolution (VHR) spatial imageries obtained from Unmanned Aerial Vehicle (UAV) has become more popular in several fields and has been utilized for various applications, due to its low cost, rapid, flexible and efficient acquisition [1]. These VHR datasets from UAV delivers detailed information on individual objects present in an image than any other medium and high resolution remote sensing datasets such as Landsat TM, ETM+, OLI/TIRS (30 m), Sentinel-2 (10 m), IKONOS (1 m panchromatic and 4 m multispectral), QuickBird (62 cm) and other HR datasets. The spatial resolution of UAV images ranges from sub-meters to few centimeters which is a major advantage of these images compared with satellite and manned aerial vehicle images. In this way, low cost UAV based VHR images can also help in updating and monitoring the growing urban landscapes which is a prerequisite for proper urban planning and smart city management. In particular, detection and extraction of buildings in an urban landscape play a key role in the estimation of the amount of urbanization, collection of property tax, smart city planning, real estate and disaster management [2–4]. Features present in the remotely sensed images are classified and detected with the help of pixel-based and object-based image classification [5, 6]. Several studies have proven the effectiveness of OBIA over the pixel-based supervised and unsupervised image classifications by comparing the classification accuracy of pixel-based image classification and object-based image classification on HR and VHR images [7, 8]. Thus, due to several limitations in the usage of pixel-based classification in classifying HR and VHR images, Geographic object-based image analysis (GEOBIA) also called as object-based image analysis (OBIA) introduced in 1970 [9], have gained popularity among researchers and have been popularly used for feature detection and extraction from HR and VHR images. The task of detection and extraction becomes more easier with ortho-mosaic image, Digital Surface Model (DSM), Digital Terrain Model (DTM) and dense point clouds generated through photogrammetric workflow of VHR UAV ortho-images which utilises structure from motion (SFM) and multi-view stereopsis (MVS) techniques for generating these layers [10, 11]. Major advantage of OBIA is that, it utilizes a group of pixels having the same properties which is clustered to form image objects or segments which are later used for image classification using a set of defined rules [12, 13]. There has been a wide range of building extraction studies reported using HR remote sensing images such as IKONOS and QuickBird using OBIA [14, 15]. A wide range of research studies has also been conducted using HR remote sensing images for assessing the accuracy of OBIA using various accuracy measurements [16–18]. However, from literature review, it is evident that only a few numbers of case studies have been reported, utilizing VHR UAV images for extraction of buildings through OBIA in a less dense urban environment, where individual buildings can be easily extracted from the surrounding objects. The extraction of buildings from complex dense urban environments present in developing countries like India and China is very difficult due to its building complexness, where differentiating one building from the other is hard through visual interpretation. Also, the presence of other urban landscapes makes the scene more complex and tedious to process. In these complex conditions, the usage of OBIA is novel and can be highly helpful for extraction of buildings in dense urban environments. The present study is aimed at the detection of buildings from a dense urban environment using object-based image classification technique and their accuracy assessment using reference buildings acquired through heads-up digitization.

2 Study Area and Datasets

The present study utilizes dense urban buildings located in the Khanjarpur area of Roorkee city situated at Haridwar district of Uttarakhand. The study area is located 31 km away from the district headquarters. It extends from 77° 54′ 19.49″ E longitude to 77° 54′ 21.34″ E longitude and from 29° 52′ 0.273″ N latitude to 29° 51′ 56.86″ N latitude, covering a total area of about 1.63 acres. The study area is present 260 m above the mean sea level (MSL). It consists of dense buildings with different heights and aerial coverage (Fig. 1).

In this study, VHR images were acquired from UAV (DJI Phantom 4pro) [19]. The UAV has inbuilt internal and external routing parameters with a high-resolution RGB camera. To generate good quality images, windless dry days were selected for performing UAV air flights. The UAV camera and flight parameters are provided in Table 1.

3 Methodology

The VHR UAV images acquired using DJI Phantom 4pro is processed through Drone2Map free trial software. In Drone2Map software, orthoimages are merged into ortho-mosaic using the obtained GCPs and orientation of acquired images whereas the projective distortions of the original images are sorted by orthorectification of UAV images. The image matching through high degree (80%) of overlap in the obtained UAV images greatly helps in acquiring dense point cloud, DSM and DTM of the study area which could be used for extraction of features. In this study, the VHR UAV processed ortho-mosaic image is used to detect and extract buildings using Object-Based Image Analysis (OBIA) technique. In the detection of buildings, normalized Digital Surface Model (nDSM) is utilised which is obtained by subtracting the DSM with the DTM of the study area through which a threshold can be applied to detect the man-made features available at different terrain heights. Thus, for the extraction of buildings, nDSM and DTM plays a significant role. The methodology carried out in the study is shown in Fig. 2.

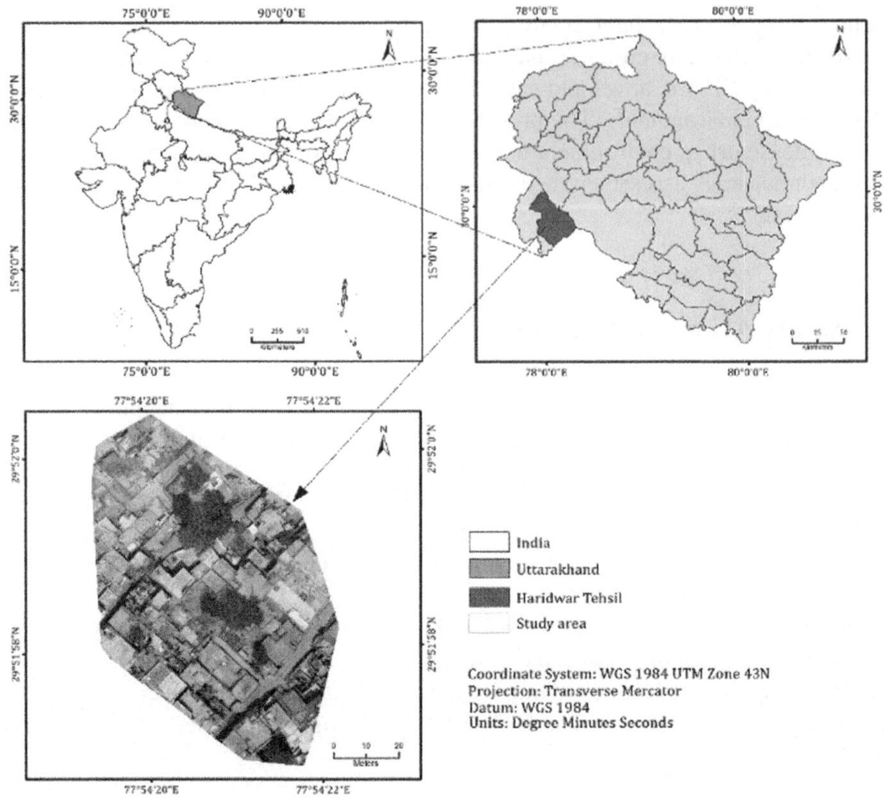

Fig. 1 Study area

Table 1 UAV camera and flight parameters

Parameters	Values
Flight altitude	150 m
Spatial resolution	1.76 cm (RGB and DSM)
Drone utilized	DJI Phantom 4pro
Camera model	FC6310_8.8_4864 × 3648 (RGB)
Sensor dimension	11.407 (mm) × 8.556 (mm)
Information	Highly dense urban
Area covered	1.63 acres
Total images	6

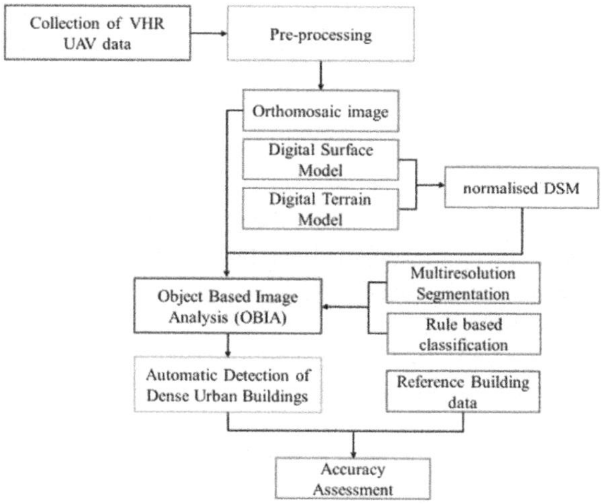

Fig. 2 Methodology flowchart

3.1 *Object Based Classification*

Object-based image analysis (OBIA) is the polygon based classification which groups pixel having identical characteristics into single object segments. OBIA mainly comprises of two processes namely segmentation and rule-based classification. OBIA has a major advantage of presenting meaningful estimation of statistical and textural information. In this study, for detection and extraction of buildings multiresolution segmentation followed by rule-based classification is performed.

3.2 *Multiresolution Segmentation*

Multiresolution segmentation is performed in e-Cognition Developer 9.1. At the initial stage, each pixel present in the VHR image is considered as seed object. After which fusion factor is estimated for surrounding pixels, where the minimum fusion factor is considered as best fitting one. If the best fitting pixel is not found, a new seed object is created from its best neighbor by considering the best candidate object. Otherwise, the calculated fusion factor is compared to the scale factor. While comparing both fusion and scale factors, if the square of the scale factor is found greater than the fusion factor, merging of two objects take place. The above process will proceed until no more possible merging is obtainable. The following equation describes both the shape and color of object features present in the fusion factor.

$$f = w_{shape} \cdot \Delta h_{shape} + w_{colour} \cdot \Delta h_{colour} \quad (1)$$

$$w_{shape} + w_{colour} = 1$$

w_{shape} and w_{colour} ranges from 0 to 1, which are the weights of shape and color heterogeneities, respectively. Δh_{colour} expresses the difference in spectral heterogeneity, which is defined as follows:

$$\Delta h_{colour} = \sum_{C} w_c \left(n_{merge} \cdot \sigma_{c,merge} - \left(n_{obj1} \cdot \sigma_c^{Obj1} + n_{Obj2} \cdot \sigma_c^{Obj2} \right) \right) \quad (2)$$

where, n_{merge}, n_{obj}, n_{obj2} are the number of pixels present in the fused objects, object 1 and object 2 respectively and whereas, $\sigma_{c,merge}$, σ_c^{Obj1}, σ_c^{Obj2} denote the standard deviation of the fused objects, object 1 and object 2 respectively. W_c is the weightage of spectral band c.

Object's compactness and smoothness are utilized for calculating the shape heterogeneity of the object.

$$\Delta h_{shape} = w_{comp} \cdot \Delta h_{comp} + w_{smooth} \cdot \Delta h_{smooth} \quad (3)$$

$$w_{comp} + w_{smooth} = 1$$

The weightage of smoothness and compactness heterogeneities, W_{comp} and W_{smooth} ranges from 0 to 1.

$$\Delta h_{smooth} = n_{merge} \cdot \frac{1_{merge}}{b_{merge}} - \left(n_{obj-1} \cdot \frac{1_{obj-1}}{b_{obj-1}} + n_{obj-2} \cdot \frac{1_{obj-2}}{b_{obj-2}} \right) \quad (4)$$

$$\Delta h_{comp} = n_{merge} \cdot \frac{1_{merge}}{\sqrt{n_{merge}}} - \left(n_{obj-1} \cdot \frac{1_{obj-1}}{\sqrt{n_{obj-1}}} + n_{obj-2} \cdot \frac{1_{obj-2}}{\sqrt{n_{obj-2}}} \right) \quad (5)$$

As given in the above Eqs. (4) and (5), the smoothness criterion (Δh_{smooth}) equals the proportion between the factual border length (l) and the perimeter (b) of the bounding box. At the same time, the compactness criterion $\left(\Delta h_{comp}\right)$ equals the ratio between the factual border length (l) and the square root of the amount of pixels present in the object. The above said equations shows that parameters comprising of the scale factor, colour heterogeneities or weights of the shape, compactness heterogeneities or weights of the smoothness and the weights of individual spectral bands are required for multi-resolution segmentation [20, 21]. Multiresolution segmentation is followed by rule-based classification.

3.3 Rule Based Classification

Rule based classification is performed based on user defined rules which utilises human knowledge. The attributes of the segmented objects is used for the creation of rule sets for mapping features. The major constrain in the rule based classification is the development of correct set of rules for mapping a relevant feature. In the present study, the main aim is to detect and extract buildings from VHR UAV images for which a set of rules is created after segmentation. In the process of creating rule set for building extraction, the Green Leaf Index (GLI) given in Eq. (6) is used for extracting vegetation, after which further rules sets are created by assigning a threshold to the nDSM in order to detect and extract the buildings.

$$GLI = \frac{2G - R - B}{2G + R + B} \tag{6}$$

where, G represents green band, R represents red band and B represents blue band respectively. After the extraction of vegetation using GLI, a rule is set where buildings are detected by providing a height threshold to nDSM. After which, another rule has been set to separate other features from building features.

4 Accuracy Assessment

Post classification accuracy assessment was carried out using the reference buildings, obtained from the orthoimages through heads-up digitization. The reference data were used as ground truth to verify the accuracy of extracted buildings in ArcGIS environment.

From the comparison of extracted and reference buildings, four components such as True Positive (TP), True Negative (TN), False Positive (FP) and False Negative (FN) were obtained. In the given Fig. 3. ABCD represents the reference building, whereas PQRS represents the extracted building from which TP denotes the features which are classified as buildings in both the reference data and the extracted data. TN refers to the features which are not classified as buildings in the reference and the resultant image. FP refers to the features which are determined as buildings in the resultant image but are not present in the reference image whereas FN denotes the features which are not classified as buildings in the resultant image, but, present in the reference image [22]. By utilizing these components, accuracy assessment of extracted object is performed. The split factor (SF), missing factor (MF), percent of building detection (PBD), correctness percent, completeness percent and quality

Fig. 3 Schematic diagram of TP, TN, FP and FN used for accuracy assessment of classified image using the reference image

percent (QP) along with omission error percent and commission error percent was evaluated for accuracy assessment using the following Eqs. (7–13) [23, 24].

$$\mathrm{SF} = \frac{TP}{FP} \tag{7}$$

$$\mathrm{MF} = \frac{FN}{TP} \tag{8}$$

$$\mathrm{Correctness}(\%) = \frac{TP}{TP + FP} \tag{9}$$

$$\mathrm{Completeness}(\%) = \frac{TP}{TP + FN} \tag{10}$$

$$\mathrm{QP} = \frac{100 \times TP}{(TP + FP + FN)} \tag{11}$$

$$\mathrm{Omission\ Error}(\%) = \frac{FN}{FN + TP} \times 100 \tag{12}$$

$$\mathrm{Commission\ Error}(\%) = \frac{FP}{FP + TP} \times 100 \tag{13}$$

5 Results and Analysis

5.1 Object Based Image Classification

The orthoimages, DSM and DTM obtained after processing through Drone2map free trial version was utilized initially in this study to obtain the nDSM. Both the ortho-mosaic image and nDSM are used for building extraction using OBIA. In the first step, mosaicked orthoimages are used for segmentation through multiresolution segmentation followed by the setting of thresholds for building extraction by creating a rule set through rule-based classification. The orthomosaic and DSM of the study area are shown in Fig. 4a, b respectively.

5.2 Multiresolution Segmentation

The multiresolution segmentation was carried out using ortho-mosaic image. In this study, after checking with different values for numerous times through trial and error method, the suitable scale, shape, and compactness parameters were set as 50, 0.5 and 0.7 respectively.

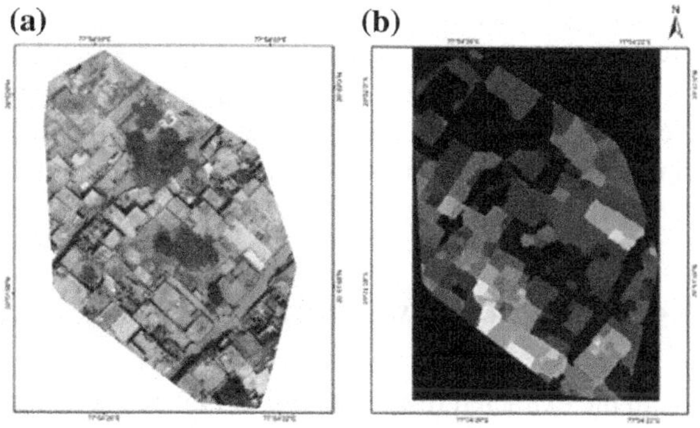

Fig. 4 **a** Orthomosaic and **b** digital surface model

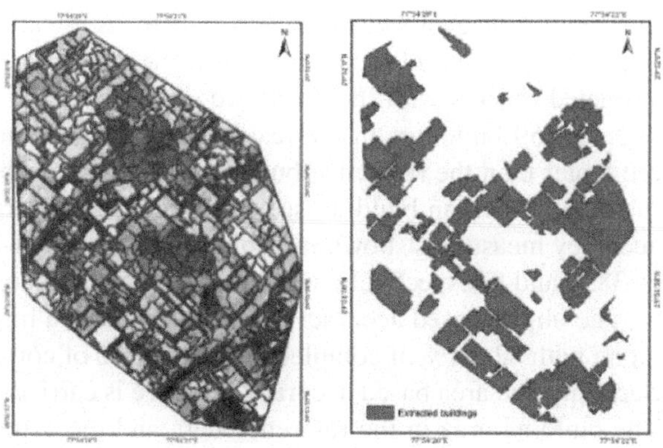

Fig. 5 Segmentation of ortho-mosaic image (left picture) and extracted buildings from object-based classification (right picture)

The image objects (IO) or segments generated from the segmentation process with above-said values of parameters, clearly indicates there is no marginal or minimal segmentation. From the segmentation shown in Fig. 5 (left picture), it is visible that image objects having same color factor are grouped in which the shape factor enhances the quality of the obtained image objects. After image segmentation, a fuzzy logic classification was applied to classify the target classes after which using nDSM, rule sets were created for the extraction of buildings.

5.3 Rule Based Classification

In the present study, the first rule set was created to separate the buildings from the vegetation for which Green Leaf Index (GLI) was executed. The threshold for GLI is given as $0 < 0.971$, within which all the image objects or segments, are classified as vegetation and above 0.971, segments are classified as buildings. This first rule set helps in differentiating the buildings and trees. The second rule set was to extract the buildings using nDSM as layer 1 whereas RGB bands of ortho-mosaic image are considered as the other three layers. In the second rule set, a threshold of ≤ 110 is given to the nDSM in order to detect and extract buildings from the study area. In this study, all the image objects other than buildings such as roads, open spaces and etc. were merged into a single feature for extracting the buildings. The rule-based classification results are shown in Fig. 5 (right picture).

5.4 Accuracy Assessment

From the extracted results shown in Fig. 5, the reference buildings (Fig. 6) obtained through heads-up digitization is compared with each other to assess the size of TP, FP and FN, both quantitatively and qualitatively. In this study, three types of accuracy measures such as object-based accuracy measure, area-based accuracy measure and object location based accuracy measure were performed to evaluate the accuracy of extracted objects with the reference objects. Totally, there are 52 reference buildings and 59 buildings in the resultant image. The numbers of extracted buildings are larger than the reference buildings which could be any misclassification due to the complex urban building structure in the study area. The result of object-based accuracy measure is shown in Table 2. The values of TP, FP and FN were estimated as 38, 6 and 8 respectively, with the help of experts opinion.

The object-based accuracy measure acquires a high overall accuracy of 88.14% along with 82. 61% of completeness, 86. 36% of correctness and 73.07% of quality. Secondly, the area based accuracy measure is carried out which is shown in Table 2. The building area in the reference data and the extracted data is calculated in m^2. The area of reference buildings and the extracted buildings along with the area of TP, FP and FN is estimated as 2543.78 m^2, 1942.92 m^2, 1396.63 m^2, 307.84 m^2 and 406.45 m^2 respectively. The object-based accuracy measure predicts the error of omission and the error of commission to about 11.94 and 13.63% respectively. However, a moderate overall accuracy of 76.37% is acquired in area-based accuracy measure along with 77.45%, 81.93% and 66.16% of completeness, correctness and

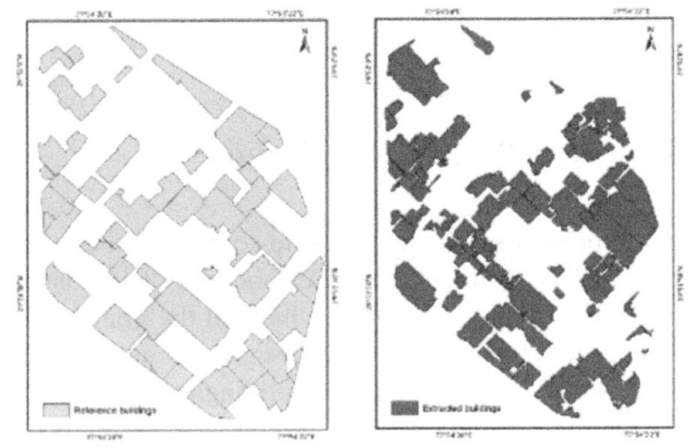

Fig. 6 Reference buildings (left picture) and extracted buildings (right picture)

Table 2 Evaluation of object based and area based accuracy measure

Accuracy measure	Object based accuracy measure	Area based accuracy measure
SF	6.33	4.53
MF	0.21	0.29
Completeness (%)	82.61	77.45
Correctness (%)	86.36	81.93
QP	73.07	66.16
Error of omission (%)	11.94	17.30
Error of commission (%)	13.63	18.06
Overall accuracy (%)	88.14	76.37

quality are acquired respectively. The error of omission and the error of commission in area-based accuracy measure is 17.30% and 18.06% respectively. After object and area-based accuracy measures, position based accuracy measure was estimated by calculating the centroid for the reference buildings and the extracted buildings from which minimum displacement, maximum displacement and mean displacement is obtained and shown in Table 2. Totally 30 samples were taken for estimating the position based accuracy of buildings in m. The mean displacement from the centroids of the reference buildings and the extracted buildings is very minimal. The overall accuracy of object-based and position based accuracy measure is high when compared to moderate accuracy of area-based accuracy measure. The mean displacement of centroid of the reference buildings and the extracted buildings is 1.53 m, whereas the maximum and minimum displacement is 4.21 m and 0.211 m respectively.

6 Discussion

In this study, an object-based hierarchical classification methodology has been established to detect and extract buildings from VHR UAV images. The semi-automatic process of multiresolution segmentation along with rule-based classification identifies buildings present in the dense urban landscape with the help of GLI for segregating the mixed up vegetation, whereas nDSM has been utilised for extraction of buildings present in the study area by providing the threshold values. After which other features other than buildings are merged into a single feature, for the easy extraction of buildings. When the extracted building polygons are compared with the reference buildings, the difference in few buildings may be due to the vegetation present at the top of the building's roof. And also the complex urban settings may also have played a large role which can be responsible for TP, FP and FN respectively.

Among all the accuracy measures, object-based accuracy measure which is based on the number of resultant objects classified as buildings have an overall accuracy of 88.14%, signifying that OBIA has nearly extracted the same amount of buildings in the reference data. However, the area based accuracy measure of extracted buildings signifies that the area of the extracted object varies moderately from the area calculated from the reference buildings. In both the cases, the error of omission and the error of commission is nearly equal whereas the quality of the extracted building is 73.07% and 66.16% respectively. The minimal value of mean displacement between the centroids of extracted and reference buildings signifies that the extracted buildings are placed in the same position as the reference buildings.

7 Conclusion

In the present study, OBIA is utilized for automatic detection and extraction of buildings from the dense and complex urban landscape using VHR UAV images. For which, automated segmentation and user defined rule sets has been provided in order to detect and extract the buildings. Accuracy of the extracted buildings has also been checked using object based, area based and position based accuracy measures, from which correctness, completeness and quality of the extracted buildings were obtained. The results of accuracy check implies that detected buildings through object based image classification is more accurate and this methodology can be used for detecting the buildings in the dense urban landscapes using VHR UAV images. In future, this study can also be developed to match the number of buildings extracted from the high dense point cloud data, and the number of buildings present in the reference data as well as in the extracted data where the classification accuracy can be calculated by utilizing, all the datasets obtained from the same VHR UAV images. From this study, it is also seen that VHR UAV images with very high GSD at centimeter level can aid in proper urban planning and smart city management than any other costly and traditional methods of surveying and photogrammetric works.

Acknowledgements The authors of this study acknowledge Dr. Kamal Jain, Professor, Indian Institute of Technology-Roorkee for providing datasets in order to carry out this study.

References

1. Awrangjeb M, Ravanbakhsh M, Fraser CS (2010) Automatic detection of residential buildings using LIDAR data and multispectral imagery. ISPRS J Photogramm Remote Sens 65:457–467. https://doi.org/10.1016/j.isprsjprs.2010.06.001
2. Li Y, Wu H, An R, Xu H, He Q, Xu J (2013) An improved building boundary extraction algorithm based on fusion of optical imagery and LIDAR data. Optik 124(22):5357–5362. https://doi.org/10.1016/j.ijleo.2013.03.045

3. Fan X, Nie G, Gao N, Deng Y, An J, Li H (2017) Building extraction from UAV remote sensing data based on photogrammetry method. In: IEEE international geoscience and remote sensing symposium (GRASS). IEEE, Fort Worth, TX, USA. https://doi.org/10.1109/igarss.2017.8127707
4. Kettig RL, Landgrebe DA (1976) Classification of multispectral image data by extraction and classification of homogeneous objects. ITGE 14:19–26
5. Feng Q, Liu J, Gong J (2015) UAV remote sensing for urban vegetation mapping using random forest and texture analysis. Remote Sens 7(1):1074–1094
6. Haralick RM, Dinstein I, Shanmugam K (1973) Textural features for image classification. IEEE Trans Syst Man Cybern 3:610–621
7. Robson BA, Nuth C, Dahl SO, Hölbling D, Strozzi T, Nielsen PR (2015) Automated classification of debris-covered glaciers combining optical, SAR and topographic data in an object-based environment. Remote Sens Environ 170:372–387
8. Khoshelham K, Nardinocchi C, Frontoni E, Mancini A, Zingaretti P (2010) Performance evaluation of automated approaches to building detection in multi-source aerial data. ISPRS J Photogramm Remote Sens 65(1):123–133. https://doi.org/10.1016/j.isprsjprs.2009.09.005
9. Gao Y, Mas JF (2008) A comparison of the performance of pixel-based and object-based classifications over images with various spatial resolutions. J Earth Sci 2(1):27–35. http://docsdrive.com/pdfs/medwelljournals/ojesci/2008/27-35.pdf
10. Blaschke T (2010) Object based image analysis for remote sensing. ISPRS J Photogramm Remote Sens 65(1):2–16
11. Carleer A, Debeir O, Wolff E (2005) Assessment of very high spatial resolution satellite image segmentations. Photogramm Eng Remote Sens 71(11):1285–1294
12. Myint WS, Gober P, Brazel A, Clarke SG, Weng Q (2011) Per-pixel vs. object-based classification of urban land cover extraction using high spatial resolution imagery. Remote Sens Environ 115:1145–1161
13. Jobin B, Labrecque S, Grenier M, Falardeau G (2008) Object-based classification as an alternative approach to the traditional pixel-based classification to identify potential habitat of the Grasshopper Sparrow. Environ Manage 41(1):20–31
14. Ettarid M, Rouchdi M, Labouab L (2008) Automatic extraction of buildings from high resolution satellite images. In: The international archives of the photogrammetry, remote sensing and spatial information sciences, vol XXXVII, Part B8, Beijing
15. Benarchid O, Raissouni N, Adib SE, Abbous A, Azyat A, Achhab BN, Lahraoua M, Chahboun A (2013) Building extraction using object-based classification and shadow information in very high resolution multispectral images, a case study: Tetuan, Morocco. Can J Image Process Comput Vis 4(1):1–8
16. Hermann JHK, Pinho C, Souza M (2006) High-resolution satellite images for urban planning. Studies in Progress at INPE (National Institute for Space Research), Brazil. ISPRS Technical Commission II Symposium, Vienna, 12–14 July 2006
17. Núñez JM, Medina S, Ávila G, Montejano J (2019) High-resolution satellite imagery classification for urban form detection. IntechOpen. https://doi.org/10.5772/intechopen.82729
18. Saba F, Zoej MJV, Mokhtarzade M (2016) Optimization of multiresolution segmentation for object-oriented road detection from high-resolution images. Can J Remote Sens 42(2)
19. Jain K (2019) Dens urban dataset for Khanjarpur area (Roorkee)
20. Yu H, Wang J, Bai Y, Yang W, Xia GS (2018) Analysis of large-scale UAV images using a multi-scale hierarchical representation. Geo-spat Inf Sci 21:33–44
21. Duarte D, Nex F, Kerle N, Vosselman G (2018) Multi-resolution feature fusion for image classification of building damages with convolutional neural networks. Remote Sens 10:1636

22. Shufelt AA, Mckeown DM (1993) Fusion of monocular cues to detect man-made structures in aerial imagery. CVGIP Image Underst 57(3):307–330. http://dx.doi.org/10.1006/ciun.1993.1021
23. Sarp G, Erener A, Duzgun S, Sahin K (2014) An approach for detection of buildings and changes in buildings using orthophotos and point clouds: a case study of Van Erriş earthquake. Eur J Remote Sens 47(1):627–642. https://doi.org/10.5721/EuJRS20144735
24. Karna BK, Bhardwaj A (2014) Integrated approach for building extraction from InSAR and optical image using object oriented analysis technique. Nepalese J Geoinf 13:16–23

Micro Level Hydrological Planning and Assessment of Tank Irrigation System

Balakumaran Ramachandran and Srinivasa Raju Kolanuvada

Abstract Dependence of Groundwater is alarmingly increasing in Tamil Nadu due to irregular rainfall patterns and improper management of surface water systems particularly, the Ground Water resources overexploited in many blocks of Tamil Nadu for domestic and irrigation use leading to seawater intrusion, depletion of water tables and water quality degradation. This study is carried out for Kolur village of Ponneri taluk, Tiruvallur district, located near the sea coast where large number of tanks are available, yet agriculture practice depend solely on groundwater due to lack of maintenance of tanks and irrigation channels. UAV drone data in this study enables automatic Micro level mapping of waterways from tanks to the fields even at the field channel level where conventional remote sensing is unusable. The maximum storage capacity of the tank has been determined using the DEM created from the UAV data. At present, paddy is the dominant crop grown in study area with the help of groundwater. The volume of water stored in the tank as a result of the runoff generated from precipitation in the catchment area during the paddy season is calculated using HEC-HMS model. Water requirement during the paddy season in the study area is calculated using evapotranspiration derived from the MODIS-MOD16A2 data and Precipitation derived from the TRMM dataset. The present study shows that tank irrigation system can supplement efficient irrigation of paddy crops in the study area by proper maintenance of the channel system, thereby decreasing the excessive use of groundwater for irrigation purposes.

Keywords UAV · Tank irrigation system · Channel system · Micro-level mapping · HEC-HMS

B. Ramachandran (✉) · S. R. Kolanuvada
Institute of Remote Sensing, Anna University, Chennai, India
e-mail: balakumaran247@gmail.com

S. R. Kolanuvada
e-mail: ksrajuirs@gmail.com

K. Jain et al. (eds.), *Proceedings of UASG 2019*, Lecture Notes in Civil Engineering 51,
https://doi.org/10.1007/978-3-030-37393-1_24

1 Introduction

All the civilizations started settling very close to the river banks due to water availability [1]. Agriculture is the main occupation of many households in Tamil Nadu, which has a declining trend due to the insufficient water availability and increase in salinity of groundwater. Tamil Nadu is popular for its cascade of tanks which were used for agriculture and drinking water sources in ancient times. The irregular rainfall patterns and improper maintenance of water bodies increased the dependence on groundwater exponentially. The Inter-Governmental Panel on Climate Change has forecasted erratic rainfall patterns during the next decade in peninsular India as a consequence of climate change [2]. Hence efficient utilization of rainfall through effective storage and distribution for agriculture is mandatory in this region. This study involves the automatic extraction of channel networks that aid in the supply of water from the tank to agricultural fields, with the help of DSM generated from the High resolution UAV data. DEMs are used to quantify basin parameters such as catchment boundaries, area, length, slope, aspect, curvature, drainage density, stream order, and Horton's ratios [3]. To extract channel networks from DEM successfully there is no perfect model developed despite rich knowledge on channel networks and their physical processes [4–7]. UAV drone data in this study enables automatic micro level mapping of waterways from tanks to fields even up to the field channels where conventional remote sensing is unusable. This study indicates that even with the available rainfall, the exploitation of groundwater can be minimized with the effective use of tank irrigation system.

2 Study Area

The study was carried out for Kolur Village of Ponneri taluk, Thiruvallur district, Tamil Nadu which has an area of 12.6 km^2 and is located between 13° 24′ 43.53″ N to 13° 27′ 20.40″ N latitude and 80° 12′ 54.52″ E to 80° 15′ 34.25″ E longitude. The village has 948 households and agriculture is the major occupation of the villagers. The average annual rainfall of Ponneri is 1479 mm with the average summer temperature above 37.9 °C and average winter temperature around 18.5 °C during December and January. The soil of the district is mostly sandy mixed with soda or other alkali or stony [8]. 4 major water tanks are located in the Kolur village serving for surface water storage mainly domestic or irrigation purposes. A water tank with an aerial extent of 126.28 acres used for surface irrigation was selected for the study. The study area for the research along with distribution of the tanks is presented in Fig. 1. The land use distribution in the Kolur village is dominated by agriculture land covering 67.8% of total geographic area followed by waterbodies covering 15.3%.

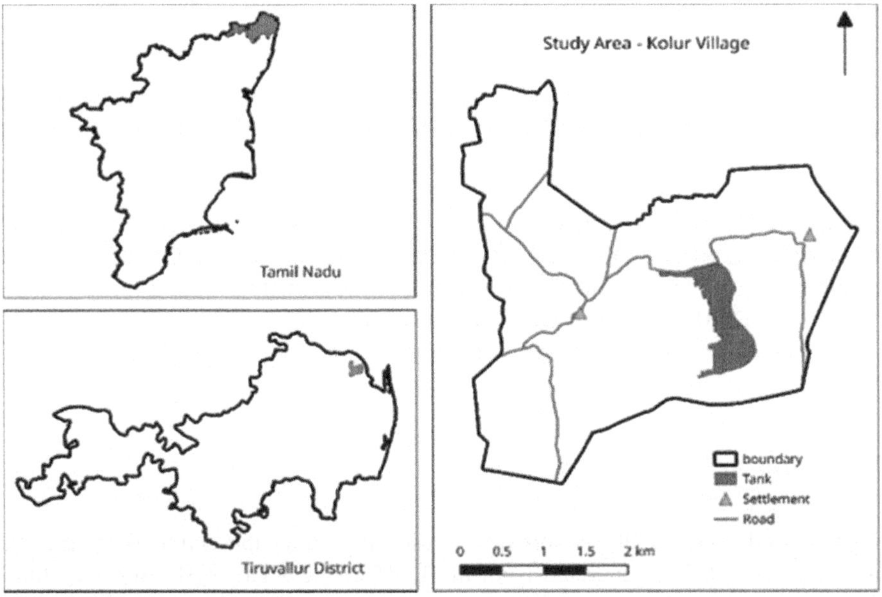

Fig. 1 Study area and its location in Tamil Nadu

3 Methodology

Based on the current cropping patterns and crop water requirements, a strategic plan for conjunctive use of surface water and groundwater in the study area is proposed. Figure 2 shows methodology for this study.

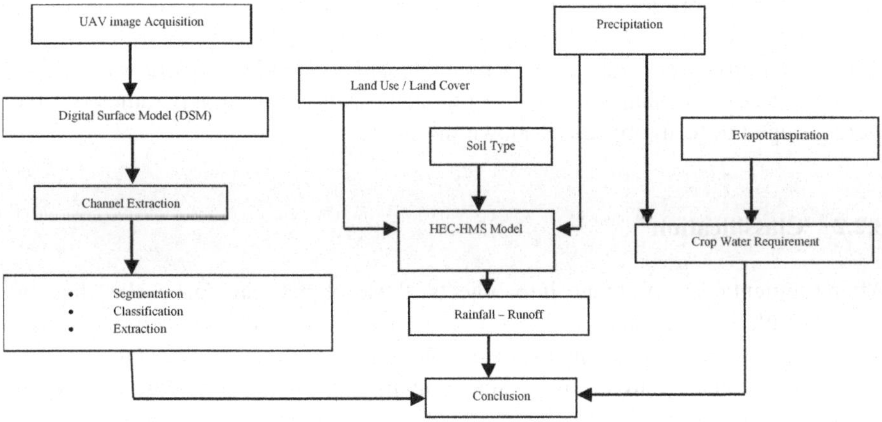

Fig. 2 Methodology

3.1 UAV Images

The drone's flight plan was prepared using the study area boundary and the quality requirements of UAV data product using eMotion software. eMotion's built-in Flight Data Manager automatically handles the georeferencing and preparation of images. The drone was flown at a flying height of 170 m with a lateral and longitudinal overlap of 70% to acquire images at 5 cm/px resolution. Images were acquired for the study area, processed using Pix4D software to generate point clouds from the acquired images. These point clouds were used to construct the digital surface model (DSM). The DSM is a 2.5 D model of the mapped area, for each (X, Y) position, the DSM has one Z value (the altitude of the highest point for this (X, Y) position).

3.2 Channel Extraction

Object-based image analysis software is used to extract the channels from 5 cm high-resolution DSM generated from the UAV images. The OBIA involves three main steps to extract the channel network from the DSM namely segmentation, classification and extraction.

3.2.1 Segmentation

The basic procedure of segmentation is to break up the image into objects representing land-based features. Segmentation is the process of grouping of pixels from data to form objects, based on the compactness and shape settings. A multi-resolution segmentation algorithm was used to break the image into segments. For breaking down the image into more meaningful objects scale, shape and compactness parameters were used. The advantage of the multi-resolution segmentation algorithm is that it can segment images of red, green or blue bands, DEM, DSM, NIR or even LiDAR. With the scale of 30, shape of 0.3 and compactness of 0.7 as input parameters, pixels were grouped to form objects as shown in Fig. 3a.

3.2.2 Classification

After segmentation of image into objects, these objects are to be classified into features with the assigned class ruleset. The classification of the objects can be made because each object has distinct statistics associated with them. The statistics of the object like geometry, area, colour, shape, texture and adjacency were used as rule sets for classification of the objects on UAV imagery. However, the use of correct statistics is of importance to classify the object. Admittedly, there is no best way to classify land cover features using Object-Based Image Analysis; hence a trial and

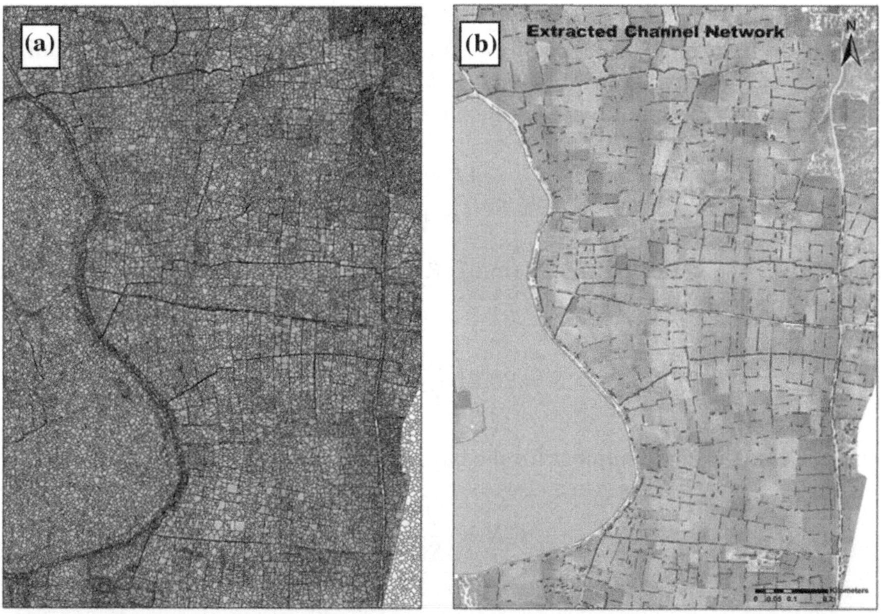

Fig. 3 **a** Multi-resolution segmentation algorithm combining pixels into objects, **b** extraction of channel network

error process of defining the characteristics of the object for classification was used to achieve better extraction of channels. The objects resulted from segmentation were classified with the rule: length/width greater than 1.5, area less than 1300 pixels, Shape index greater than or equal to 2 and mean brightness less than 200.

3.2.3 Extraction

A separate layer is derived with the objects that are classified as per requirement. The extracted layer can also be exported as vector file with the help of export vector layer rule set. This information will also help to assess capacity and cultivable area catered by each segment of channel. The objects classified as channel network is exported as vector layer (Fig. 3b).

3.3 Rainfall-Runoff Modelling

The runoff generated as a result of precipitation that gets stored in the tank is estimated based on the HEC-HMS model. HEC-HMS is a free modeling tool, which is useful for complete modeling of hydrological processes and includes many traditional hydrologic analysis procedures. Surface runoff is predicted for the rainfall by using the SCS curve number method [9]. When the rainfall is greater than the initial abstraction, surface runoff occurs which is the key concept in the SCS method.

$$Q = \frac{[R - 0.2S]^2}{[R + 0.8S]}$$

where, Q is the collected runoff (mm), R is the precipitation (mm) and S is the retention parameter.

$$S = 25.4\left(\frac{1000}{CN} - 10\right)$$

where, CN is the curve number for the day.

$$CN = \frac{25{,}400}{S + 254}$$

3.4 Crop Water Requirement

The irrigation water requirement basically represents the difference between the crop water requirement and effective precipitation [10].

$$IWR = ETp - P$$

The potential evapotranspiration (PET) is derived from the MOD16A2 Version 6 evapotranspiration/latent heat flux product, an 8-day composite product produced at 500 m pixel resolution. The algorithm used for the MOD16 data product collection is based on the logic of the Penman-Monteith equation, which includes inputs of daily meteorological re-analysis data along with MODIS remotely sensed data products [11]. The daily precipitation data for the study area is obtained from the TRMM dataset with a resolution of 0.25 arc degrees. This dataset merges microwave data from multiple satellites, including SSMI, SSMIS, MHS, AMSU-B and AMSR-E, each inter-calibrated to the TRMM Combined Instrument.

4 Results and Discussion

4.1 Channel Extraction

The channel network is extracted using object-based image analysis of UAV high-resolution data. The result Fig. 3b indicates the existence of the channel networks from the tank, which are discontinuous due to obstructions like trees, bushes, debris dumped and silting of channels. Wherever, siltation has occurred, the discontinuity of channels is verified using morphological filtering of the outcome by checking continuity of linear features. Some of the field bunds were also classified as channels because of their linearity leading to reduced accuracy in channel extraction.

4.2 Rainfall-Runoff Modelling

HEC-HMS simulation is done for the Kharif season to obtain the surface runoff collected into the tank. The basin model with the hydrological elements like sub basin, junction and sink connected together in a dendritic network enables breakup of watershed into manageable pieces (Fig. 4).

Estimation of surface runoff using SCS curve number requires CN value for the hydrological element, sub basin. Major soil type is clay loam categorized as Group C of the Hydrologic Soil Groups (HSG) with moderate runoff potential. The weighted curve number value of the study area was determined based on the land use/land cover map and the soil data, was given as input to the basin model [12]. The daily precipitation data during the paddy season is provided to the meteorological model through the time-series data. The result from the simulation run shows that a volume of 335,712 m^3 of surface run off drains into the tank during the paddy season as a result of precipitation.

4.3 Crop Water Requirement

The PET from the MODIS data was obtained as a cumulative sum of 8 days from which monthly PET is derived by interpolation to obtain crop water requirement. The precipitation value (mm/h) from the TRMM dataset is used to calculate accumulated rainfall for the entire month. The precipitation is then subtracted from the PET to calculate the irrigation water requirement. The area of agricultural land irrigated by the channel network from the study tank is 567.8 acres. The volume of water required for the crops during the kharif season for irrigation from the tank is 711,142 m^3 calculated based on the understanding from AQUASTAT—FAO [13]. Figure 5 shows the graphical representation of potential evapotranspiration and precipitation for the kharif season in the study area.

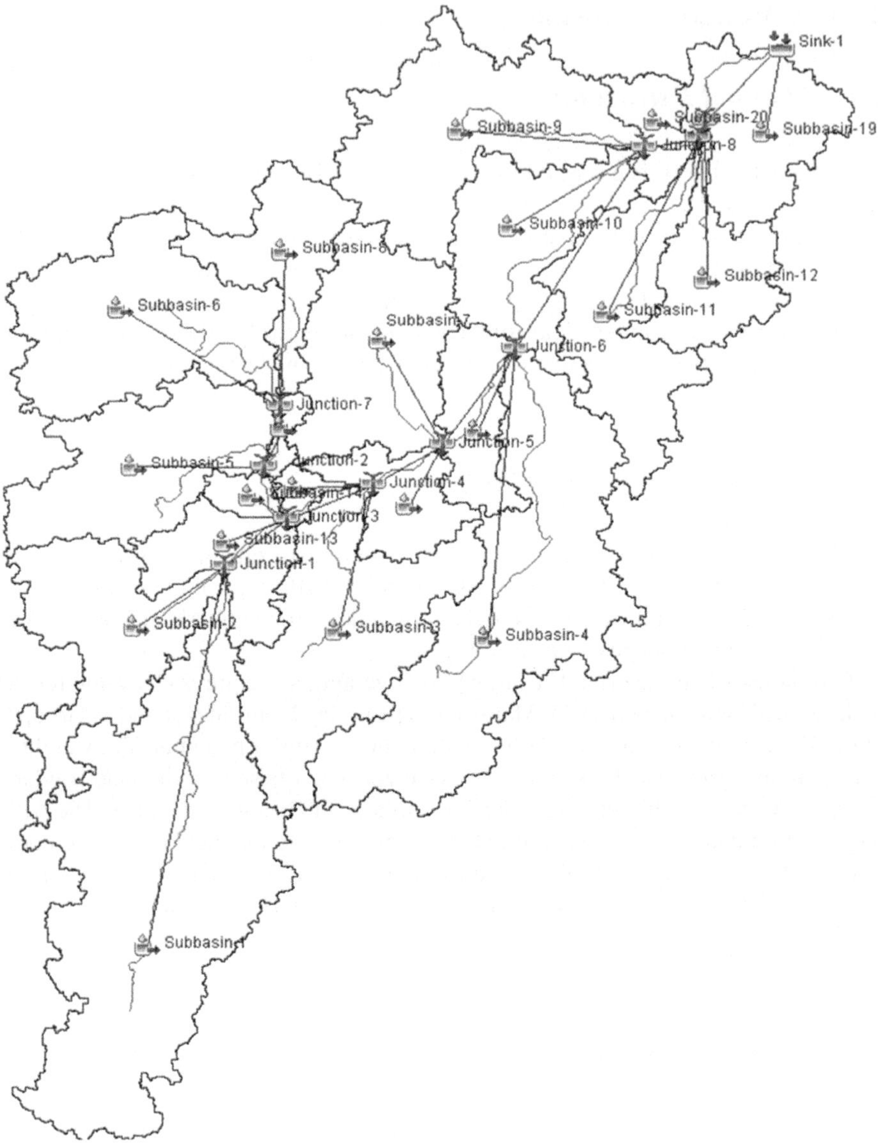

Fig. 4 Dendritic network of hydrological elements

The tank has area of 511,037.72 m^2 and maximum storage capacity calculated by the generation of Triangulated irregular network (TIN) model is 3,043,145.25 m^3. Figure 6 show the TIN model generated for the calculation of storage capacity and also the agricultural land that can possibly be cultivated with the help of the channel network.

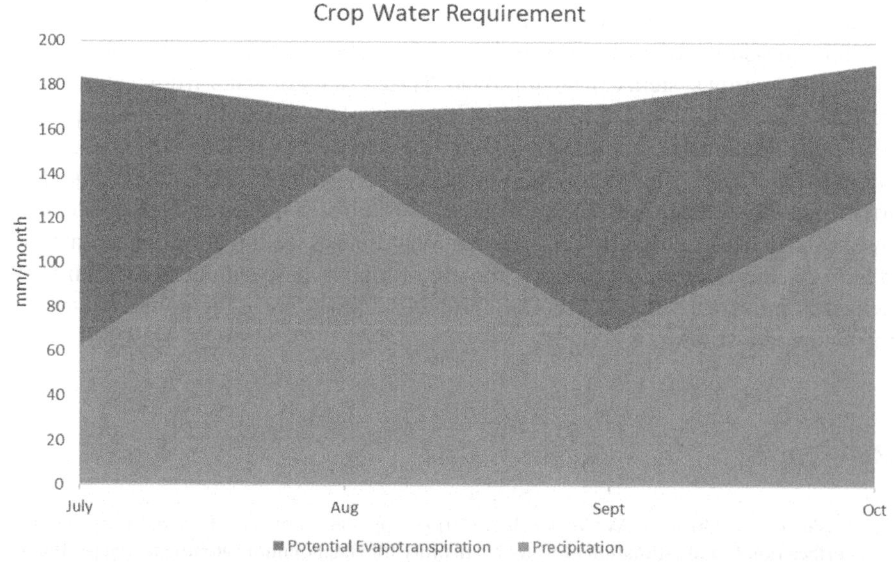

Fig. 5 Potential evapotranspiration and precipitation for the Kharif season in study area for crop water requirement

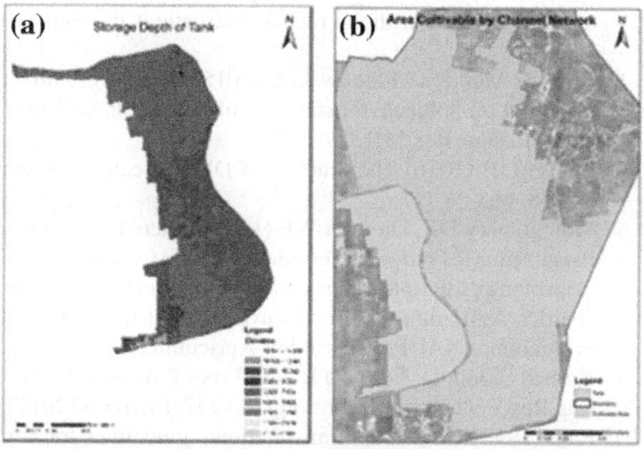

Fig. 6 **a** Triangulated irregular network (TIN) model to calculate storage capacity of tank, **b** agricultural land cultivable by channel network

5 Conclusion

The study indicates the efficiency of UAV imagery for extraction of channel network which is in existence and neglected due to improper maintenance. The DEM thus derived from UAV stereodata enabled quantification of storage capacity of the tank and helped in hydrological modelling of input to the tank through possible precipitation in the area. Understanding the inflows to tank through precipitation helped in deciding the necessity of desiltation. Currently, the agricultural lands around the tank were irrigated using the tank water by pumping which resulted in lower efficiency. With the measures for proper maintenance of the channel network, the efficiency of the tank irrigation system can be improved. The capacity of water from the precipitation can be completely met by the storage capacity of the tank. Though the crop water requirement cannot be completely met by the capacity of the tank, the tank water can satisfy 47.2% of the crop water requirement and can be a major supplement for irrigation of paddy crops in the study area. This reduces the extensive stress caused by excessive use of groundwater for irrigation purposes leading to salinization of the fragile groundwater system. Hence, the use of topographic data derived from UAV, helped in improving the planning of agricultural practices in the command area of the tank under study.

References

1. Kasheebai B, Shivapur AV, Venktesh B (2017) Application of SWAT model for generating surface runoff and estimation of water availability for Balehonnuru catchment area for Badhra River Basin. Int Res J Eng Technol (IRJET) 04(08)
2. IPCC (2014): Climate Change (2014) Synthesis report. Contribution of working groups I, II and III to the fifth assessment report of the intergovernmental panel on climate change [Core Writing Team, RK Pachauri, LA Meyer (eds)]. IPCC, Geneva, Switzerland, 151pp
3. Srinivasa Raju K, Kavya Lakshmi P, Sivakumar S (2019) Multi-criteria-based approach for optimal siting of artificial recharge structures through hydrological modeling. Arab J Geosci 12:190
4. Clubb F, Mudd S, Milodowski D (2015) Reply to comment by P. Passalacqua and E. Foufoula-Georgiou on “objective extraction of channel heads from high-resolution topographic data”. Water Resour Res 51:1377–1379
5. Lindsay JB (2016) The practice of DEM stream burning revisited. Earth Surf Process Landf 41:658–668
6. Montgomery DR, Dietrich WE (1988) Where do channels begin? Nature 336:232–234
7. Passalacqua P, Foufoula-Georgiou E (2015) Comment on “objective extraction of channel heads from high-resolution topographic data” by Fiona J. Clubb et al. Water Resour Res 51:1372–1376
8. District Agriculture Plan Thiruvallur District (2008) National Agricultural Development Programme (NADP), Centre For Agricultural And Rural Development Studies (CARDS)
9. Kurbah S, Jain DM (2017) Rainfall-runoff modeling of a river basin using SWAT model. Int J Eng Res V6(12). https://doi.org/10.17577/ijertv6is120111
10. FAO (1998) Crop evapotranspiration—guidelines for computing crop water requirements—FAO irrigation and drainage paper 56
11. Running S, Mu Q, Zhao M (2017) MOD16A2 MODIS/terra net evapotranspiration 8-day L4 global 500 m SIN Grid V006. 2017. Distributed By NASA EOSDIS Land Processes DAAC. https://doi.org/10.5067/MODIS/MOD16A2.006
12. S. Satheeshkumar, S. Venkateswaran, R. Kannan (2017) Model. Earth Syst. Environ. 3:24. https://doi.org/10.1007/s40808-017-0301-4
13. AQUASTAT—FAO’s Information System on Water and Agriculture (2012). Fao.Org. http://www.fao.org/nr/water/aquastat/water_use_agr/index3.stm

Cost-Effective Real-Time Aerial Surveillance System Using Edge Computing

Md. Shahzad Alam and Sujit Kumar Gupta

Abstract Nowadays there is an emerging need for surveillance in order to maintain the public places more secure and ensure the safety and security of the people. Many government agencies require some autonomous system for surveillance of the large areas which can give them precise and real-time information like number of vehicles, people, and other objects. An aerial surveillance system will be very effective in this scenario and platform like Unmanned Aerial vehicle (UAV) will be very reliable and cost-effective option for this task. To make the system fully autonomous, we require real-time object detection that is computationally complex and time consuming due to the heavy load on the limited processing and payload capacity of low-cost UAV. In this paper, we propose a cost-effective approach for aerial surveillance in which we move the heavy computation tasks to the cloud while keeping limited computation on-board of UAV system using Edge computing technique. Further this will maintain the minimum communication between UAV and the cloud thus proposed system will reduce the network traffic and also delay. Proposed system is based on the state-of-art technique YOLO (You Look Only Once) for real time object detection.

Keywords Real-time object detection · Surveillance system · Real-time · Unmanned aerial vehicle · Edge computing

1 Introduction

In the past, aerial surveillance is mainly used by military and large enterprise and carried out by aircraft and helicopters which are costly and require large infrastructure. Introduction of Unmanned Aerial Vehicle (UAV) has enabled many applications in areas like surveillance, surveying, search and rescue. But high end surveillance UAV mainly used by military is still costly for civilian use and require skilled person to operate. But due to recent advancement, UAV has become cheap, light weight and miniaturize and easily available to anyone. Due to this many small government agencies like Police Department as well as private agencies require these platform for surveillance as it is become very difficult to monitor large areas with just CCTV

Md. Shahzad Alam · S. K. Gupta (✉)
National Institute of Technology Karnataka, Manglore, Surathkal, India

K. Jain et al. (eds.), *Proceedings of UASG 2019*, Lecture Notes in Civil Engineering 51,
https://doi.org/10.1007/978-3-030-37393-1_25

as there position are fixed and adding more CCTV camera will increase the cost of infrastructure. Introducing UAV as surveillance platform will overcome these deficiency as one UAV can cover large area and it is mobile and can be controlled remotely by any person. Now a days even low cost UAV are equipped with senors, on-board limited processing capacity and their payload can be also changed with the requirement.

Currently surveillance is done manually by monitoring every feed of CCTV camera which makes the work more complex and susceptible to human error. So surveillance system should be autonomous and should take some decision on their own to reduce human dependence. By integrating computer vision techniques like object detection with video feed received from UAV, we can make whole system autonomous as it can detect object without any human intervention. Low cost drone have limited processing capability and object detection are computationally complex so it is not feasible to apply object detection technique on-board. So applying object detection on cloud rather than on-board UAV could be feasible option in this scenario. In order to get the information in real-time the system should be optimized to avoid any network delay so leveraging Edge computing technique in this scenario, we can reduce the network delay by moving some computation from cloud to on-board UAV to make system real-time.

In this paper, we propose an autonomous real-time surveillance system with low cost drone by integrating object detection with high definition video of UAV using limited resources on-board and moving some computation to cloud. The main contributions of this paper are summarized as follows:

1. To develop a real-time object detection system using aerial surveillance system platform as a case study.
2. To design the entire system architecture for UAV surveillance and also to optimize the communication between drone and the cloud server.
3. To create web-based application for the end-user to visualize the analyzed output data. The rest of this paper is organized as follows; related work is discussed in Sect. 2. The proposed methodology is explained in Sect. 3. Experimental results and analysis is discussed in Sect. 4. Finally, Sect. 5 concludes the paper with future directions.

2 Related Work

Several researchers in the past have used UAV as platform for surveillance in many applications like tracking and detecting moving object. Huang et al. [6] proposed visual-inertial drone system for real-time motion detection, it helps in tracking moving object in a frame. It uses on-board processor which has limited capacity, but it is expensive. To make the system more low-cost another approach by Sadeghi et al. [10] has proposed an autonomous detection and tracking of moving object in which it does not uses any on-board computing but directly sending video feed from UAV

Table 1 Summary of existing works

Authors	Methodology	Advantages	Limitations
Lee et al. [8]	It uses BING algorithm for objectness estimator on-board and R-CNN object detection technique on cloud	It uses hybrid approach by moving object detection at cloud and keeping low-level computation on-board	It's near to real-time as it uses R-CNN which still has low frame per second than YOLO
Chen et al. [1]	It proposed object tracking using Euclidean space equations and uses GPS to calculate relative position	It's onboard processor can achieve real-time performance in the drone for object detection and tracking	Processing capacity is limited to on-board snapdragon processor and also makes it costly
Dick et al. [2]	It proposed the high speed object tracking using server on the network edge to optimize communication between drone and cloud	It leverages edge computing platform to decrease the latency between drone and cloud server	All the processing is done on cloud server and no computations are done locally on drone
Kim et al. [7]	It proposed dynamic offloading from drone to cloud server decision algorithm for tracking moving object	It reduces the drone battery consumption as well as execution time using offloading algorithm	The offloading algorithm is based only on the relative velocity of object betweens the frames

to system. They used camera estimation technique to overcome the limitation of feature tracking using a moving camera mounted on UAV. But they did not consider the time to compute the complex computation of object detection on server and the latency it will generate while sending video feed to the server from UAV.

Lee et al. [8] proposed the cloud based object detection for UAV in which it moves the highly complex computational task to the cloud while keeping the low-level task on-board UAV. Summary of the existing works is shown in Table 1.

3 Proposed Methodology

3.1 System Overview

Our proposed system consists of three main components, first component is local one in which all the processes are done locally i.e. on-baord UAV as it has limited processing capability. Second component is cloud level in which we have applied cloud based object detection technique YOLO (You Look Only Once) [11] and the last component is at the user level in which we process all the detected objects and present it to the user in a useful information like number of person, vehicle and other objects in real-time. Figure 1 shows the entire proposed system flow in which we first capture video from on-board high definition camera of UAV and then apply motion detection technique in on-board Raspberry Pi. If it is detected the motion then it will send the captured image to the cloud where we apply YOLO object detection technique and get the count of person and vehicle and send it to the end user. If motion is not detected then it will send the previous detected objects count to the user (Figs. 2 and 3).

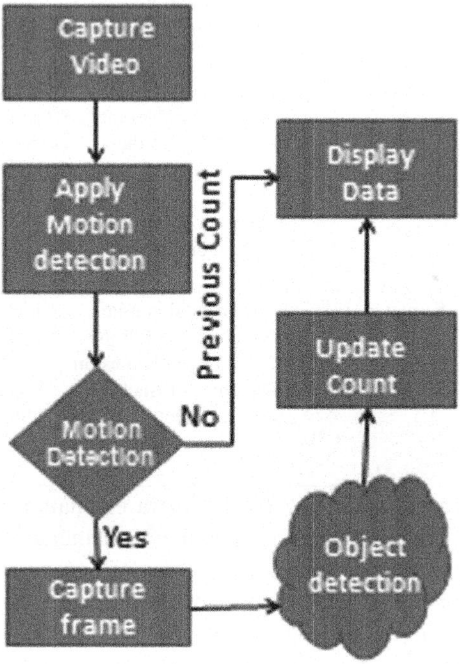

Fig. 1 Proposed detection framework

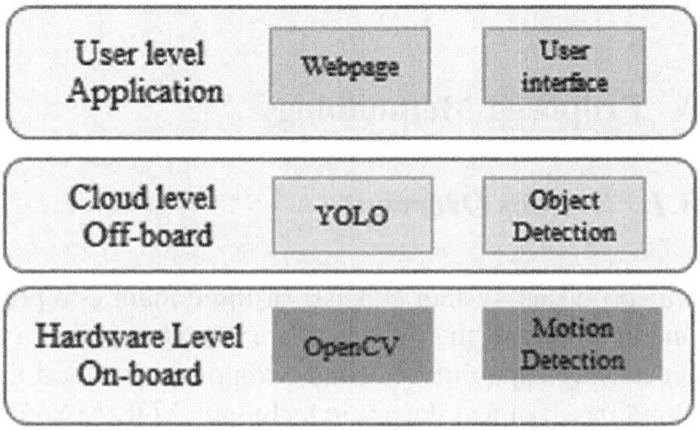

Fig. 2 System architecture

Fig. 3 Motion detection in target area

3.2 Edge Computing On-Board Motion Detection

Since we applied object detection technique in the cloud and system is in real-time so we need to decrease the network delay and also it should not send the same frame which contain same number of objects again to cloud. So to overcome this problem, we leveraged the advantage of edge computing technique. In this scenario if there is no movement in the target area that means there is no new object entered in the target area so the count of object is same as in the previous frame. If any motion is detected then there may be a new object entered that target area so we captured new frame again and send it to cloud for further processing. To detect the motion on-board we mounted Raspberry PI 3 on UAV in which we applied Open CV technique as it is lightweight and work in real-time to detect any motion in that target area as shown in Figure.

3.3 Cloud Based Object Detection

To detect the object in the cloud we applied real-time object detection technique YOLO (You Look Only Once) which is faster as compared to other state-of-art techniques like R-CNN [3], Fast-CNN [4] and Faster R-CNN [12] as these techniques

Table 2 Dataset for training model

Altitude	10 m	13 m	16 m	19 m
Image frames	50	50	50	50

need more than one convolution network. But YOLO considers object detection as a single regression problem, straight from image pixels to bounding box coordinates and class probabilities. As generally top-view of object is captured by UAV so we need to train the model to detect the objects from its top view so we trained the model from dataset of 20 videos of 2 min duration in different altitudes as the object size changes with the change in altitude of UAV. For collecting training videos we consider the target area as a parking area which consists of vehicle and person at altitude difference of 3 m and the details of dataset are shown in Table 2. We then converted the videos into frames and split the frames altitude wise and trained our model with all the interested objects. To extract the useful information like number of person, vehicle, and other objects from the output image of classifier we counted the bounding box of labeled object. After getting all the count object-wise we sent it to the end user.

3.4 User Interface

It is very necessary to represent the detected objects data in a meaningful way to the user where this information can be used for the different applications like in case police department they may interest in knowing the estimated number of people and vehicle which can further help them for adequate police personnel deployment on the basis of count and further can also helps in traffic management. So we developed a webpage which can be accessed anywhere and provided a interactive user interface in which they will get all information at one place. Further we processed that information and provided the analysis of that data like in case of parking area if number of detected object is near to maximum capacity of parking area than we can show the user directly the full parking area.

4 Experimental Result and Analysis

4.1 Experimental Setup

For the experiment we used custom built quad copter as used in [4] which we built by assembling the quadcopter from parts as shown in Figs. 4 and 5. It is mounted with Raspberry PI 3 and high definition camera.

Fig. 4 Parts of UAV

Fig. 5 Assembled quadcopter

It's camera are down facing with 80° coverage of target area and its camera resolution is 1280 × 720 at 30 fps. For cloud server we used Intel Xeon processor and for GPU we used Nvidia Tesla G40 GPU with 64 GB RAM. The whole hardware configuration is shown in Table 3.

Table 3 Harware configuration

	On-board (UAV)	Off-board (cloud)
CPU	1.2 GHz quad-core ARMv8	Intel Xeon processor
RAM	1 GB	64 GB
GPU	NA	16 GB Nvidia Tesla G40
OS	Rasbian	Ubuntu 16.04

4.2 Results and Analysis

Evaluation of On-Board Motion Detection We deployed the motion detection on-board UAV on Raspberry Pi 3 as a light-weight computational device. To detect the motion we used Open CV in which we implemented the background subtraction algorithm which is less computation complex and suitable for this type of task. Figure 6 shows actual image and background subtracted image of a person in-front of drone.

Evaluation of Cloud Based Object Detection For evaluation of object detection accuracy on cloud we collected sample frames and evaluated SSD [11] and YOLO object detection in which we got 71% accuracy for YOLO and 76.4% accuracy for SSD. But SSD works at 13 FPS where as YOLO works at 30 FPS. So to achieve the result in real-time we trade-off the accuracy with speed. Figure 7 shows the screenshot of our object detection in which we applied YOLO object detection using our own trained model.

To evaluate the object detection accuracy effectively, we captured frames at different altitude as we know the objects appear smaller as we increased the altitude of

Fig. 6 Background subtraction output

Fig. 7 Sample image frame snapshot from UAV in left and detected/identified object at right

Table 4 Accuracy at different altitude

Object	Detected object	Ground truth	Height
Person	5	5	5
Car	3	3	5
Person	5	6	10
Car	3	3	10
Person	8	9	15
Car	3	3	15

UAV. So considered three sets of altitude 5, 10 and 15 to evaluate our system. Table 4 shows the accuracy of system at different altitudes and we observed that our system works correctly even by increasing the altitude of UAV. Further, as altitude increases the target area of UAV also increases thus it sometime also decreases the accuracy of system.

Performance Evaluation w.r.t Network Delay We processed the image at two ends first one locally using on-board hardware and other one at cloud. The on-board hardware act as an edge device in which we deploy motion detection technique, since this task is processed in milliseconds, it has almost zero network delay. But when we send the image frame to the cloud over the network and apply object detection and then send the results to the user, this entire process will have considerable network latency as our system is working in real-time. To evaluate the performance of our proposed system we used iFogSim [5] simulator in which we considered two test cases; first one is with only cloud where we are processing videos directly in cloud as shown in Fig. 8 without using any edge device. In the second case, we considered with edge device where we are filtering frames using motion detector and sent only those frames in which objects are available as shown in Fig. 9. To get network delay

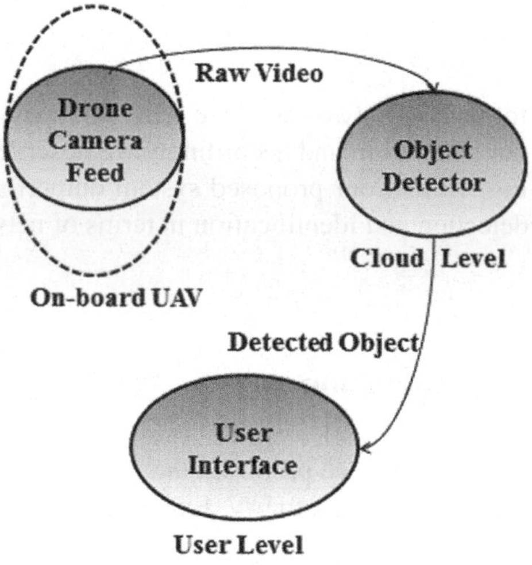

Fig. 8 Case 1: application with cloud only

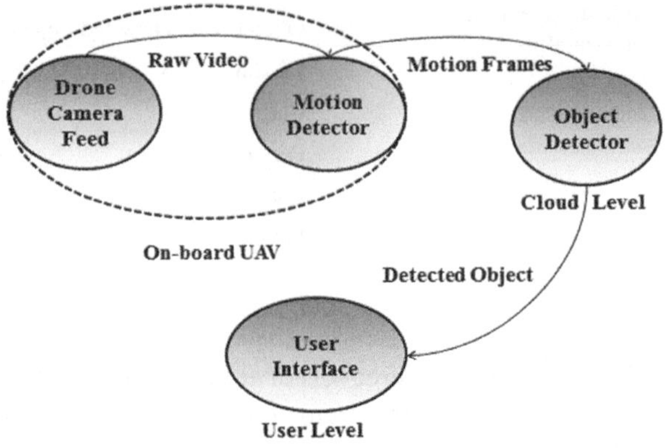

Fig. 9 Case 2: application with edge as well as cloud

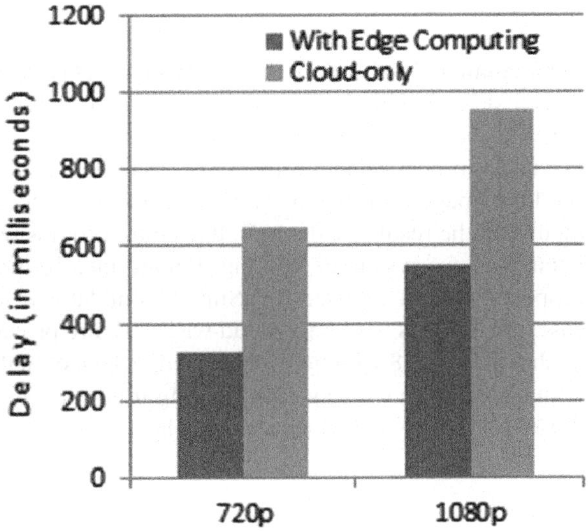

Fig. 10 Network delay of different frame size

for the above two cases, we sent two types of frame resolutions, 720p and 1080p for our simulation and accordingly we observed the network delay. It is observed from Fig. 10 that our proposed system outperforms the cloud based system with object detection and identification in terms of network delay.

5 Conclusion

In this paper, we proposed the autonomous real-time surveillance system architecture using low cost UAV. Experimental results demonstrated that we over-came the problem of on-board limited processing capacity of UAV by moving complex computation to cloud. It also shows that by using edge computing technique we can decrease the network delay to make the system work in real-time. The proposed system may generate network delay if the object is moving at faster rate then each time motion will be detected and frame will be sent to the server each time for further processing thus it will create long queue at server site and may takes more time to process. So to address this problem the proposed model can be further optimized by introducing other technique for edge computing and for object detection we can further increase its accuracy by adding more object classes and number of images during the training of model.

References

1. Chen P, Dang Y, Liang R, Zhu W, He X (2018) Real-time object tracking on a drone with multi-inertial sensing data. IEEE Trans Intell Transp Syst 19(1):131–139
2. Dick J, Phillips C, Mortazavi SH, de Lara E (2017) High speed object tracking using edge computing
3. Girshick R, Donahue J, Darrell T, Malik J (2014) Rich feature hierarchies for accurate object detection and semantic segmentation. In: Proceedings of the IEEE conference on computer vision and pattern recognition, pp 580–587
4. Girshick R (2015) Fast r-cnn. In: Proceedings of the IEEE international conference on computer vision, pp 1440–1448
5. Gupta H, Vahid Dastjerdi A, Ghosh SK, Buyya R (2017) iFogSim: a toolkit for modeling and simulation of resource management techniques in the internet of things, edge and fog computing environments. Softw Pract Exp 47(9):1275–1296
6. Huang C, Chen P, Yang X, Cheng KTT (2017) REDBEE: a visual-inertial drone system for real-time moving object detection. arXiv preprint arXiv:1712.09162
7. Kim B, Min H, Heo J, Jung J (2016) Dynamic offloading algorithm for drone computation. In: Proceedings of the international conference on research in adaptive and convergent systems, pp 123–124. ACM, 2016
8. Lee J, Wang J, Crandall D, Sabanovi S, Fox G (2017) Real-time, cloud-based object detection for unmanned aerial vehicles. In: IEEE international conference on robotic computing (IRC), pp 36–43. IEEE, 2017
9. Liu W, Anguelov D, Erhan D, Szegedy C, Reed S, Fu CY, Berg AC (2016) Ssd: single shot multibox detector. In: European conference on computer vision. Springer, Cham, pp 21–37
10. Sadeghi-Tehran P, Clarke C, Angelov P (2014) A real-time approach for autonomous detection and tracking of moving objects from UAV. In: 2014 IEEE symposium on evolving and autonomous learning systems (EALS). Orlando, FL, pp 43–49
11. Redmon J, Farhadi A (2016) YOLO9000: better faster, stronger. arXiv preprint
12. Ren S, He K, Girshick R, Sun J (2015) Faster R-CNN: towards real-time object detection with region proposal networks. In: Advances in neural information processing systems, 91–99

The Potential of UAV Based Remote Sensing for Monitoring Hindu Kush Himalayan Glaciers

Aman Rai, Aayushi Pandey, Prabuddh Kumar Mishra and Kailash Chandra Tiwari

Abstract Remote sensing acts as a valuable tool for the regular mapping and monitoring of glaciers. The spatial resolution of satellite data used for mapping and monitoring is often too coarse to monitor the small changes in average sized glaciers. The high cost of high-resolution LIDAR and airborne datasets along with the problem cloud cover and shadow of peaks are major drawbacks of satellite datasets. To tackle these problems, the use of unmanned aerial vehicles (UAV) has grown significantly for monitoring the changes in glaciers over the last decade, however; the use of UAVs for monitoring Himalayan glaciers is nonexistent. Hindu Kush Himalaya (HKH) constitutes over 54,000 glaciers with an average size of 1.1 km^2. Having more than 6000 km^3 ice reserves, HKH glaciers influence both global and local climate. Meltwater from these glaciers feed Asia's ten largest river systems which together provide water to more than 1.3 billion people. Thus, monitoring the changes in Himalayan glaciers is vital for assessment of climate change and hydrology of the region. Majority of the Himalayan glaciers are unexplored due to the remoteness and harsh environment. Hence, there is a great need for highly accurate and systematic observation of HKH glaciers with the high spatial and temporal resolution to further advance our understanding of the climate-glacier system in the region. This paper attempts to assess the potential of UAVs for monitoring glaciers in HKH region.

Keywords The himalayas · Glaciers · Unmanned aerial vehicles · Remote sensing · Glacier monitoring

A. Rai (✉) · P. K. Mishra
Department of Geography, Shivaji College, University of Delhi, New Delhi, India
e-mail: amanrai.ar29@gmail.com

P. K. Mishra
e-mail: prabuddh@shivaji.du.ac.in

A. Pandey · K. C. Tiwari
Department of Civil Engineering, Delhi Technological University, New Delhi, India
e-mail: aayushipandey50@gmail.com

K. C. Tiwari
e-mail: kcchtphd@gmail.com

K. Jain et al. (eds.), *Proceedings of UASG 2019*, Lecture Notes in Civil Engineering 51,
https://doi.org/10.1007/978-3-030-37393-1_26

1 Introduction

The Hindu Kush Himalayan region covers the largest concentration of glaciated area outside the polar region [49, 50]. The region encompasses a total hill and mountain area of 4,192,000 km^2 in eight countries (Afghanistan, Bangladesh, Bhutan, China, India, Myanmar, Nepal, Pakistan) covering the total glaciated area of 60,054 km^2 that consist of 54,252 individual glaciers with a total ice reserve of estimated 6127 km^3 [3]. The region is referred as the third pole and freshwater tower of Asia [22, 81]. Such a large extent of glaciers in the area influence both local and global climate and hydrology of the region. Ten largest river systems in south and central Asia i.e. The Amu Darya, Brahmaputra, Ganges, Indus, Irrawaddy, Mekong, Salween, Tarim, Yangtze, and Yellow River receive meltwater from the glaciers [39, 46], which together provides support to about 1.3 billion people in their respective basins [3]. Due to the climate change, glaciers are shrinking and retreating [4], monitoring the continuously evolving and changing surface features of the glaciers provide insight into the structure, mass balance and internal dynamics of the glacier. Changes in glaciers have a profound impact on downstream communities, ecology, human livelihoods, and food security, on the other hand, it also alters downstream hydrology, hydropower potential [39, 37, 63, 71], and increases the volume and number of unstable proglacial lakes causing potential glacier lake outburst floods (GLOFs) [3]. Therefore, monitoring glacier extents, mass balances and surface velocity are essential to understand how climate change impacts the status of glaciers [60]. Accurate mapping of glaciers can increase the accuracy of hydrological modeling, prediction of avalanche and mass balance studies of glaciers [76]. Despite the importance of glaciers of the Himalayan region, there is limited data regarding the dimensions of these glaciers and their behaviour in response to the climate change, this is due to the remoteness, high elevation, the complex political situation [12] and rough weather conditions.

Taking into consideration the extent and isolation of glaciers, remote sensing acts as a valuable tool for the regular mapping and monitoring of glaciers [8, 13, 60]. Remote sensing data provides a synoptic view for mapping and monitoring the dynamics of glaciers. Remote sensing data have been used for glacier mapping since the late 1970s [1, 15, 25]. Resolution and sensor capabilities of satellites improved in recent years and many medium spatial resolution satellites were launched in the past decades. Various studies have been conducted on glaciers all over the world using freely available satellite data [11, 13, 21, 25], and several attempts were made to produce glacier inventory of the Himalayan region by global glacier inventory initiatives [31, 58, 64]; Pfeffer et al. [61]. These satellite remote sensing based inventories focused on larger areas. To monitor the broader regions satellite remote sensing may prevail over the difficulties of monitoring through the field, but the spatial (30 m/pixel or 23.5 m/pixel) and fixed temporal resolution is often too coarse for detailed and frequent investigations. The average size of individual glaciers in the Himalayan region is relatively small (1.1 km^2) [3], and monitoring small glaciers is challenging task as the spatial resolution of conventional remote sensing satellites data is too coarse for small glaciers [88]. Use of traditional aerial remote sensing is also constrained by the higher cost of the data along with the notable issues of mountain regions such as cloud cover, the shadow of the peaks and safety of the platform [88].

The spatial and temporal resolution of aerial remote sensing data is also a limitation for accurate and frequent monitoring. Thus, there is a great need for highly accurate and systematic observation of glaciers with the high spatial and temporal resolution to further advance our understanding of the climate-glacier system. UAVs provide the required capabilities to bridge the gap between extensive fieldwork [38] and coarse resolution satellite imagery. The recent development of UAV technology has enabled glaciologists to carry out their aerial surveys at reasonable costs with high resolution, automated, rapid and multispectral sensor equipped UAVs [9, 38, 43, 48, 47, 67–70, 74, 85, 86]. Majority of the UAV based studies have used the imagery in the visual spectrum and structure-from-motion technique (SfM) to recreate glacier surfaces and track motion. The number of glacier studies using the UAVs has grown globally with the advancement of UAV technology in recent years. This paper is an attempt to review the potential of UAVs and its applications for monitoring the Himalayan glaciers.

2 Literature Review

Monitoring glacier surface has been an important task for researchers from early stakes measurements in the nineteenth century to in situ topographic surveys of present days. Data derived from various remote sensing platforms such as ground-based devices [30, 62], aircraft [5, 52] or satellites [7, 33] have been used to monitor the status of glaciers worldwide. Advancements in UAV technology have accelerated the use of UAVs in scientific research of various disciplines [42]. Among the diverse applications of UAVs, glaciology is gaining pace due to its potential to carry out rapid, cost-efficient and automated surveys [9, 38, 43, 68, 86]. To analyze the current and background status of UAVs for glacier monitoring we extracted the articles published till February 2019 on the application of UAVs for Glaciers from Elsevier Scopus Database using the keywords "unmanned aerial systems" or "UAS" or "unmanned aerial vehicles" or "UAV" or "Drones" and "Glacier". Total numbers of publications on the applications of UAVs in glaciology were found to be 110 till February 2019 (Fig. 1), which is far less as compared to other applications of UAVs such as Agricultural science (2529) and Environmental monitoring (1874). Studies in various alpine, as well as continental glaciers, were found in the published work on the application of UAVs in glaciology.

Existing studies covered the parts of the Arctic [41], Beaufort/Chukchi Sea region in Alaska [80], Fountain glacier in Canadian Arctic [87], French Alps [82], Greenland [20], Southern Ocean in Antarctic [29], Eden Valley drumlin field in England [17], Rhone Glacier in Swiss Alps [27] and Lirung glacier and Langtang Khola in Nepal Himalaya [38, 48, 55]. UAV based studies explored various applications of UAVs in glaciology such as monitoring glacier motion, measuring seasonal or multi-year

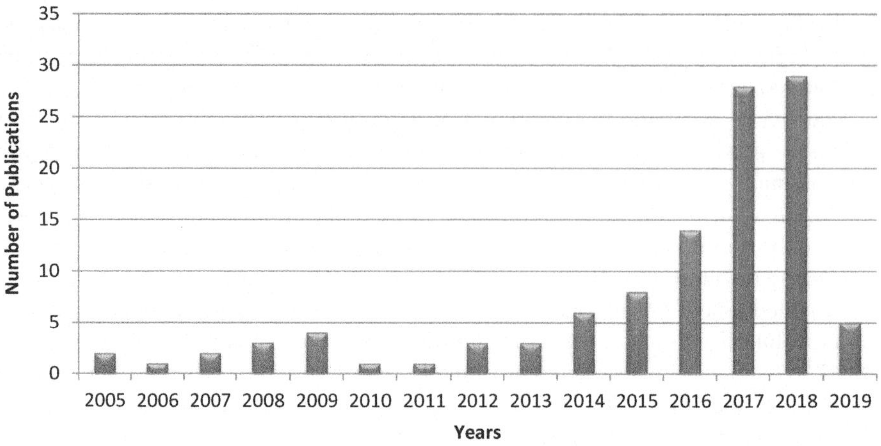

Fig. 1 Number of publication on the use of UAVs in glaciology extracted from Elsevier Scopus database

melt, analyzing calving and crevasse dynamics and mapping of surface characteristics such as drainage networks, albedo, debris cover, and cryoconite holes [19, 38, 42, 48, 47, 67, 69, 70, 74, 85]. The available literature indicated that the use of UAVs for monitoring glaciers is still in its initial phase in the Himalayan region, it has only been tested in the central Himalayas, i.e. Nepal. However, all these studies indicate the development of UAVs offers excellent potential for accurate, rapid, cost-effective and frequent monitoring of small scale glaciers.

3 Changes in Himalayan Glacier and the Need for UAV Based Studies

According to the studies based on satellite remote sensing data, glaciers in many locations all over the globe are retreating, and in recent decades, the rate of retreat has accelerated [90]. Glacier changes across the HKH region cannot be generalized; however, the general trend in the HKH region appears to be negative. The glaciers in the region are shrinking due to changing climate [28, 44, 45, 73, 89]. Based on the observation of individual glaciers, it is indicated that the annual retreat rates vary from basin to basin. In some cases, it was observed that in recent years, the retreat rate has doubled compared to the early seventies [4]. However, some studies report advances in glaciers in the Karakoram region [35, 34, 73]. The significant variations in glacier response to climatic changes are subjected to the variation in elevation because glaciers at higher elevations remain below freezing temperature even during the presence of a warmer climate. The glaciers of the HKH region are situated above 3200 msl and out of total glaciated area 60,054 km^2 in HKH region, total 60% of the glaciated area lies between the elevation range of 5000–6000 msl [3]. According to Bajracharya et al. [2], the glaciers below 5700 msl are mainly sensitive to climate change unless they are covered by thick debris. Small clean ice glaciers at lower elevation are the most sensitive to climate change [3]. 79, 60 and 77% total glacier area of the three largest river basins Indus, Ganga and Brahmaputra respectively lie below this critical elevation [3].

The valley glaciers and small glaciers are retreating faster with the formation and expansion of glacial lakes [2]. Monitoring the changes in the low altitude average sized glacier in this region is vital for the assessment of rates of retreat among different glaciated regions or basins. A better understanding of the changing glaciers will also help to manage its impact on downstream hydrology, predicting potential GLOFs and avalanches. However, due to the harsh environment, remoteness, inaccessible terrain and hazardous conditions for field investigations, the majority of the glaciers in the Himalayan region remain relatively unstudied. Until now, researchers have relied on the traditional methods of data acquisition, i.e. satellite and aerial photography for monitoring glaciers in the HKH region. The coarse spatial and fixed temporal resolution along with the higher cost of the traditional methods is limiting factors for accurate monitoring. The increasing spatial and temporal resolution also increases the cost; additionally, cloud cover and the safety of aerial platforms in high mountains are also an issue of concern. The development of task-specific and fully automated UAVs with all the necessary equipment and capabilities such as ultra-high resolution, onboard Global Navigation Satellite System (GNSS), availability of a wide range of light weighted sensors, software-designed flight planning for optimal land coverage, onboard autopilot to follow a pre-determined course and the ability of simultaneous acquisitions of images from different sensors makes it more than other remote sensing platforms.

UAV technology offers great potential to monitor small Himalayan glaciers in lower elevation range, and their limited size allows high-resolution coverage at centimeter spatial resolution. Integrating UAV based studies with satellite and aerial remote sensing-based studies will enhance understanding of small scale processes of glaciers by providing new insights. It will also help to improve models and predictions of future glacier change. The use of UAVs in glaciology is currently limited [86], and in the greater Himalayas nonexistent. However, the application of UAVs for glaciological monitoring has the potential to prevail over many of the difficulties related to present day extensive field-based and geodetic methods.

4 Applications of UAV for Monitoring Himalayan Glacier

4.1 Mass Balance Analysis

Accurate temporal monitoring of glacier mass balance is necessary to understand the dynamics of glacier mass. Studies in the past have been based on the sparse data point and significant generalizations [26], however recently developed UAVs and Structure from motion photogrammetry are being wildly used to improve the understanding of glacier changes [77, 87]. Using UAVs for obtaining temporal DEMs can be useful in mass balance studies of glaciers. The cost-effective and highly accurate temporal DEMs are the most significant advantages of UAVs over other methods of acquiring DEMs, which requires higher cost and has fixed temporal resolution. UAV based glacier mass balance studies are suitable for thick debris covered glaciers of Himalayas where field observations are challenging. However, high-resolution DEM differencing saves huge capital and time while providing high accuracy equivalents to millions of stakes in the field [38]. In 2013 a UAV based study in conducted on Lirung glacier in central Himalayas successfully acquired high-resolution imagery and DEM that could be used for flow and mass balance analyses [38]. Though carrying out mass balance studies requires UAVs to fly over the entire catchment, which can be a major task for large Himalayan glaciers.

4.2 Monitoring Debris-Covered Glaciers

In the Himalayas, about 24% of the total glacier area is covered by debris [75]. Debris covers a large proportion of glacial tongues in the region. About 40% of the glacier ice mass in the ablation zone is covered by debris [47]. In the Himalayan glaciers, a large part of the mass loss is determined by the presence of debris cover [65, 66, 73] as debris thickness plays a vital role in reducing the melt rates [32, 54, 56, 57]. UAVs offer great potential to measure the surface of debris-covered glaciers in high spatial and temporal resolution. Many recent studies have already tested the capabilities of UAVs for debris mapping [38, 48, 83]. It can be used for routine monitoring of the Himalayan debris-covered glaciers where field visits are often risky and physically challenging. Using thermal sensor equipped UAVs will allow the detailed mapping of debris surface temperature, information of surface temperature can provide insight into the debris properties. This information can also be used to quantify the melt process and estimation of debris thickness. Similarly, multispectral and hyperspectral sensors could also be used to estimate the distribution of debris thickness [72]. Development of more reliable algorithms for calculating land surface temperature (LST) of glaciers can also be done by using UAVs based thermal data to downscale the coarse resolution satellite thermal remote sensing data [9]. Thermal images can also be used to automate the mapping of debris-covered glacier [8, 10, 14]. Using UAVs for debris-covered glacier mapping and monitoring may revolutionize the traditional methods of research on debris-covered glaciers as it has the potential to overcome the limitations of satellite remote sensing and field-based studies. A UAV based study on debris-covered Lirung Glacier in the Central Himalaya reported conducting three separate missions in one day to map changing surface temperatures of the debris-covered glacier, thus, proving UAVs as a better alternative on the contrary to time-consuming traditional field-based study.

4.3 GLOF Studies

Snow and glacier melt of shrinking glaciers are accompanied by the development and expansion of proglacial lakes, causing the increase of potential glacial lake outburst flood (GLOF) hazard. More than 50 events of glacial lake outburst have been recorded in the HKH region; however, the records are available only for parts of China, Nepal, Pakistan and Bhutan [51]. Newly formed lakes due to shrinking glaciers in the HKH have increased the risk of GLOFs. A study revealed that there are nearly 25,614 glacial lakes in five major river basins of HKH region (Amu Darya, Indus, Ganges, Brahmaputra, and Irrawaddy basin) covering an area of 1444 km^2, out of which 79% of lakes have the total area of less than 0.05 km^2 [51]. High-resolution UAVs provides excellent opportunities to monitor the rise and fall in the water level of these lakes as well as the UAV data products such as DEM can be used to extract high-resolution terrain parameters, i.e. elevation, slope, aspect and curvature for GLOF modeling. Apart from this, UAVs can also help to find vulnerable sites to restrict development plans and high elevated risk-free zones for planning and preparedness at the local level in the events of the flood.

4.4 Temporal Change Analysis

Aerial coverage of satellite remote sensing is larger than UAVs; however, the spatial and temporal resolution is much poorer as compared to UAVs. The flexibility to acquire data on demand is one of the most significant advantages of UAVs for temporal analysis of daily, seasonal or annual variation. On the other hand, satellites are hampered by clouds and shadows that reduce the accuracy of the results while performing temporal change analysis. Highly detailed and accurate ortho-mosaics and DEMs obtained from overlapping multiview images allow accurate assessments of glacier surface and height changes. Changes in surface properties, like ice cliffs and lakes, can also be assessed using DEM differencing techniques to improve our understanding of their role in glacier melt further. Apart from the change detection in spatial distribution, high-resolution UAV data can also be used for analysis of annual and seasonal changes in the thickness, changes in glacier surface, terminus position, seasonal melt impact on upstream dynamics and thinning, changes in the depth, density, orientation and nature of crevassing and their impact on calving rate. Multispectral sensors, temporal flexibility and ultra-high spatial resolution of the UAV make it capable of monitoring overall glacier health and detecting small changes which may improve our understanding of the spatiotemporal dynamics of Himalayan glaciers. The risks in UAV based studies are also smaller compared with direct measurements as the scientists do not need to be on the glacier for prolonged periods.

4.5 Geomorphological Mapping

Rapid mapping and geomorphological characterization of proglacial terrain are essential for a proper understanding of the glacier landform dynamics. Small UAVs can be used to acquire ultra high-resolution low-altitude images. These images can subsequently be processed using SfM technique to generate detailed orthoimages and DEMs. Extracting terrain parameter, i.e. slope, aspect, elevation and curvature from high accuracy DEMs can be used for high-resolution mapping of various glacial depositional, erosional landforms and lakes, i.e. arêtes, horns, hanging glaciers, tarn lakes, drumlins, kames, moraine-dammed lakes. Some recent studies used UAVs in the mapping of small fragments of proglacial areas [6, 16, 23, 53, 79, 84], apart from the mapping of glacial geomorphological features modeling of avalanche prone areas can also be done using UAV based DEMs along with high-resolution aerial images [10, 78].

4.6 Other Applications

Other applications include high resolution DEM and ortho-mosaic generated from UAVs to quantify ice surface displacements and monitoring high resolution seasonal surface velocities and flow patterns. This is not possible with other remote-sensing platforms as satellite imagery does not allow detailed analysis of glacier surface velocity and availability of images depend on the fixed satellite overpass compared to the on-demand deployment of the UAV system. High-resolution UAVs allow continuous values of the surface velocities of a single season to be obtained. The detailed knowledge of the smaller scale variations in flow, both spatially and temporally also helps to understand the bigger picture of heterogeneous mass wasting and distribution of surface features found on debris-covered glaciers [38]. Immerzeel et al. [38] and Kraaijenbrink et al. [48] successfully used a small fixed-wing UAV to map glacier changes and surface velocities at 5000 m altitude in the Himalaya. Monitoring the dynamics of the supra-glacial lake can also be done using UAVs. The high-resolution UAV-acquired imagery can be used for both quantitative and qualitative analysis by multitemporal visual inspections and automated mapping of the surface features. Monitoring ice surface elevation and energy-balance modelling of supraglacial lakes can also be done using UAV based DEM, on the other hand, spectral libraries for different types of snow and glacier facies and debris cover can also be developed by using hyperspectral sensors in UAVs.

5 Benefits and Challenges

The potential of accurate mapping and monitoring the changes in small glaciers are growing with the development of low cost, multispectral, easy to use and task-specific UAVs. While the conventional satellite data plays a significant role in monitoring larger regions, UAV based studies for monitoring glaciers are gaining pace in recent years. UAVs are capable of providing centimeter-resolution imagery that can provide unique insights into the spatiotemporal dynamics of the glaciers. The flexibility to deploy on demand and cost-effectiveness makes it ideal for small-scale studies in high mountain regions. UAVs are especially suited to study the Himalayan glaciers where steep slope and thick debris cover makes field investigations huge task and frequent cloud cover and shadow of higher peaks limit the use of satellite imagery. Moreover, it is a higher spatial and temporal resolution, and its ability to survey potentially dangerous areas of the glacier surface, i.e. avalanche prone areas and crevasses is a significant advantage. Correct deployment and attainment of accurate ground control data result in the higher accuracy and precision of Orthoimages and DEMs generated through SfM are better than satellite imagery and aerial LIDAR [24, 36]. Annual monitoring on benchmark glaciers through UAVs could play an important role to understand climate change impacts on debris-covered Himalayan glaciers, which is still very uncertain [18]. The application of UAVs to monitor Himalayan glaciers has great potential; however, there are a few significant limitations, such as ground control points (GCPs) must be installed and surveyed to generate high accuracy DEMs. Installing GCPs is a physically challenging, labour intensive and highly dangerous task on debris-covered glaciers. GCPs distributed across the glacier surface to help to minimize distortions in the SfM processing workflow. Other alternatives of reducing DEM error is through increased overlap and sidelap, which further increases flight line density and thus total survey time. Flying the UAVs in High Mountains where air pressure decreases with increasing height is a challenging task as it flies much faster and is less stable. Sometime to get enough lift during take-off is also challenging, yet the highest flights at an altitude of 5700 m above mean sea level has been successfully conducted in the central Himalayan region [40], which seems to be the upper limit for currently existing UAVs. However, according to Bajracharya et al. [3], glaciers most sensitive to climate change lies below the 5700 m of altitude. Local wind and flight altitude also plays a major role in the stability and successful survey. Choosing a suitable platform (Fixed-wing or multirotors) is also an important task which should be carefully considered based on the survey area, local conditions and objective of the survey.

6 Conclusion

Currently, the use of UAVs for glacier monitoring in the Himalayan region is almost nonexistent except in the parts Nepal. However, the use of UAVs for different glaciological applications is gaining pace around the world. UAVs have tremendous future potential to overcome the challenges of other remote sensing platforms for monitoring small Himalayan glaciers which covers the larger part of the HKH region. UAV technology has the potential to revolutionize classical field-based methods that are labour intensive and costly. The ultra-high spatial resolution, flexibility, multispectral sensors, ability to generate high resolution and accurate DEMs and techniques such as SFM offer great opportunities for glaciologists to carry out studies related to glacier mass balance, GLOF, surface structure mapping, seasonal analysis, surface velocity monitoring, debris-covered glacier monitoring at relatively faster than other remote sensing and field methodologies. However, UAVs has its limitations such as short battery backup which is insufficient for large glacier monitoring, distortion caused by wind and flying in low air pressure also creates hurdle in UAV based studies. Complications in large data processing and handling is also a major limitation along with restrictions and regulations of using UAVs however we hope these problems will be tackled in near future and entry of low-cost and task-specific UAVs in recent years will smooth the progress of UAVs applications for monitoring Himalayan glaciers.

References

1. Andersen T (1982) Operational snow mapping by satellites. Paper read at hydrological aspects of alpine and high mountain areas, at Exeter, UK
2. Bajracharya SR, Maharjan SB, Shrestha F (2014) The status and decadal change of glaciers in Bhutan from 1980s to 2010 based on the satellite data. Ann Glaciol 55:159–166. https://doi.org/10.3189/2014AoG66A125
3. Bajracharya SR, Maharjan SB, Shrestha F, Guo W, Liu S, Immerzeel W, Shrestha B (2015) The glaciers of the Hindu Kush himalayas: current status and observed changes from the 1980s to 2010. Int J Water Resour Dev 31(2):161–173. https://doi.org/10.1080/07900627.2015.1005731
4. Bajracharya SR, Mool PK, Shrestha BR (2007) Impact of climate change on Himalayan glaciers and glacial lakes: case studies on GLOF and associated hazards in Nepal and Bhutan. ICIMOD, Kathmandu
5. Baltsavias EP, Favey E, Bauder A, Bösch H, Pateraki M (2001) Digital surface modelling by airborne laser scanning and digital photogrammetry for glacier monitoring. Photogramm Rec 17:243–273
6. Bernard É, Friedt JM, Tolle F, Marlin C, Griselin M (2016) Using a small COTS UAV to quantify moraine dynamics induced by climate shift in Arctic environments. Int J Remote Sens 38:2480–2494
7. Berthier E, Vincent C, Magnússon E, Gunnlaugsson A, Pitte P, Le Meur E, Masiokas M, Ruiz L, Pálsson F, Belart JMC, Wagnon P (2016) Glacier topography and elevation changes derived from Pléiades sub-meter stereo images. Cryosphere 8:2275–2291
8. Bhambri R, Bolch T, Chaujar RK (2011) Mapping of debris covered glaciers in the Garhwal himalayas using ASTER DEMs and thermal data. Int J Remote Sens 32:8095–8119
9. Bhardwaj A, Sam L, Martín-Torres FJ, Kumar R (2016) UAVs as remote sensing platform in glaciology: present applications and future prospects. Remote Sens Environ 175:196–204
10. Bhardwaj A, Joshi PK, Snehmani, Singh MK, Sam L, Gupta RD (2014) Mapping debris-covered glaciers and identifying factors affecting the accuracy. Cold Reg Sci Technol 106–107:161–174. https://dx.doi.org/10.1016/j.coldregions

11. Bhardwaj A, Joshi PK, Snehmani, Sam L, Singh MK, Singh S, Kumar R (2015) Applicability of Landsat 8 data for characterising glacier facies and supraglacial debris. Int J Appl Earth Obs Geoinf 38:51–64. https://dx.doi.org/10.1016/j.jag.2014.12.011
12. Bolch T, Kulkarni A, Kaab A, Huggel C, Paul F, Cogley JG, Stoffel M (2012) The state and fate of himalayan glaciers. Science 336:310–314. https://doi.org/10.1126/science.1215828
13. Bolch T, Menounos B, Wheate R (2010) Landsat-based glacier inventory of western Canada, 1985–2005. Remote Sens Environ 114:127–137. https://doi.org/10.1016/j.rse.2009.08.015
14. Bolch T, Buchroithner MF, Kunert A, Kamp U (2007) Automated delineation of debris-covered glaciers based on ASTER data. In: Geoinformation in Europe proceedings of 27th EARSeL symposium, pp 403–410, Bozen, Italy, 04–07 June 2007
15. Campbell WJ, Weeks WF, Ramseier RO, Gloersen P (1975) Geophysical studies of floating ice by remote sensing. J Glaciol 15(73):305–328
16. Chandler BMP, Evans DJA, Roberts DH, Ewertowski MW, Clayton AI (2016) Glacial geomorphology of the Skálafellsjökull foreland, Iceland: a case study of 'annual' moraines. J Maps 12:904–916
17. Clayton AI (2012) Remote sensing of subglacial bedforms from the British ice sheet using an unmanned aerial system (UAS): problems and potential. Durham E-thesis Durham University. https://etheses.dur.ac.uk/5907/1/MScR_Thesis_AClayton.pdf
18. Cogley JG (2011) Himalayan glaciers now and in 2035. In: Singh VP, Singh P, Haritashya (eds) Encyclopaedia of snow, ice and glaciers. Springer, in press, UK
19. Cook JM, Sweet M, Cavalli O, Taggart A, Edwards A (2018) Topographic shading influences cryoconite morphodynamics and carbon exchange. Arct Antarct Alp Res 50:S100014
20. Crocker RI, Maslanik JA, Adler JJ, Palo SE, Herzfeld UC, Emery WJ (2011) A sensor package for ice surface observations using small unmanned aircraft systems. IEEE Trans Geosci Remote Sens 50(4):1033–1047. https://doi.org/10.1109/TGRS.2011.2167339
21. Dozier J (1984) Snow reflectance from Landsat-4 Thematic Mapper. IEEE Trans Geosci Remote Sens 22(3):323–328
22. Dyhrenfurth GO (1955) The third pole—the history of the high himalaya, 1st UK edn. Ex Libris, Werner Laurie, London
23. Ewertowski MW, Evans DJA, Roberts DH, Tomczyk AM (2016) Glacial geomorphology of the terrestrial margins of the tidewater glacier, Nordenskiöldbreen, Svalbard. J Maps 12:476–487
24. Fonstad MA, Dietrich JT, Courville BC, Jensen JL, Carbonneau PE (2013) Topographfonic structure from motion: a new development in photogrammetric measurement. Earth Surf Proc Land 38:421–430. https://doi.org/10.1002/esp.3366
25. Foster J, Schultz D, Dallam WC (1978) Ice conditions on the Chesapeake Bay as observed from landsat during the winter of 1977. In: Paper read at 35th Eastern snow conference, at Hanover, New Hampshire
26. Fountain AG, Vecchia A (1999) How many stakes are required to measure the mass balance of a glacier? Geogr Ann Ser A Phys Geogr 81:563–573
27. Friedli E (2013) Photogrammetric methods for the reconstruction and monitoring of glaciers. Master thesis ETH Zurich. https://www.igp.ethz.ch/photogrammetry/education/oldstudentsprojects/MA_Report_EFriedli_final.pdf
28. Fujita K, Nuimura T (2011) Spatially heterogeneous wastage of himalayan glaciers. Proc Natl Acad Sci 108:14011–14014. https://doi.org/10.1073/pnas.1106242108
29. Funaki M, Higashino SI, Sakanaka S, Iwata N, Nakamura N, Hirasawa N, Kuwabara M (2014) Small unmanned aerial vehicles for aeromagnetic surveys and their flights in the South Shetland islands. Antarct Polar Sci 8(4):342–356. https://doi.org/10.1016/j.polar.2014.07.001
30. Gabbud C, Micheletti N, Lane SN (2015) Lidar measurement of surface melt for a temperate alpine glacier at the seasonal and hourly scales. J Glaciol 61:963–974
31. Haeberli W, Bosch H, Sherler K, Østrem G, Wallen CC (eds) (1989) World glacier inventory status 1988. IAHS, Wallingford
32. Hagg WJ, Mayer C, Lambrecht A, Helm A (2008) Sub-debris melt rates on southern inylchek glacier, central tian shan. Geogr Ann Ser Phys Geogr 90(1):55–63

33. Herman F, Anderson B, Leprince S (2011) Mountain glacier velocity variation during a retreat/advance cycle quantified using sub-pixel analysis of ASTER images. J Glaciol 57:197–207
34. Hewitt K (2011) Glacier change, concentration, and elevation effects in the Karakoram himalaya, upper Indus Basin. Mt Res Dev 31:188–200. https://doi.org/10.1659/MRDJOURNAL-D-11-00020.1
35. Hewitt K (2005) The Karakoram anomaly? Glacier expansion and the "elevation effect," Karakoram himalaya. Mt Res Dev 25:332–340. https://doi.org/10.1659/0276-4741(2005)025%5b0332:tkagea%5d2.0.co;2
36. Hugenholtz CH, Whitehead K, Brown OW, Barchyn TE, Moorman BJ, LeClair A, Riddell K, Hamilton T (2013) Geomorphological mapping with a small unmanned aircraft system (UAS): feature detection and accuracy assessment of a photogrammetrically-derived digital terrain model. Geomorphology 194:16–24. https://doi.org/10.1016/j.geomorph.2013.03.023
37. Immerzeel WW, Van Beek LPH, Bierkens MFP (2010) Climate change will affect the Asian water towers. Science 328:1382–1385. https://doi.org/10.1126/science.1183188
38. Immerzeel WW, Kraaijenbrink PDA, Shea JM, Shrestha AB, Pellicciotti F, Bierkens MFP, de Jong SM (2014) High-resolution monitoring of himalayan glacier dynamics using unmanned aerial vehicles. Remote Sens Environ 150:93–103. https://doi.org/10.1016/j.rse.2014.04.025
39. Immerzeel WW, Pellicciotti F, Bierkens MFP (2013) Rising river flows throughout the twenty-first century in two himalayan glacierized watersheds. Nat Geosci 6:742–745. https://doi.org/10.1038/NGEO1896
40. Immerzeel WW, Kraaijenbrink PDA, Shea JM, Shrestha AB, Pellicciotti F, Bierkens MFP, de Jong SM (2016) Using unmanned aerial vehicles for glacier monitoring in the himalayas. University of Utrecht, Utrecht, Netherlands, pp 21
41. Inoue J, Curry J, Maslanik J (2008) Application of aerosondes to melt-pond observations over Arctic sea ice. J Atmos Ocean Technol 25(2):327–334
42. James MR, Robson S (2012) Straight forward reconstruction of 3D surfaces and topography with a camera: accuracy and geoscience application. J Geophys Res 117
43. Jouvet G, Weidmann Y, Seguinot J, Funk M, Abe T, Sakakibara D, Seddik H, Sugiyama S (2017) Initiation of a major calving event on the Bowdoin glacier captured by UAV photogrammetry. Cryosphere 11:911–921
44. Kargel JS, Cogley JG, Leonard GJ, Haritashya U, Byers A (2011) Himalayan glaciers: the big picture is a montage. Proc Natl Acad Sci 108:14709–14710. https://doi.org/10.1073/pnas.1111663108
45. Kaser G, Cogley JG, Dyurgerov MB, Meier MF, Ohmura A (2006) Mass balance of glaciers and ice caps: consensus estimates for 1961–2004. Geophys Res Lett 33(19):1–5. https://doi.org/10.1029/2006GL027511
46. Kaser G, Großhauser M, Marzeion B (2010) Contribution potential of glaciers to water availability in different climate regimes. Proc Natl Acad Sci USA 107:20223–20227. https://doi.org/10.1073/pnas.1008162107
47. Kraaijenbrink P, Shea JM, Litt M, Steiner JF, Treichler D, Koch I, Immerzeel WW (2018) Mapping surface temperatures on a debris-covered glacier with an unmanned aerial vehicle. Front Earth Sci 6:64. https://doi.org/10.3389/feart.2018.00064
48. Kraaijenbrink P, Meijer SW, Shea JM, Pellicciotti F, Jong SMD, Immerzeel WW (2016) Seasonal surface velocities of a himalayan glacier derived by automated correlation of unmanned aerial vehicle imagery. Ann Glaciol 57
49. Kulkarni AV (1991) Glacier inventory in Himachal Pradesh using satellite data. J Indian Soc Remote Sens 19(3):195–203
50. Kulkarni AV (1994) A conceptual model to assess effect of climatic variations on distribution of himalayan glaciers. In: Global change studies: scientific results from ISRO geosphere biosphere programme (ISRO-GBP-SR-42-94), pp 322–326. Indian Space Research Organisation, Bangalore, India
51. Maharjan SB, Mool PK, Lizong W, Xiao G, Shrestha F, Shrestha RB, Khanal NR, Bajracharya SR, Joshi S, Shai S, Baral P (2018) The status of glacial lakes in the Hindu Kush himalaya. ICIMOD Research Report 2018/1. ICIMOD, Kathmandu

52. Mertes JR, Gulley JD, Benn DI, Thompson SS, Nicholson LI (2017) Using structure-from-motion to create glacier DEMs and orthoimagery from historical terrestrial and oblique aerial imagery. Earth Surf Proc Land 42:2350–2364
53. Midgley NG, Tonkin TN, Graham DJ, Cook SJ (2018) Evolution of high-Arctic glacial landforms during deglaciation. Geomorphology 311:63–75
54. Mihalcea C, Mayer C, Diolaiuti G, Lambrecht A, Smiraglia C, Tartari G (2006) Ice ablation and meteorological conditions on the debris-covered area of Baltoro glacier, Karakoram, Pakistan. Ann Glaciol 43(1):292–300
55. Miles ES, Pellicciotti F, Willis IC, Steiner JF, Buri P, Arnold NS (2016) Refined energy-balance modelling of a supraglacial pond, Langtang Khola, Nepal. Ann Glaciol 57(71):29–40. https://doi.org/10.3189/2016AoG71A421
56. Nicholson L, Benn DI (2006) Calculating ice melt beneath a debris layer using meteorological data. J Glaciol 52(178):463–470
57. Östrem G (1959) Ice melting under a thin layer of moraine, and the existence of ice cores in moraine ridges. Geogr Ann 41(4):228–230
58. Paul F, Barry RG, Cogley JG, Frey H, Haeberli W, Ohmura A, Zemp M (2010) Recommendations for the compilation of glacier inventory data from digital sources. Ann Glaciol 50:119–126. https://doi.org/10.3189/172756410790595778
59. Paul F, Hasty M (2014) retreat of glaciers in northern Patagonia from 1985–2011. J Glaciol 60(224):1033–1043. https://doi.org/10.3189/2014JoG14J104
60. Paul F, Bolch T, Kääb A, Nagler T, Nuth C, Scharrer K et al (2014) The glaciers climate change initiative: methods for creating glacier area, elevation change and velocity products. Remote Sens Environ
61. Pfeffer WT, Arendt AA, Bliss A, Bolch T, Cogley JG, Gardner AS, Sharp MJ, Randolph Consortium (2014) The Randolph glacier inventory: a globally complete inventory of glaciers. J Glaciol 60:537–552. https://doi.org/10.3189/2014JoG13J176
62. Piermattei L, Carturan L, Guarnieri A (2015) Use of terrestrial photogrammetry based on structure-from-motion for mass balance estimation of a small glacier in the Italian alps. Earth Surf Proc Land 40:1791–1802
63. Rankl M, Kienholz C, Braun M (2014) Glacier changes in the Karakoram region mapped by multimission satellite imagery. Cryosphere 8:977–989. https://doi.org/10.5194/tc-8-977-2014
64. Raup B, Kaab A, Kargel J, Bishop M, Hamilton G, Lee E, Khalsa S (2007) Remote sensing and GIS technology in the global land ice measurements from space (GLIMS) project. Comput Geosci 33:104–125. https://doi.org/10.1016/j.cageo.2006.05.015
65. Reid TD, Brock BW (2010) An energy-balance model for debris-covered glaciers including heat conduction through the debris layer. J Glaciol 56(199):903–916
66. Reid TD, Carenzo M, Pellicciotti F, Brock BW (2012) Including debris cover effects in a distributed model of glacier ablation. J Geophys Res 117(D18):D18105
67. Rippin DM, Pomfret A, King N (2015) High resolution mapping of supra-glacial drainage pathways reveals link between micro-channel drainage density, surface roughness and surface reflectance. Earth Surf Process Landf 40:1279–1290
68. Rossini M, Di Mauro B, Garzonio R, Baccolo G, Cavallini G, Mattavelli M, De Amicis M, Colombo R (2018) Rapid melting dynamics of an alpine glacier with repeated UAV photogrammetry. Geomorphology 304:159–172
69. Ryan JC, Hubbard A, Box JE, Todd J, Christoffersen P, Carr JR, Holt TO, Snooke N (2015) UAV photogrammetry and structure from motion to assess calving dynamics at store glacier, a large outlet draining the Greenland ice sheet. Cryosphere 9:1–11
70. Ryan JC, Hubbard A, Stibal M, Irvine-Fynn TD, Cook J, Smith LC, Cameron K, Box J (2018) Dark zone of the Greenland Ice Sheet controlled by distributed biologically-active impurities. Nat Commun 9:1065
71. Schaner N, Voisin N, Nijssen B, Lettenmaier DP (2012) The contribution of glacier melt to streamflow. Environ Res Lett 7(034029):1–8. https://doi.org/10.1088/1748-9326/7/3/034029
72. Schauwecker S, Rohrer M, Huggel C, Kulkarni A, Ramanathan AL, Salzmann N, Stoffel M, Brock B (2015) Remotely sensed debris thickness mapping of Bara Shigri glacier, Indian himalaya. J Glaciol 61:675–688. https://doi.org/10.3189/2015JoG14J102

73. Scherler D, Bookhagen B, Strecker MR (2011) Spatially variable response of himalayan glaciers to climate change affected by debris cover. Nat Geosci 4(1):1. https://doi.org/10.1038/ngeo1059
74. Seier G, Kellerer-Pirklbauer A, Wecht M, Hirschmann S, Kaufmann V, Lieb GK, Sulzer W (2017) UAS-based change detection of the glacial and proglacial transition zone at Pasterze Glacier, Austria. Remote Sens 9:549
75. Sharma AK, Singh SK, Kulkarni AV, Ajai (2013) Glacier inventory in Indus, Ganga and Brahmputra basins of the himalaya. Natl Acad Sci Lett. 10.1007/s40009-013-0167-6
76. Shukla A, Ali I (2016) A hierarchical knowledge-based classification for glacier terrain mapping: a case study from Kolahoi Glacier, Kashmir himalaya. Ann Glaciol 57(71):1
77. Smith M, Carrivick J, Quincey D (2016) Structure from motion photogrammetry in physical geography. Prog Phys Geogr 40:247–275
78. Snehmani, Bhardwaj A, Pandit A, Ganju A (2013) Demarcation of potential avalanche sites using remote sensing and ground observations: a case study of Gangotri glacier. Geocarto Int 29. https://doi.org/10.1080/10106049.2013.807304
79. Tonkin TN, Midgley N, Cook SJ, Graham DJ (2016) Ice-cored moraine degradation mapped and quantified using an unmanned aerial vehicle: a case study from a polythermal glacier in Svalbard. Geomorphology 258:1–10
80. Tschudi MA, Maslanik JA, Perovich DK (2008) Derivation of melt pond coverage on Arctic sea ice using MODIS observations. Remote Sens Environ 112(5):2605–2614. https://doi.org/10.1016/j.rse.2007.12.009
81. Viviroli D (2004) The hydrological significance of mountains: from regional to global scale. Hydrol Earth Syst Sci 8(6):1016–1029
82. Walter M, Niethammer U, Rothmund S, Joswig M (2009) Joint analysis of the super-sauze (French alps) mudslide by nanoseismic monitoring and UAV-based remote sensing. First Break 27(8):53–60
83. Westoby MJ, Brasington J, Glasser NF, Hambrey MJ, Reynolds JM (2012) Structure-from-Motion photogrammetry: a low-cost, effective tool for geoscience applications. Geomorphology 179:300–314
84. Westoby MJ, Dunning SA, Woodward J, Hein AS, Marrero SM, Winter K, Sugden DE (2017) Sedimentological characterization of Antarctic moraines using UAVs and structure-from-motion photogrammetry. J Glaciol 61:1088–1102
85. Westoby MJ, Dunning SA, Hein AS, Marrero SM, Sugden DE (2016) Interannual surface evolution of an Antarctic blue-ice moraine using multi-temporal DEMs. Earth Surf Dyn 4:515–529. Remote Sensing 2018, 10, 1547 16 of 17
86. Whitehead K, Moorman BJ, Hugenholz CH (2013) Brief communication: low-cost, on-demand aerial photogrammetry for glaciological measurement. Cryosphere 7:1879–1884
87. Whitehead K (2013) An integrated approach to determining short-term and long-term patterns of surface change and flow characteristics for a polythermal Arctic glacier. Ph.D. thesis, University of Calgary, Canada. https://hdl.handle.net/11023/812
88. Wigmore O, Mark B (2017) Monitoring tropical debris-covered glacier dynamics from high-resolution unmanned aerial vehicle photogrammetry, Cordillera Blanca, Peru. Cryosphere 11:2463–2480
89. Zemp M, Hoelzle M, Haeberli W (2009) Six decades of glacier mass-balance observations: a review of the worldwide monitoring network. Ann Glaciol 50:101–111. https://doi.org/10.3189/172756409787769591
90. Zemp M, Roer I, Kääb A, Hoelzle M, Paul F, Haeberli W (2008) Global glacier changes: facts and figures. World Glacier Monitoring service, Geneva, UNEP/Zurich. https://www.zora.uzh.ch/id/eprint/4173/

A Review of UAV Regulations and Policies in India

Saurabh Srivastava, Saurabh Gupta, Onkar Dikshit and Syam Nair

Abstract The emergence of Unmanned Aerial Vehicle (UAV) is driving a paradigm shift in research and development. Initially, UAVs were limited for defense purposes only. Now, these are getting popular and are widely used in different sectors such as infrastructure, mining, and media. With the increase in usage of UAVs, legal and policy complexities of drone laws have also increased. However, the current legal framework that regulates UAVs, needs to be reviewed due to the rapid increase in UAV's market in India. The introduction of national drone policy from Director General of Civil Aviation (DGCA) which is UAVs regulatory body in India, reduces the ambiguities in the system. However, some bottlenecks still exist. The present policy legalizes the use and operation of drones in India which needs to be analyzed carefully. This paper examines the existing UAVs regulations and issues like safety, security, and privacy associated with the use of drones. The complexities in the existing law and major policy gaps in India are also analyzed. Moreover, a comparative analysis of existing national and international drone laws is presented with emphasis on the registration of aircraft and important guidelines in regard to the qualification and license for pilots.

Keywords UAVs · Drones · Regulation · Policies · Legal framework · DGCA

1 Introduction

Unmanned Aerial Vehicle (UAV), also referred to as remotely piloted aircraft, is an aircraft, which are operated with no pilot on board [1]. Initially, UAVs were used in military contexts. Now, these are widely used in different sectors such as surveying/mapping, infrastructure, mining, aerial cinematography, and industrial inspection. UAV technology is one of the most dynamic growth sectors in the aerospace industry. A report by the Ministry of Civil Aviation, Government of India, estimated UAVs market in India to touch US$886 million by 2021, while the global market is

S. Srivastava (✉) · S. Gupta · O. Dikshit · S. Nair
Department of Civil Engineering, Indian Institute of Technology Kanpur, Kanpur, Uttar Pradesh, India
e-mail: saura@iitk.ac.in

K. Jain et al. (eds.), *Proceedings of UASG 2019*, Lecture Notes in Civil Engineering 51,
https://doi.org/10.1007/978-3-030-37393-1_27

likely to touch US$21.47 billion [2]. As the global market for drones has grown, legal and policy complexities of drone laws have also increased. There has been significant interest in the regulations and policy gaps in India associated with Unmanned Aerial Vehicles (UAVs).

This paper examines the existing UAVs regulations and issues like safety, security, and privacy associated with the use of drones. The complexities in the existing law and major policy gaps in India are also analyzed. This paper presents a comparative analysis of existing national and international drone laws.

2 Background

In India, UAVs were not allowed to be used due to the lack of proper regulations and security concerns. The first true notification regarding drones came from the Office of the Director General of Civil Aviation (DGCA), India's civil aviation regulator, on October 7, 2014. Director General of Civil Aviation (DGCA) which is UAVs regulatory body in India, released the much awaited National Drone Policy on 27th August 2018. The policy came into effect from 1st December 2018. This regulation succeeds two other draft regulations that were issued by the DGCA in April 2016 and November 2017. A task force has also been set up to provide for further recommendations and may even modify the current regulation. Ministry of Civil Aviation released a Drone Ecosystem Policy Roadmap (also known as Drone Policy 2.0) in January 2019 [6].

The regulations provide a very detailed framework for licensing and other requirements for operation of UAV in India. The process of registering initializations and submissions of applications and clearances can be carried out through an online platform called DigitalSky. It implements "no permission, no take-off" (NPNT) regime [5].

3 Methodology

The paper presents a comparison of various data sources related to UAV regulations and policies. It includes a comparative analysis of existing national and international drone laws. This analysis encircles international guidelines and national regulatory frameworks which are analyzed in a comparative manner. UAV regulation of nine countries has been analyzed on the basis of 9 parameters.

Table 1 Online links to national and international UAV regulations

Web source	Content
http://dgca.nic.in/ [6]	Documentation for India
http://jarus-rpas.org/regulations [4]	Documentation for the regulation of UAV for 29 countries
https://www.legislation.gov.au/Details/F2017C00742 [8]	Documentation for Australia
http://dronelawjapan.com/ [10]	Drone law for Japan
https://digitalsky.dgca.gov.in/ [7]	Indian online portal for UAVs registration
https://uavcoach.com [11]	Free resource guides, weekly digest, online training courses

4 Database

Our database includes all the regulatory documents available online on the websites of UAV regulatory bodies of different countries. UAV regulations of 9 countries are included in this paper. Table 1 presents a list of known sources that provide links to national and international UAV regulations and their content.

5 International Context: UAV Regulation Across the World

The International Civil Aviation Organization (ICAO) has been the primary platform leading the global governance efforts. The ICAO is a UN specialized agency that codifies the principles and techniques of international air navigation [3]. ICAO published an online toolkit in 2016, which delivers general guidance for regulators and operators. The ICAO further issued recommendations to the safe integration of UAVs into controlled airspace [1].

Joint Authorities for Rulemaking on Unmanned Systems (JARUS) is a group of 59 national authorities and experts gathering regulatory expertise from all around the world [4]. The purpose of JARUS is to recommend a single set of technical, safety and operational requirements for all aspects linked to the safe operation of the UAVs.

6 Analysis

UAV regulation of nine countries has been analyzed. These regulations firstly classify UAVs by the mass. Moreover, rules are based on these categories only. UK, Australia, Austria, and the USA used similar mass categories ranges in 5, 25 kg, and more than

25 kg while Canada and Japan classification are similar ranging from less than 25 kg and more than 25 kg. For India, DGCA classifies under Nano (0.25 kg), Micro (0.25–2 kg), Mini (2–25 kg), Large (more than 25 kg) subheads. The maximum height for civil drone one can fly 400 ft for almost every country. However, for USA and China this is limited to 300 ft. We also compared nations policy for lateral distances for UAVs. Many countries have suggested a mainly visual line of sight (VLOS). USA and India presently do not have any regulations about it while the UK and Austria suggested 500 m. The process of application for flying UAV is different for different countries. For example Belgium, Japan, and USA do not require any prior permission, while other countries require prior permission before going to the field. In India, DGCA recently launched Digital Sky portal for online registration and approval for the use of UAV. The registration number is also not required for every country except Austria, China, and India. There is also no prior licensing for any of the country, but it is expected to have prior knowledge of drone's operation and laws associated with it. Hence, many countries demand certification course. All countries are lacking in definite policies for Data Protection and Privacy. As a guideline, it is suggested to respect one's privacy. In view of ease of registering and flying, the DGCA took a great step reducing the paperwork with the online portal which is not available in most of the countries. Table 2 presents a comparative analysis of 9 countries UAV regulations.

7 National Context: UAV Regulation in India

Director General of Civil Aviation (DGCA) which is UAVs regulatory body in India, implemented the much awaited National Drone Policy on 1st December 2018 [6]. The following subsections describe the current and proposed regulations in India.

8 Current Regulations (Drone Policy 1.0)

In the current drone policy, UAVs are termed as Remotely Piloted Aircraft (RPA). The DGCA has categorized RPA in five different categories. An RPA is classified into 'Nano' category that has a weight less than or equal to 250 g. There is no need to register a Nano RPA, but the rest would have to be registered and issued a unique identification number (UIN). RPAs that weigh between 250 and 2 kg are classified as 'Micro'. If an RPA weighs between 2 and 25 kg, it'll be classified as 'Small'. The 'Medium' category RPA weighs between 25 and 150 kg. RPA that weigh greater than 150 kg is classified as 'Large'. A permit named Unmanned Aircraft Operator Permit (UAOP) is required by the owners of the RPA to fly them. It can be obtained from the DGCA (Director General of Civil Aviation). However, this permit isn't required if a Nano RPA is operating in uncontrolled airspace (below 50 ft). A Micro RPA operating below 200 ft does not require permission, but the user must inform the local police

Table 2 Comparative analysis of 9 countries UAV regulations

Country	Mass (in kg)	Maximum height	Lateral distance	Application	Registration of UAV	License	Pilot training	Data protection	Privacy
USA	>0.25 0.25–25 25–150 <150	300 ft height above ground level	Currently, no maximum lateral distance from pilot but must remain VLOS	None	Registration number	Certification under supervision or online training courses	Certification	NA	Refers related to law
UK	>20 20–150 <150	>7 kg 400 ft	500 m or VLOS limited	Various approval requirements for different flight operations	NA	NA	Pilot competency	Refers to data protection act, CCTV code of practice	NA
Australia	>2 2–25 25–150	>7 kg must be able to be seen adequately for VLOS	VLOS	For 2–25 kg required	NA	>20 kg or BVLOS 20 kg or less (VLOS)—pilot competency assessment required if requesting a permission	License >2 kg	Advice to respect personal privacy	Advice to respect personal privacy

(continued)

Table 2 (continued)

Country	Mass (in kg)	Maximum height	Lateral distance	Application	Registration of UAV	License	Pilot training	Data protection	Privacy
Austria	>5 5–25 25–150	400 ft	500 m	General permission single approval for risky operations	Needed	For riskier categories gradual increase of pilot qualifications. Varies from operator responsibility, via operator declaration to ACG approval	Depending on the scenario	NA	NA

(continued)

Table 2 (continued)

Country	Mass (in kg)	Maximum height	Lateral distance	Application	Registration of UAV	License	Pilot training	Data protection	Privacy
Belgium	<1 <5 >5	300 ft	NA	No, but prior authorization for high-risk class 1 operations and declaration from the operator for low-risk class 1 operations. Class 2 operations need no authorization	NA	Yes, remote pilot license including medical for all class 1 activities. For class 2 activities only practical examination with certificate from examiner when applicant passed successfully, without medical	NA	NA	NA
Canada	>25 No category above that	Above 400 ft with approval and relevant conditions	Related to the distance between pilot and observer	Yes, special flight operations certificate (SFOC). Assessed on a case-by-case basis	NA	No pilot license	Pilot competency	Advice to respect personal privacy	Advice to respect personal privacy

(continued)

Table 2 (continued)

Country	Mass (in kg)	Maximum height	Lateral distance	Application	Registration of UAV	License	Pilot training	Data protection	Privacy
Japan	Up to 25 excluding lighter than 0.2	150 m	NA	No, but prior approval for high-risk operations	NA	NA	NA	NA	NA
China	0.25 or more	Above 300 ft AGL only possible with derogation to rules of the air	VLOS (without approval)	Flight authorization and operational certificate	Registration	Different licensing criteria under different weight segments	Certification	NA	NA
India	Nano (up to 0.25) Micro (0.25–2) Mini (2–25) Small (25–150) Large (150 or more)	No height limit. The standard specifies 400 ft; however, heights above 400 ft can be conducted provided the operator can mitigate risks	NA	Via DGCA can be done online	Required UIN (unique identification number)	No, as of now but need to undergo minimum training requirements specified in the civil aviation requirements on “requirements for operation of civil RPAS”	One needs to undergo minimum training requirements specified in the civil aviation requirements on “requirements for operation of civil RPAS”	Feedback and review mechanism proposed	No specific rules presently

NA: Indicates that the parameter is not addressed by the UAV regulations office. RPA owned and operated by the National Technical Research Organisation (NTRO), Aviation Research Centre (ARC) and Central Intelligence Agencies do not require permission. However, these agencies must intimate the concerned authorities prior to the conduct of actual operations.

The UAOP will have to be issued by DGCA within seven working days of submission of the necessary documents. These UAOPs are not transferrable and shall be applicable for not more than five years. All RPA operations will have to be approved by Digital Sky Platform. The Digital Sky Platform is a unique unmanned traffic management (UTM) system which facilitates registration and licensing of RPA and operators in addition to giving instant (online) clearances to operators for every flight. RPAs will only be allowed to fly during the day time and within the "visual line of sight". One cannot fly RPA near "permanent or temporary Prohibited, Restricted, and Danger Areas" and eco-sensitive zones.

As far as safety/security requirements are concerned, the operator is responsible for the safe custody of the RPAS. In case of loss of RPA, the owner/operator must report immediately to the local police office. The operator shall be responsible for notifying any incident. If an RPA is damaged and cannot be restored to original condition, the same shall be notified to DGCA by the owner/operator for cancellation of UIN. The RPAS operator must ensure that all security measures as mentioned in the security program are in place before operation of each flight. Any changes in the contact details specified in UIN shall be immediately notified to DGCA and all other concerned agencies [5].

9 Proposed Regulations (Drone Policy 2.0)

The implementation of national drone policy and the launch of the DigitalSky Platform under Civil Aviation Regulations ("CAR") 1.0 laid down the foundation for UAS operations in India. Under the chairmanship of Hon'ble Minister of State for Civil Aviation, Jayant Sinha, Ministry of Civil Aviation had constituted a task-force called the *drone task force* to provide further recommendations for Civil Aviation Regulations ("CAR") 2.0. In January 2019, the committee has presented a drone ecosystem policy roadmap on CAR 2.0. This drone ecosystem policy roadmap would allow commercial usage of drones in India. Explaining the emerging scenario, the then Minister of State for Civil Aviation said, "India is set to become a global leader as far as the drone ecosystem is concerned. It is important for us to have a policy road map and regulations that support the growth of the drone ecosystem" [2].

Various recommendations are given in drone ecosystem policy roadmap cover following areas/aspects:

- Beyond the visual line of sight operations
- Autonomous Operations
- Drone Corridor
- Airworthiness

- UAS Traffic Management
- DigitalSky Service Providers
- Pilot Training
- Droneports
- Payload/Cargo
- Make in India
- Insurance.

Drone Policy 2.0 recommends the expansion of UAS operations beyond the visual line of sight (BVLOS). It will open the barrier for the commercial use of drones. Drone Policy 2.0 proposes infrastructure like droneports, drone corridors and UAS Traffic Management (UTM). The policy seeks to establish **droneports, an area dedicated to facilitating take-off and landing of the drones. Policy proposes** drone Corridor, segregated airspace, to keep commercial drone operations away from manned aircraft airspace. **UAS Traffic Management (UTM)** should be established which would be responsible for managing UAS induced traffic, especially in drone corridors. The draft policy proposes 100% foreign direct investment (FDI) under automatic route in RPAS-based commercial civil aviation services. Under Drone Policy 1.0, there is no mention of FDI [7].

10 Future Trends and Challenges

The present state of UAV regulation and policies are good enough to take care of present needs, as DGCA proposes many developments in policies. The demand of UAVs in India will increase in the coming year as predicted need further development in policies will be necessary such as pilot licensing, air traffic management, rules on privacy and data protection needs to be further developed. Also, there is some lack of awareness needs to encounter with some awareness and training workshops. Although DGCA in their CAR 2.0 mentioned about restricted areas. Hence this definition does not give guidelines for identifying the area in cities, outskirt of cities in term of minimum/maximum heights, lateral distances.

There are also some other commercial uses of drone introduced by various startups like logistic service, delivering food and beverages and many more, that need to be properly taken care of because as market increases there will be increased in the number of players which requires policies too.

Policies for damage to physical objects due to UAVs crashes, injuries to people, bird, and animals also need to be reviewed. A collision between different UAVs needs to be taken into account.

The government needs to be accountable and spread awareness about the advantages of UAVs among people. Including some basics about UAVs in the curriculum, and drafting rules and policies in different local languages will help people to learn faster.

11 Conclusion

Between the introduction of draft regulation by DGCA in April 2016 and the implementation of National Drone Policy from 1st December 2018, there was a great deal of uncertainty about the future of UAV use in India. Now that the DGCA has implemented the National Drone Policy, it will help in shaping the technological future of Indian society.

Considering India's varied topographic and demographic profile, it is important that the policy focus on the commercial use of UAS remains sensitive to the factors unique in the Indian context. This would not only help in effective implementation of the proposed policy but also help in shaping the technological future of the country. The proposed regulations (Drone Policy 2.0) should be integrated with Drone Policy 1.0 as the current policy is a general framework while the proposed one focuses on commercial application/use of drones.

References

1. Stöcker C, Bennett R, Nex F, Gerke M, Zevenbergen J (2017) Review of the current state of UAV regulations. Remote Sens 9:459. https://doi.org/10.3390/rs9050459
2. MoCA (2019) Drone ecosystem policy roadmap. In: Global Aviation Summit, Mumbai, India. 15–16 Jan 2019
3. International Civil Aviation Organization. https://www.icao.int/Pages/default.aspx. Accessed on 10 June 2019
4. JARUS (n.d.) Regulations. http://jarus-rpas.org/regulations/. Accessed on 14 June 2019
5. DGCA (2018) Requirements for operation of civil Remotely Piloted Aircraft System (RPAS). File no. 05-13/2014-AED, office of the Director General of civil aviation. http://dgca.nic.in/cars/D3X-X1.pdf. Accessed on 17 May 2019
6. Directorate General of Civil Aviation. http://www.dgca.nic.in/. Accessed on 11 June 2019
7. MoCA (2018) Digital sky platform launched. press releases: Digital Sky. https://digitalsky.dgca.gov.in/. Accessed on 12 June 2019
8. Federal Register of Legislation (2019) Civil aviation safety regulations 1998 vol. 1. Available at https://www.legislation.gov.au/Details/F2019C00314. Accessed on 11 May 2019
9. Herrmann M (2017) A comparison of Unmanned Aerial Vehicle regulations in the United States and Europe. In: Proceedings of 53rd ASC Annual International Conference, Seattle, Washington, pp 299–306
10. Current Law on Drones and UAV's in Japan. http://dronelawjapan.com/. Accessed on 11 May 2019
11. Drone Training, Industry News, and Free Resources. https://uavcoach.com/. Accessed on 12 June 2019

Multi Frequency Polarimetric Decomposition of UAVSAR Data

Udit Asopa and Shashi Kumar

Abstract UAVSAR is an airborne SAR System which operates in various frequencies of microwave. The UAVSAR is developed by NASA/JPL. UAVSAR data has quad pol capability. Quad pol SAR data or fully polarimetric SAR data (having polarization channels as HH, HV, VH and VV) has capability of distinguishing the geographical features. Different geographical features behave differently for different wavelengths such as for P Band, L Band, S Band, etc. This study is focused on the differentiation of the scattering behavior of different object under different wavelengths. In this research PolSAR Data have been used to characterize the scattering behavior of the objects. It has been found that model-based decomposition techniques provide result of different scatterers according to the mathematical model used in the approach. This causes variation in the scattering values for the same features using decomposition modelling. To overcome this problem, roll invariant parameters have shown their potential over decomposition model to get unique scattering characteristics of the targets. In this research we have worked upon the data of P band and L band dataset of the UAVSAR for the same geographical location located near Candle Lake in Canada. The feature like water body is clearly visible in both the dataset while the vegetation i.e. forest patch is clearly visible in the L band due to the less penetration of the L Band EM wave compared to P Band and the sub-canopy features are better distinguishable in the P Band dataset because of its ability to penetrate through the canopy of the tree.

Keywords SAR · Polarimetric decomposition · Roll invariant parameters · H alpha decomposition

U. Asopa (✉) · S. Kumar
Indian Institute of Remote Sensing, Dehradun 248001, India

S. Kumar
e-mail: shashi@iirs.gov.in

K. Jain et al. (eds.), *Proceedings of UASG 2019*, Lecture Notes in Civil Engineering 51,
https://doi.org/10.1007/978-3-030-37393-1_28

1 Introduction

Remote sensing is done to analyze and map the geographical features, monitor their change in state. Microwave remote sensing is a mode of remote sensing which uses employs the use of microwaves for the target mapping. The microwave backscattering is dependent on the dielectric constant of the target [1]. The targets like vegetation, waterbody, soil, etc. have different dielectric constant values due to which they have different amount of backscattering. This phenomena of different type of target giving different type of return is the basis of polarimetric decomposition [2]. The SAR data having all channels of polarization available is represented through the scattering matrix [2]. Eigenvector decomposition of a covariance matrix [C] and coherency matrix [T] were introduced in [3]. The method to decompose the target based on the scattering model of three component was given by Freeman and Durden [4] and 4 component method was provided by Yamaguchi et al. in [5]. The Polarimetric decomposition can be utilized in every sector such as forest characterization, geo-hazard study [6], ice and water study, etc.

With time the new platform for SAR data acquisition has been developed which is the Airborne SAR i.e. the sensor is flown over the area in a confined aircraft. Various space agencies use and organizations employs this platform for data acquisition. In this article, the dataset used are acquired with the UAVSAR (Unmanned Aerial Vehicle—Synthetic Aperture Radar) of NASA-JPL.

2 Study Area

The study area chosen to carry out the research is located in the Saskatchewan State of Canada Country. The Study area chosen, is surrounding the candle lake. The candle lake is located at the 53° 49′ 30″ N, 105° 17′ 40″ W.

Figure 1 displays the extent of the employed UAVSAR datasets laid over the sentinel 2 FCC satellite image of same area. Out of the full extent, the study site is displayed in Fig. 2 which lies in both the datasets and is near the tooth-shaped lake in between the co-ordinates (53° 51′ 56.47″ N, 105° 19′ 50.60″ W) and (53° 47′ 36.05″ N, 105° 11′ 57.65″ W). This whole area is in Saskatchewan Province. This province has abundant forest resources and approximately 44% of area is classified as forest [7]. Primarily trees of this area are used in the timber industry. The main tree species present in this area are "*Populus tremuloides Michx.*", "*Pinus banksiana Lamb.*", "*Picea glauca (Moench) Voss*", "*Picea Mariana (Mill.) B.S.P.*", etc. The forest of this region can be classified as displayed in Fig. 3.

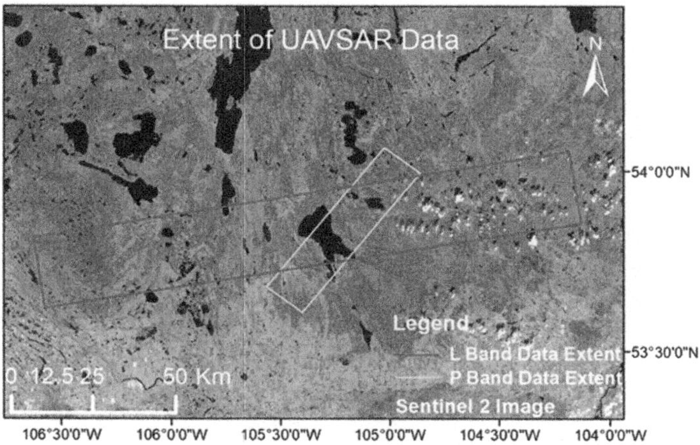

Fig. 1 The extent of UAVSAR L band and P band data laid over the Sentinel 2 satellite image

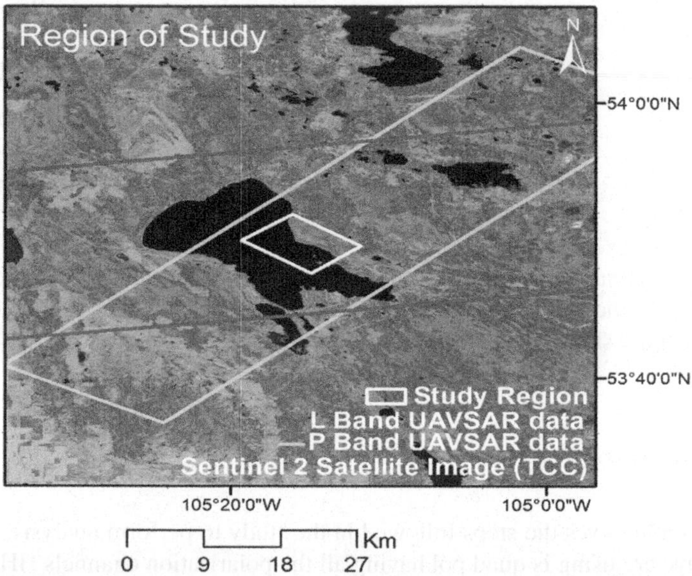

Fig. 2 Location of study region

3 Dataset

In this study, the employed dataset is acquired with the NASA JPL's UAVSAR in 2017. Two of the acquired datasets are used in this study, one of L Band and other is of P Band. Table 1 displays the dataset properties such as the frequency, date of acquisition, flight heading direction, etc.

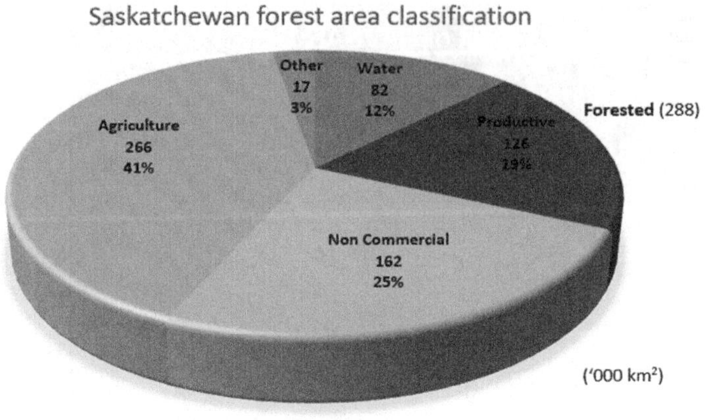

Fig. 3 The forest area classification of Saskatchewan province. *Source* [7]

Table 1 Properties of datasets

Property	Dataset 1	Dataset 2
Band	L	P
Frequency	1260.16 MHz	430 MHz
Wavelength	23.79 cm	68 cm
Date of acquisition	8 Sep, 2017	7 Aug, 2017
Polarization channels	Quad Pol	Quad Pol

Figure 4 displays the L Band dataset of the full extent and Fig. 5 shows the P band image of the full extent (area). The displayed images are pauli decomposed. Figures. 6 and 7 shows the subset image of both dataset of the study area.

4 Methodology

This section involves the steps followed in the study to perform analysis. The SAR data that we are using is quad pol having all the polarisation channels HH, HV, VH and VV (Fig. 8).

Since all the polarization channels are present, the [S] matrix of each pixel of the employed SAR data can be made such as

$$S = \begin{bmatrix} S_{HH} & S_{HV} \\ S_{VH} & S_{VV} \end{bmatrix} \tag{1}$$

In a monostatic system of acquisition, the cross pol components of the [S] matrix are same i.e. "$S_{HV} = S_{VH}$" [8]. Due to this condition, only three distinct complex

Fig. 4 L band UAVSAR data (Pauli decomposed)

Fig. 5 P band UAVSAR data (Pauli decomposed)

Fig. 6 Subset of study region of L band dataset

Fig. 7 Subset of study region of P band dataset

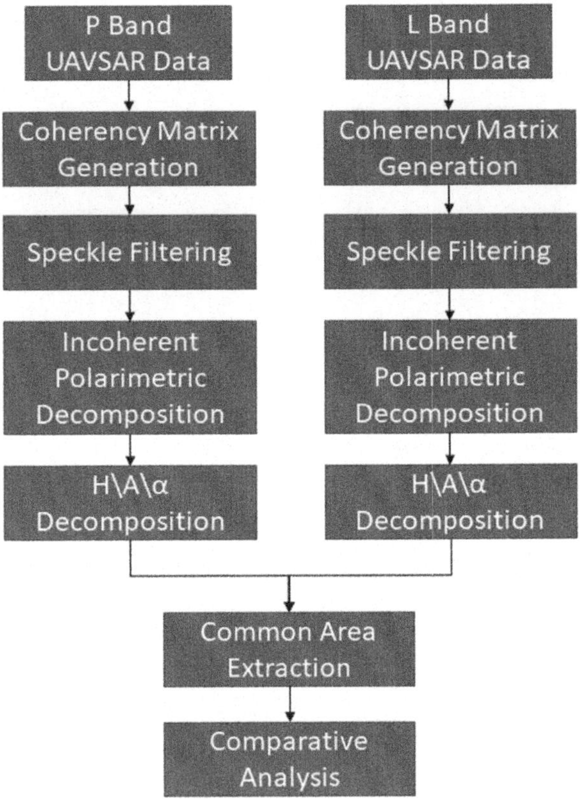

Fig. 8 Methodology chart

components of the S matrix. Thus making a scattering vector of coherent type of these elements using the linear polarization basis.

$$\vec{K} = \begin{bmatrix} S_{HH} \\ \sqrt{2}\, S_{HV} \\ S_{VV} \end{bmatrix} \tag{2}$$

The multilooked 3 × 3 hermitian covariance matrix for this vector is

$$C = \left\langle \vec{K} \vec{K}^{\mathrm{T}} \right\rangle \tag{3}$$

Here, < > denotes the multilook averaging and ^{T} denotes the transpose of K matrix and complex conjugate. The $\sqrt{2}$ is applied here to check the consistency in the total power calculation [2]. The pauli basis is selected for the making of the coherent scattering vectors.

$$\vec{K}_P = \frac{1}{\sqrt{2}} \begin{bmatrix} S_{HH} + S_{VV} \\ S_{HH} - S_{VV} \\ 2 * S_{HV} \end{bmatrix} \tag{4}$$

The coherency matrix can be made as

$$\langle [T] \rangle = \left\langle \vec{K}_P \vec{K}_P^{\mathrm{T}} \right\rangle \tag{5}$$

The eigenvectors and eigenvalues can be employed to generate the diagonal form of coherency matrix, thus calculating them is important. The eigenvalues of the coherency matrix have physical significance [9]. The coherency matrix can be written in form of

$$\langle [T] \rangle = [U_3][E][U_3]^{-1} = \sum_{i=1}^{i=3} \lambda_i u_i u_i^{*\mathrm{T}} \tag{6}$$

Here, [U] is the eigenvector matrix and [E] is eigenvalue matrix. [U] = [u_1 u_2 u_3] is the unitary matrix in which u_1, u_2 and u_3 are Eigen-vectors. The matrix [E] is the diagonal matrix of elements λ_1, λ_2 and λ_3. The matrix [U] is

$$[U_3] = \begin{bmatrix} \cos\alpha_1 & \cos\alpha_2 & \cos\alpha_3 \\ \sin\alpha_1 \cos\beta_1 e^{i\delta_1} & \sin\alpha_2 \cos\beta_2 e^{i\delta_2} & \sin\alpha_3 \cos\beta_3 e^{i\delta_3} \\ \sin\alpha_1 \cos\beta_1 e^{i\gamma_1} & \sin\alpha_2 \cos\beta_2 e^{i\gamma_2} & \sin\alpha_3 \cos\beta_3 e^{i\gamma_3} \end{bmatrix} \tag{7}$$

There are four variables α, β, δ and γ. Out of the 4 variables two are from eigenvalues and two are from eigenvectors. From eigenvalues the variables obtained are anisotropy (A), entropy (H) and from eigenvectors the variables obtained are angles

α and β. The angle *alpha* (α) is the measure of scattering mechanism which ranges between 0° and 90° while angle *beta* (β) is the measure of physical orientation of the object about the line of sight and it falls between 0° and 180° [10]. The value of these parameters can be calculated as follows:

$$\alpha = \sum_{i=1}^{i=3} P_i \alpha_i \tag{8}$$

$$\beta = \sum_{i=1}^{i=3} P_i \beta_i \tag{9}$$

here, P_i is the probabilities obtained from the eigenvalues λ_i with values ranging between 0 and 1.

$$P_i = \frac{\lambda_i}{\sum_{i=1}^{i=3} \lambda_i} \tag{10}$$

The Entropy is defined from the logarithmic sum of eigenvalues of the coherency matrix:

$$H = -\sum_{i=1}^{i=3} P_i log_3(P_i) \tag{11}$$

The randomness of the scattering is represented by the entropy H. Low value of H represents the "single scattering mechanism" (isotropic) while high value represents the "random mixture of scattering mechanisms with equal probability" and hence a depolarizing target. The parameter A (anisotropy) is complementary to the H (entropy). The high value of anisotropy represents the "two dominant scattering mechanism" with even probability and a low significant remaining mechanism, while low value means two non-negligible secondary mechanisms and dominant first scattering mechanism with equal importance.

$$A = \frac{\lambda_2 - \lambda_3}{\lambda_2 + \lambda_3} \tag{12}$$

5 Results

For Polarimetric Decomposition H-A/Alpha decomposition was selected, which consists roll invariant parameters. These parameters were calculated in the process. Entropy-alpha (H-α) plane was made via plotting the scatter plot of calculated alpha across the entropy. The Entropy-Alpha plane consists of 9 zones which denotes various type of scattering mechanism found on the ground. Figure 9 shows the Entropy-Alpha plane of the P band dataset and Fig. 10 shows the Entropy-Alpha plane of L band dataset.

Since in Figs. 9 and 10 we can easily visualize the difference between the datasets. The alpha value employed in the making of scatter plot is displayed in Figs. 11 and 12 respectively of dataset P band and L band.

The value of alpha for any pixel gives the information of its scattering mechanism thus by evaluating the alpha value we can estimate the type of scattering. As explained earlier higher value of alpha (alpha > 60) means the double bounce type of scattering, value between 30 and 60 refers to the volume scattering kind of mechanism and rest falls into the category of surface scattering.

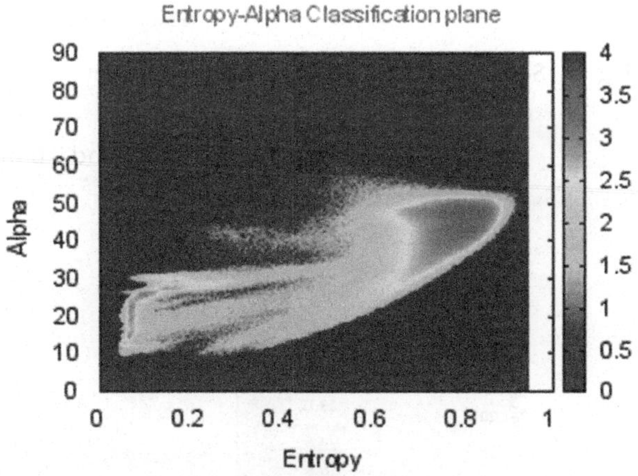

Fig. 9 H-alpha plane of P band dataset

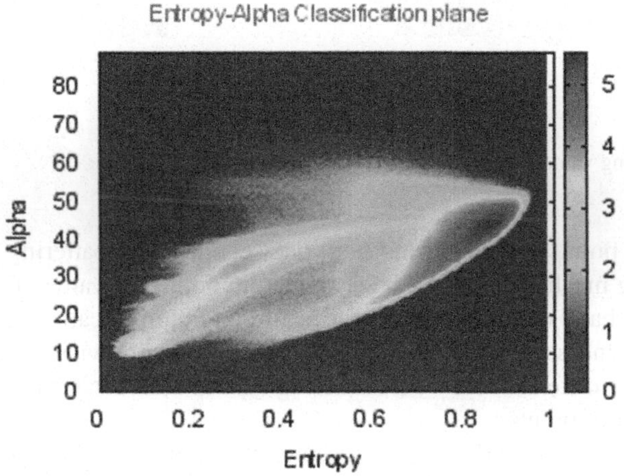

Fig. 10 H-alpha plane of L band dataset

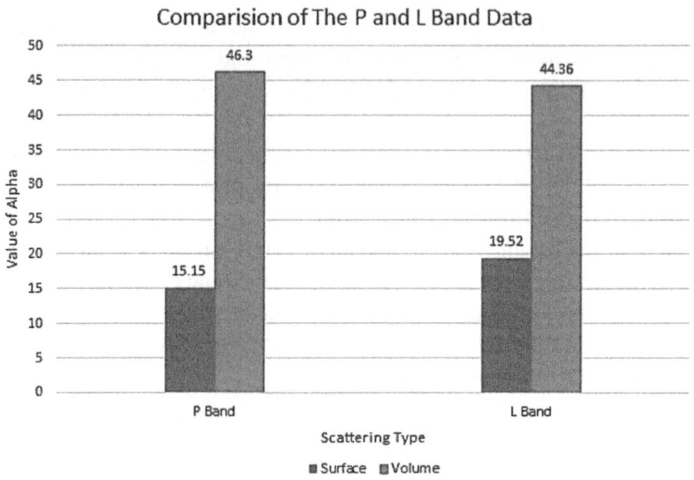

Fig. 11 Scattering value scatterplot for 1st set of point

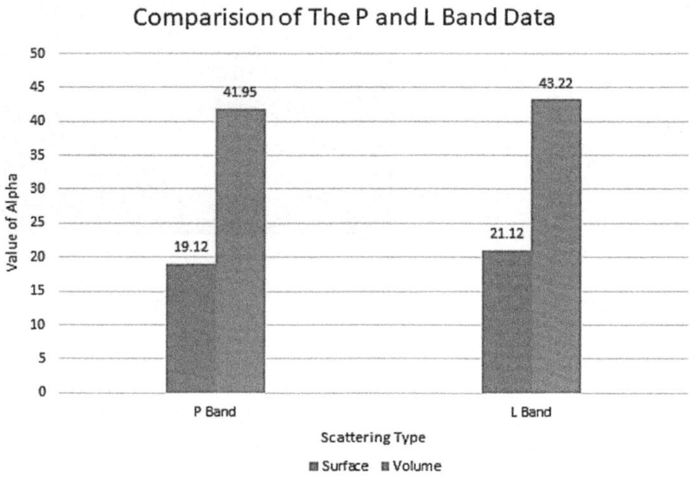

Fig. 12 Scattering value scatterplot for 2nd set of points

Two set of points are taken into both datasets and the scattering is analyzed in both scattering mechanism. For 1st set of feature point, the value of surface feature is low in the P band than L band and for feature of volume scattering type the value is higher. For 2nd set of feature point, the same pattern is visible. Tables 2 and 3 shows the value of feature for both scattering type in both dataset for 1st set of point and for 2nd set of points.

The above Chart were developed based on the value of alpha. From these charts it can easily be seen that the P band is giving better response for surface type of

Table 2 Scattering mechanism wise alpha value for both dataset for 1st point set

Scattering type	P band	L band
Surface	15.15	19.52
Volume	46.3	44.36

Table 3 Scattering mechanism wise alpha value for both dataset for 2nd point set

Scattering type	P band	L band
Surface	19.12	21.12
Volume	41.35	43.22

scattering mechanism while L band is giving good response for volume type of scattering mechanism.

Acknowledgements For this study authors are grateful to the Indian Institute of Remote Sensing for providing the Lab facility. The authors are deeply grateful to the NASA JPL for providing the dataset for research purpose.

References

1. Singh RP (1984) Dielectric properties and microwave remote sensing. Adv Sp Res 4(11):97–101
2. Boerner WM, Mott H, Luneburg E (2002) Polarimetry in remote sensing: basic and applied concepts, pp 1401–1403
3. Cloude R, Pottier E (1997) An entropy based Classification scheme for.pdf 35(1):68–78
4. Freeman A, Durden SL (1998) A three-component scattering model for polarimetric SAR data. IEEE Trans Geosci Remote Sens 36(3):963–973
5. Yamaguchi Y, Moriyama T, Ishido M, Yamada H (2005) Four-component scattering model for polarimetric SAR image decomposition. IEEE Trans Geosci Remote Sens 43(8):1699–1706
6. Yonezawa C, Watanabe M, Saito G (2012) Polarimetric decomposition analysis of ALOS PALSAR observation data before and after a landslide event. Remote Sens 4(8):2314–2328
7. Canada-Saskatchewan Partnership Agreement in Forestry (1994) Saskatchewan's forests, p 15
8. Ulaby FT, Elachi C (1990) Radar polarimetry for geoscience applications. Artech House
9. Cloude SR (1996) A review of target decomposition theorems in radar polarimetry. IEEE Trans Geosci Remote Sens 34(2):498–518
10. Ouarzeddine M, Souissi B, Belhadj-Aissa A (2006) Target detection and characterization using H/alpha decomposition and polarimetric signatures (1):395–400

Analyzing the Effect of Distribution Pattern and Number of GCPs on Overall Accuracy of UAV Photogrammetric Results

Basant Awasthi, Shashank Karki, Pratikshya Regmi, Deepak Singh Dhami, Shangharsha Thapa and Uma Shankar Panday

Abstract Unmanned Aerial Vehicle (UAV) based photogrammetry is becoming a valuable source of data for topographic mapping, volume calculations, terrain mapping, and generating 3D models. However, the use of UAVs for any purpose requires basic knowledge of various flight settings. The number and distribution of Ground Control Points (GCPs) are the most crucial, therefore, the number of GCPs should be used economically. This paper discusses the accuracy of UAV based photogrammetric products in Corridor mapping and area with Undulating terrain for different sets of flight settings. Influence of number of GCPs and their distribution patterns are assessed for optimal accuracy. For the accuracy assessment of GCP distribution, various configurations of GCPs were tested. The acquired accuracy was then compared for each of these configurations and the most suitable ones were determined for each terrain type. In corridor mapping, the distribution of GCPs depends upon the length of an area with GCPs alternating each side of the linear feature, separated by an offset distance along with the feature. In our study, the optimum number of GCPs was found to be four, along with the feature being mapped. Similarly, in the area with undulating terrain, the GCPs should be established in places covering all elevations with a minimum of five GCPs in shape of a die. Our results show that distribution and number of GCP used during UAV based survey play a major role in the accuracy of Digital Surface Model (DSM) and orthomosaics.

B. Awasthi (✉) · S. Karki · P. Regmi · D. S. Dhami · S. Thapa · U. S. Panday
Department of Civil and Geomatics Engineering, Kathmandu University,
P.O. Box 6250, Dhulikhel, Kavre, Nepal

S. Karki
e-mail: shashank.karki480@gmail.com

P. Regmi
e-mail: pratiregmi78@gmail.com

D. S. Dhami
e-mail: deepakdhami18@gmail.com

S. Thapa
e-mail: shangharsha.thapa@gmail.com

U. S. Panday
e-mail: uspanday@ku.edu.np

K. Jain et al. (eds.), *Proceedings of UASG 2019*, Lecture Notes in Civil Engineering 51,
https://doi.org/10.1007/978-3-030-37393-1_29

The accuracy not only depends upon the number of GCPs but also on its distribution pattern. Therefore, the choice of suitable pattern and number of GCPs for a particular mission can help obtain results with desired accuracy as well as economic feasibility.

Keywords UAV · Photogrammetry · GCP · Accuracy · Economy

1 Introduction

Unmanned Aerial Vehicle (UAV) technology is becoming more enthusiastically employed by land surveyors for a variety of applications. The introduction of drones as a low-cost and easily accessible alternative used for conducting measurements from the air has increased its popularity in the aerial survey. Recent developments in sensor technologies and flying platforms have significantly increased UAV applications spanning over volume calculations, creating orthophotos, generating 3D models, acquiring data for Geographic Information Systems (GIS), conducting construction inspections, general mapping of terrain and much more. The UAV technology is in constant development, and new applications are arriving over time. These low-cost drones, if used properly, can generate sufficiently accurate products however it is mostly dependent upon how the missions and surveys are conducted in an optimal to achieve the possible highest accuracy.

Traditional surveying practices use instruments such as Total Station, Auto Level and DGPS. The most accurate among these is DGPS that gives accuracy up to 10 cm. To achieve the same accuracy using only the GPS and IMU technology of the UAV is not possible. Most of the UAVs are equipped with a commercial grade GPS that provides accuracy up to 2 m. Hence, each photo was taken, as well as the final output by the UAV and photogrammetric processing, will have a maximum error of 2 m [1]. This accuracy is not acceptable for most of the projects done in photogrammetric practices. To accommodate this accuracy, Ground Control Points (GCP) are used. GCPs are simple markers distributed throughout the working site, which are geographically referenced using survey instruments such as DGPS. These points are used during the processing of the images as geo-referencing points to obtain survey grade accuracies in the final product. But the question of a suitable number of GCPs for a certain land cover type and its distribution is still unanswered. In this study, we find the optimum number of GCPs and its distribution pattern for a particular area based on the errors obtained in each of the GCP configurations suitable for the project.

2 Related Work

The classical approach of bundle block adjustment for image orientation is transitioning slowly with the introduction of adjustment with GPS projection centers and inertial measurement units. These add-ons are able to determine the full exterior orientation. The combination of bundle block adjustment along with the GPS and IMUs can be used to solve various problems regarding image orientation as discussed in [2]. However, the result obtained is totally dependent upon the quality of the units used.

The combination of GPS/INS in UAV increased the accuracy of UAV to determine the point of measurement as discussed in [3]. For direct georeferencing, onboard GPS data are used while ground checkpoints are being used for evaluation, as mentioned in [4]. By doing so horizontal accuracy can be fulfilled but may not the vertical accuracy [3]. It shows the importance of indirect georeferencing, i.e. establishing the Ground Control Points (GCPs). A minimum of three GCPs is required but increasing the number of GCPs will lead to higher accuracy of the final results [5]. The optimum number of GCPs and their distribution pattern vary with the area and the type of feature to be mapped as introduced by [4, 6]. The research on area with undulating terrain and corridor mapping for the GCPs has undergone but not for the common purpose. So, it seems essential to perform research on these two sites.

3 Methodology

In this study, the methodology is divided into several phases, namely, data preparation, data collection, preprocessing and result and discussion. Figure 1 shows the workflow of the methodology used in the study.

Among the different parameters involved in image acquisition using a UAV, a set of configurations that are to be analyzed are prepared. When a parameter, for example, the number of GCPs is to be varied, all other parameters such as image overlap, UAV velocity, the height of flight etc. are kept the same. The analysis is done over different land cover types. Therefore two different sites are taken, and the acquisition of images, processing, and analysis for each is done exclusively. For the acquisition of the image, during flight planning, the flight parameters viz. overlap, the height of the flight, camera velocity, etc. are properly set such that it covers all permutations of a number of GCPs and distribution configurations. The images are acquired such that appropriate Ground Sampling Distance is obtained. The position of Ground Control Points is to be determined using any of the Ground-based survey methods. These GCPs are to be used later for geo-referencing the images. The images are then processed in either of the commercial or open software available in the market. These software match tie points, make a mosaic of the photos, geo-reference the photo, and finally produce output in the form of orthomosaic or Digital Surface Model. The quality reports for each configuration are also obtained, which are later used for analysis.

The Root Mean Square Errors (RMSE) in the obtained accuracy reports are evaluated and visualized using graphs. Also, the point clouds obtained from the processing software are subjected to analysis using image processing software to obtain a qualitative representation of the differences among different point clouds.

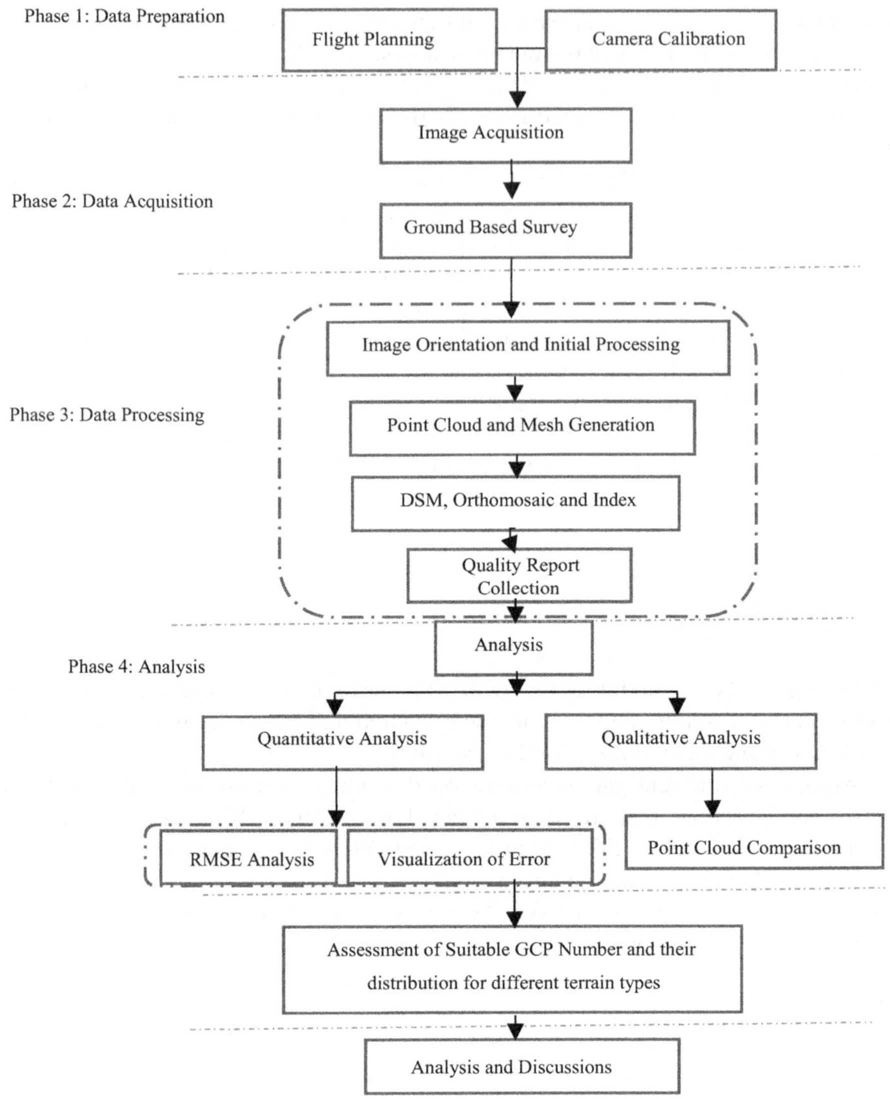

Fig. 1 Research methodology

4 Workflow

The major land cover types for aerial photogrammetric mapping to be concerned with were found as Corridor Mapping and Mapping area with undulating terrain (Table 1).

Table 1 Sites used for study

Site	Location	Area (m^2)
Site 1: Corridor	Punyamata river, Panauti	108,400
Site 2: Area with Undulating terrain	Kathmandu university, Dhulikhel	368,500

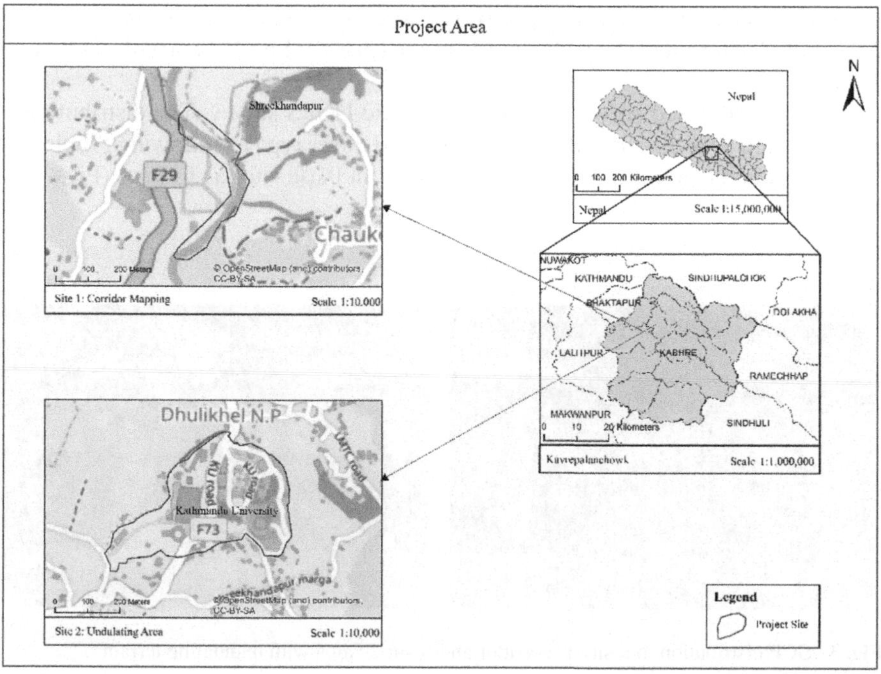

Fig. 2 Project area* (*scales are appropriate for A4 (8.27″X11.69″) paper only)

Project Area
See Fig. 2.

4.1 Number of GCPs and their Distribution Pattern

Map Pilot application was used for flight planning during image acquisition. The height of flight, overlap, flight mission and area were manually set in the application for each site. DJI Phantom 3 Advanced was used for image acquisition. The camera lens used was of FOV 94° 20 mm (35 mm format equivalent) f/2.8 focus at ∞ and the image size was 4000 × 3000. The images were acquired with an overlap of 80/60 for both the sites, height of flight was 40 and 50 m for Site 1: Corridor and Site 2:

Table 2 Number of control points: ground control points and check points

Site	Total ground stations	Ground control points	Check points
Site 1: corridor	15	3–9	6
Site 2: area with undulating terrain	20	3–9	11

Area with Undulating terrain respectively and Ground Sampling Distance (GSD) of 2.03 cm/0.8 in for Site 1: Corridor and 4.08 cm/1.6 in for Site 2: Area with Undulating Terrain.

The Ground-based survey was done by static DGPS survey. The distribution of control points was distributed according to land cover type as following (Table 2).

The Ground Control points were distributed in these sites as follows (Figs. 3, 4, and 5).

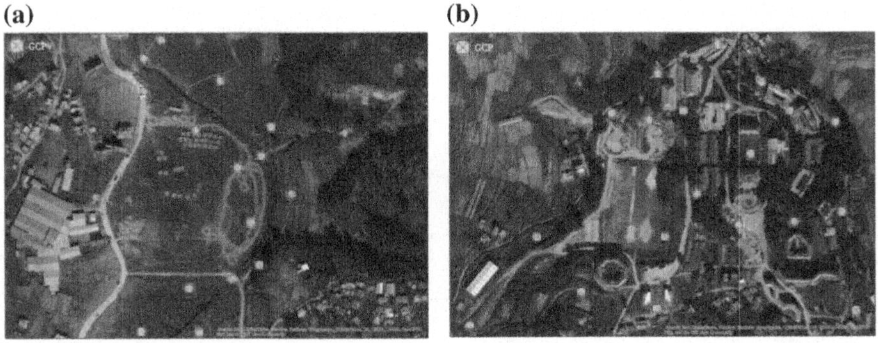

Fig. 3 GCP distribution in **a** site 1: corridor and **b** site 2: area with undulating terrain

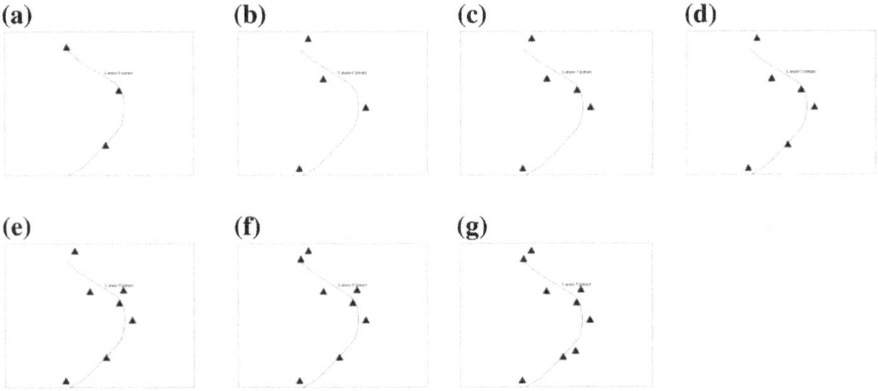

Fig. 4 Set of different GCP configurations used for (a) site 1: corridor **a** 3 GCP, **b** 4 GCP, **c** 5 GCP, **d** 6 GCP, **e** 7 GCP, **f** 8 GCP, **g** 9 GCP

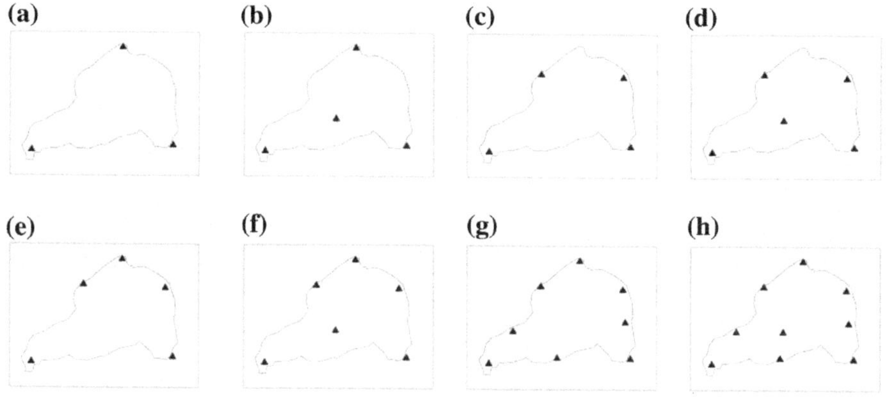

Fig. 5 Set of different GCP configurations used for (b) site 2: area with undulating terrain **a** 3 GCP, **b** 3 GCP with 1 at center, **c** 4 GCP, **d** 4 GCP with 1 at center, **e** 5 GCP, **f** 5 GCP with 1 at center, **g** 8 GCP, **h** 8 GCP with 1 at center

Furthermore, the different GCP configuration used in each of these areas were as following.

5 Preprocessing

The coordinates of these control points were determined by static DGPS survey using Stonex S8 Plus. GNSS Solutions was used for processing the raw data files. The coordinates of the control points were used as the true value for accuracy assessment.

All acquired images were processed using commercial photogrammetric software known as Pix4DMapper. The Control Points were registered and referenced using the co-ordinates obtained by DGPS survey during exterior orientation. For each of the sites, the configuration of these control points as different sets of GCPs and Check points were selected. Then, the images were processed for each of the configurations for both the sites. As, we know the maximum accuracy is obtained when the maximum number of GCPs are used. So, the Orthomosaic and Digital Surface Model for each of the site were processed with the configuration having the maximum number of GCPs. The orthomosaics and digital surface model for the sites with the configuration with the maximum number of GCPs were obtained are as shown in Figs. 6 and 7 respectively.

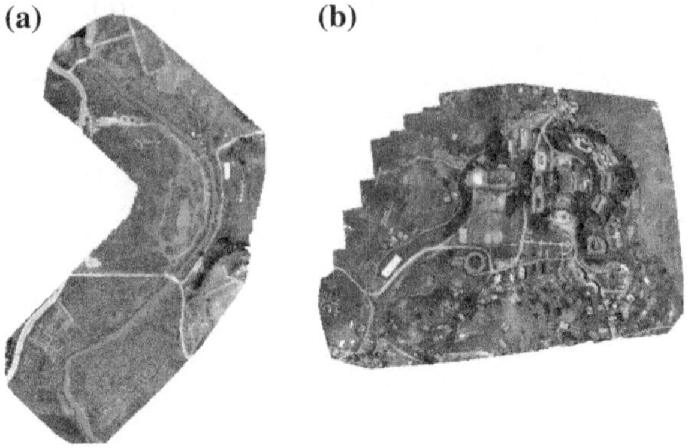

Fig. 6 Orthomosaic* of site 1: corridor (GCP configuration with 9 GCPs: 6 alternating the feature separated by an offset and 3 along the feature) and area with undulating terrain (GCP configuration with 9 GCPs: 8 along the boundary and 1 at the center) respectively

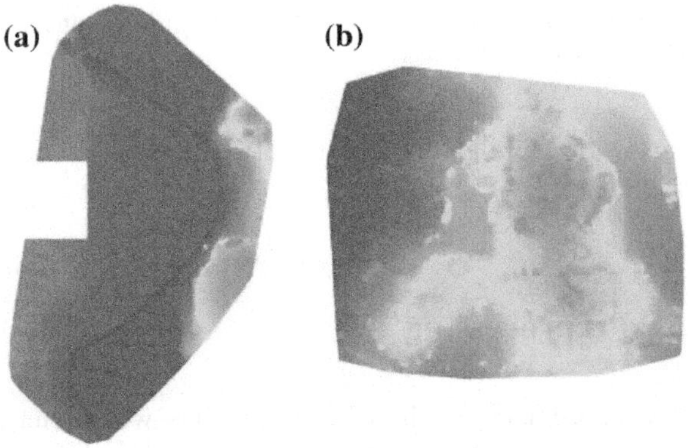

Fig. 7 Digital Surface Models* of site 1: corridor (GCP configuration with 9 GCPs: 6 alternating the feature separated by an offset and 3 along the feature) and area with undulating terrain (GCP configuration with 9 GCPs: 8 along the boundary and 1 at the center) respectively

6 Results and Discussions

This study utilized a UAV system for analysis of the accuracy of photogrammetric products when the parameters involved varied in large scale mapping of two different land cover types. The analysis of the results from the processed images was done in quantitative as well as qualitative terms.

6.1 *Quantitative Analysis*

Residual errors for each platform were examined to analyze the accuracy of the photogrammetric products. An equation to calculate the errors in the photogrammetric product is shown in the Eq. (1).

$$\epsilon = yi - \mu \tag{1}$$

where

ϵ Error
yi Ground Truth Value
μ Estimated Value.

Corridor Mapping

The site was covered with a total of 15 control points; all of which were used as ground control points and checkpoints. The errors in *X* and *Y* and *X*, *Y* and *Z* were found as follows (Table 3).

The errors in *X*, *Y* and *Z* in the above patterns were found to range from a total of 18–11 cm and errors in *X* and *Y* were found to range from 11.6 to 8.2 cm. The reduction in error was found to be slow but steady. For mapping such a linear area, the densification of GCPs increased the accuracy of the product. The addition of GCPs along the length of the linear feature increased the accuracy of the product which showed evidence in terms of decrement in RMSE value at an almost constant rate. The use of GCPs is therefore seemed dependent upon the accuracy required for a particular project. The pattern found most suitable for such a site was GCPs distribution in an alternating way on both side of the linear feature and if possible, along the body of the feature. Increasing the number of GCPs sideways of the feature was not observed to increase the accuracy by a significant amount.

Also, we could see an enhanced accuracy in *X* and *Y* after the usage of 7th GCP onwards, which might be due to the fact that the horizontal accuracy of the 7th GCP might be erroneous. However, when viewed in overall positioning (horizontal and vertical), the accuracy pattern can be seen to follow the previous pattern, further strengthening the fact that addition of GCPs resulted in an increment in accuracy.

The analysis of the different GCP configuration suggested that the accuracy of photogrammetric works increases as the number and overlap increase. However, the increment in accuracy over increment in number of GCPs showed saturation after certain limit (Figs. 8 and 9).

From the graph shown in Figs. 8 and 9, we can see the comparable RMS Errors in case of *X* and *Y*. However, it was found to have significant difference in RMS Errors in case of *Z*. With 3 GCPs along the feature, there was about 27 cm error and it

Table 3 Root mean square error (RMSE) in *X* and *Y* and in *X*, *Y* and *Z* for different GCP configuration in site 1: corridor

GCP configuration	RMS error in *X* and *Y* (m)	RMS error in *X*, *Y*, and *Z* (m)
3 GCPs	0.111	0.182
4 GCPs	0.109	0.160
5 GCPs	0.109	0.159
6 GCPs	0.082	0.136
7 GCPs	0.116	0.129
8 GCPs	0.112	0.119
9 GCPs	0.113	0.113

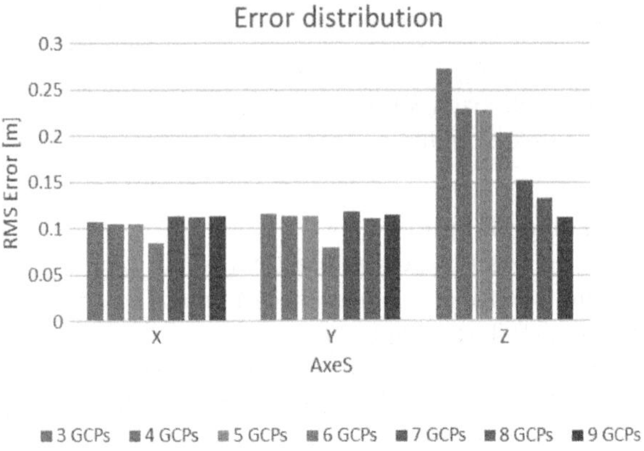

Fig. 8 Error distribution graphs generated using RMSE values for different GCP distribution in site 1: Corridor

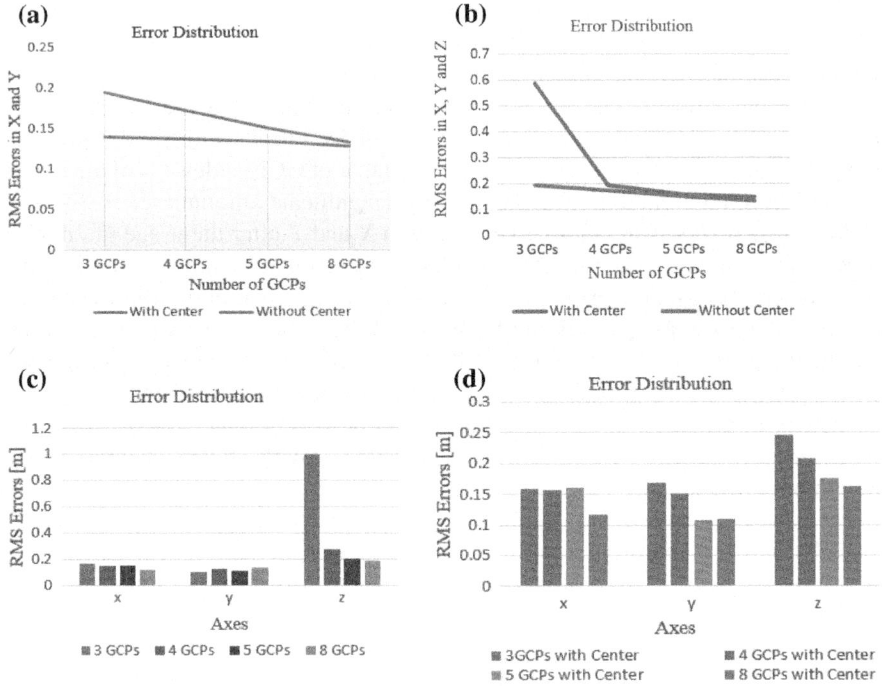

Fig. 9 Error distribution graphs generated using RMSE values for different GCP distribution in site 2: Area with Undulating terrain

goes on decreasing with increase in number of GCPs and was minimum for 9 GCPs. Therefore, the major concern for error was coined in *Z* i.e. elevation values. Even for *Z*, the decrement of error was considerable for up to a certain point, in our case 4 GCPs. On increasing the number of GCPs to 5 the decrease in error was minimal and

Table 4 Root mean square error (RMSE) in X and Y and in X, Y and Z for different GCP configuration in site 2: Area with undulating terrain

GCP configuration	RMS error in X and Y (m)	RMS error in X, Y, and Z (m)
3 GCPs	0.140	0.586
3 GCPs with a central GCP	0.139	0.194
4 GCPs	0.138	0.195
4 GCPs with a central GCP	0.134	0.156
5 GCPs	0.154	0.173
5 GCPs with a central GCP	0.137	0.151
8 GCPs	0.127	0.148
8 GCPs with a central GCP	0.114	0.132

this trend continued. Hence, for the site, 4 GCPs seemed to be adequate for obtaining accuracy comparable to the accuracy obtained by the use of higher number of GCPs.

Therefore, we could say, the optimum number of GCPs for a similar linear feature is 4 GCPs along the length of the feature in case there is a single turning along the linear feature. An additional GCP will be required for mapping a feature with an additional turning as the values of X and Y will be affected by turnings.

Area with Undulating Terrain

The site was covered with a total of 20 control points; all of which were used as ground control points and check points. The errors in X and Y and X, Y and Z were found as following (Table 4).

The errors in X, Y and Z in the above patterns were found to range from a total of 58–15 cm and errors in X and Y were found to range from 16 to 12 cm. The decrease in error follows a similar pattern of saturation of decrement of errors as the GCPs increase. On increasing the number of GCPs the pattern continued through 3 GCPs to 5 GCPs with an additional GCP at the center (6 GCPs), so the trend would be the same for higher number of GCPs. As we have a total of 9 GCPs, for selecting the reference configuration, we selected the configuration of 8 GCPs with a central GCP. The following graphs showed error distribution in all cases.

The above graphs showed the importance of positioning a GCP in the central region of the area. There was sudden decrease of error for every set of patterns when an additional GCP was added in the central area. From the Fig. 9a we could clearly see the decrease in error in Z from approx. from a unit meter to quarter of it in case of 3 GCPs with one at Center. The errors in only X and Y tend to decrease in a steady manner, however when the errors in Z were also considered we could see the steepest slope in decrease in RMS Errors with only 3 GCPs configuration in Fig. 9b. Also, from Fig. 9b showing error in X, Y and Z comparing the patterns of 3 GCP with center and 4 GCP without center, we could observe more accuracy with the inclusion of GCP in central position despite constant number of GCP. This clearly revealed the importance of central GCP in distribution pattern. The finding from it was that the inclusion of central GCP decreases error gradually on increasing the number of GCPs. The optimum accuracy could be meet by placing more GCPs in the boundary if in case the placement of central GCP seems impossible. However, there would be fewer GCP requirement if we could place the central GCP.

The increment in number of GCPs would decrease the error magnitude to certain extent which after that reached the saturation stage meaning that the errors would be almost constant as shown in Fig. 9c, d. This could be interpreted as "Increasing GCPs after a certain extent has minimal effect of the accuracy of photogrammetric product". The configuration with 4 GCPs along the boundary with an additional central GCP (altogether 5 GCPs) was found to be the most efficient configuration both in terms of economy and accuracy. Hence, based on the observation, it would be wise to state that the optimum number of GCPs would be 5 assuming one being placed at the central location of the study area.

6.2 Qualitative Analysis

The configuration giving the highest accuracy was selected as the reference configuration. The point clouds processed from all the other configuration were compared with the reference configuration for each land cover type. The distance between the point clouds were calculated using open source Cloud Compare software and shown as an absolute distance map. It computes the distances of each of its points relatively to the reference cloud.

Corridor Mapping

The configuration with 9 GCPs (i.e. 6 were on the two sides of the feature such that the adjacent GCPs were alternating the linear feature separated by an offset and 3 were along the feature) was chosen as reference. The extremities of the area were found to be the major zones of error due to absence of referencing line. The resulted distance maps are represented as following.

The Fig. 10 shows absolute distance map for GCP configuration

Area with Undulating Terrain

The configuration with 8 GCPs along the boundary and an additional one at the center was chosen as reference. The man-made features of Kathmandu University at high altitudes can be seen as the major zones of error. The distance maps are shown in Fig. 11.

The green zones are erroneous zones that represent the deviation from the reference point cloud. These were mostly visible in the 3 GCP configuration mainly in the central region. This could be due to the absence of any referencing line in the central region as the line joining GCPs did not cross over this region. This was readily accommodated by an additional GCP at the center. Furthermore, usage of

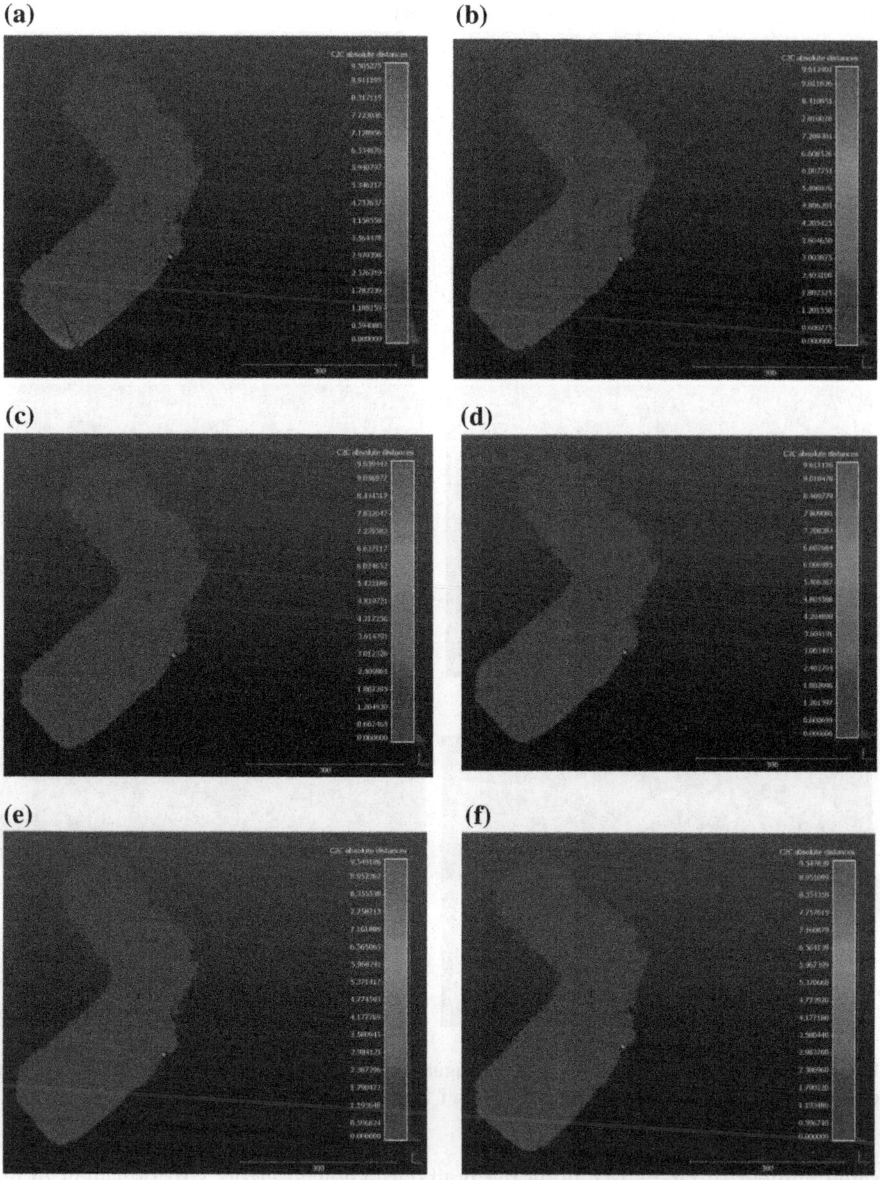

Fig. 10 Absolute distance map for GCP configuration with **a** 3 GCPs, **b** 4 GCPs, **c** 5 GCPs, **d** 6 GCPs, **e** 7 GCPs, **f** 8 GCPs along the feature

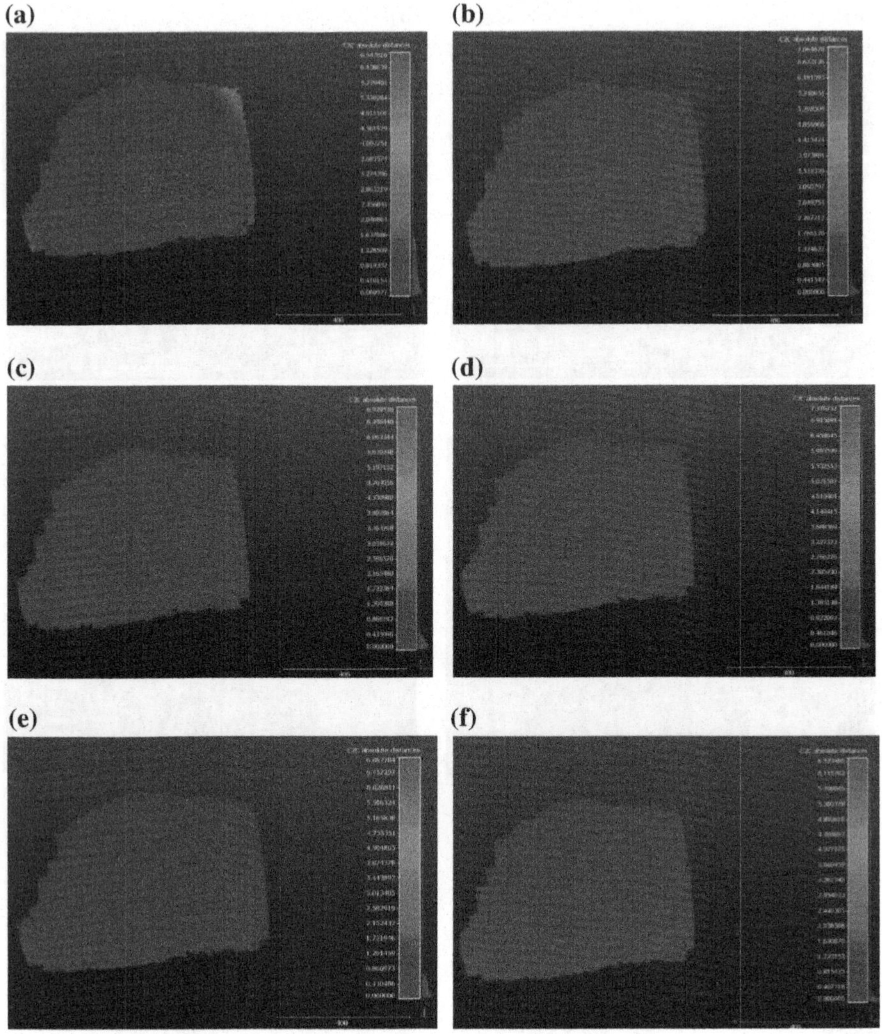

Fig. 11 Absolute distance map for GCP configuration with **a** 3 GCPs, **b** 3 GCPs and 1 at center, **c** 4 GCPs, **d** 4 GCPs and 1 at center, **e** 5 GCPs, **f** 5 GCPs and 1 at center

same number of GCPs (4) along the boundary showed errors still prevalent in the central regions. This phenomenon suggested GCP location to be more significant than its quantity. The increase in accuracy on increasing the number of GCPs could be clearly seen as the light blue zones dominated other colors in the maps with larger number of GCPs. The saturation of accuracy while adding GCPs could be seen in these maps too.

7 Conclusion

The analysis of different results obtained from various configurations of numbers and patterns and GCPs were successfully carried out. This study has demonstrated the effects of variation in accuracy of photogrammetric products for three different sites when different sets of GCPs were used in different patterns.

The GCP arrangement made in such a way that they are alternating each side of the linear feature being mapped separated by an offset distance alongside it was observed to be the appropriate one in case of corridor mapping. The inclusion of GCPs along the feature showed better suit in the case but this context might not be similar for all types of linear features such as rivers. The conducted research for this site showed a gradual increase in accuracy with the addition of a greater number of GCPs. There should be consideration of project budget and accuracy while devising the optimum number of GCPs to be used for a given area. For the case of our study area, it was noticed that the optimum number of GCPs was four along the feature. The optimum number of GCPs for any corridor mapping is the same with a couple of additional GCP for each mission with certain deviations from the preceding mission placed exactly on the mission overlap area.

The best fit arrangement of GCPs for undulating terrain followed a die shape (i.e. GCP at the central region and other GCPs well distributed along the boundary). On following this arrangement, one should be assured about the fact that the GCPs in the chosen pattern covered all elevations range.

The plot showing accuracy variations with different GCP pattern practice showed saturation on the increment in accuracy with increased number of GCPs. The saturation was seen on accuracy for a pattern with four GCPs along the boundary and one at the center. This indicated the optimum number of GCPs to be used is five (with one GCP at center and rest along the boundary) and GCPs arrangement must be made in die shape for such type of terrain.

References

1. Snavely N (2008) Scene reconstruction and visualization from internet photo collections. IPSJ Trans Comput Vis Appl 3
2. Oniga VE, Breaban AI, Statescu F (2018) Determining the optimum number of ground control points for obtaining high precision results based on UAS images. Multidiscip Digit Publ Inst Proc 2(7):352
3. Cesetti A, Frontoni E, Mancini A, Ascani A, Zingaretti P, Longhi S (2011) A visual global positioning system for unmanned aerial vehicles used in photogrammetric applications. J Intell Rob Syst 61(1–4):157–168
4. Tahar K (2013) An evaluation on different number of ground control points in unmanned aerial vehicle photogrammetric block. Int Arch Photogramm, Remote Sens Spat Inf Sci XL-2/W2(40):93–98
5. Skarlatos D, Procopiou E, Stavrou G, Gregoriou M (2013) Accuracy assessment of minimum control points for UAV photography and georeferencing. In: First international conference on remote sensing and geoinformation of the environment (RSCy2013)

6. Prajwal M, Rishab J, Vaibhav S, Karthik KS (2016) Optimal number of ground control points for a UAV based corridor mapping. Int J Innov Res Sci Eng Technol 5(9):28–32
7. Tihar KN, Ahmad A, Akib WAAWM, Mohd WMNW (2012) Assessment on ground control points in unmanned aerial system image processing for slope mapping studies. Int J Sci Eng Res 3(11):1–10

CityGML Based 3D Modeling of Urban Area Using UAV Dataset for Estimation of Solar Potential

Harikesh, Sachchidanand Singh, Vaibhav Shrivastava and Vishal Sharma

Abstract 3D GIS modelling is the latest trend in Remote sensing for urban planning, utility mapping and many more applications. Mapping of an urban area using UAV (unmanned aerial vehicle) remote sensing, gives high accuracy which was not possible through traditional sensors. Urban area, which is continuously expanding, have an urgent need to find alternative energy sources to facilitate their increasing power demand. In this regard, solar energy can prove to be a vital source. In this work, high-resolution DTM (Digital Terrain Model), prepared from RGB point clouds for the Roorkee urban area using UAV survey, is used for building height extraction and 3D visualization, shadow analysis of the buildings and solar potential estimation. The CityGML based 3D city model is generated using UAV cloud dataset. CityGML, which is based on Geographic Markup language, is an open data model for the storage and it facilitates interoperability of virtual 3D city model. 3D GIS model is prepared using Computer Generated Architectural Rule Technique and various other tools such as ESRI City Engine, ArcGIS and FME software. The building height obtained from the 3D model, was validated from ground survey and the solar potential was validated from National Renewable Energy Laboratory, solar maps obtained from https://maps.nrel.gov/nsrdb-viewer/ website. The result depicted that the present status of Roorkee urban area has a strong potential to reduce the electricity load of the city via harnessing the solar energy, thus leading to a sustainable future.

Keywords 3D modelling · DEM · UAV · Geographic markup language · CityGML

1 Introduction

The modelling of cities continued in the last decade to flourish from 2D drawings to complex 3D models and to even n-dimension. Even the unique complexities of the cities and the different interrelationships between the objects within it could be

Harikesh (✉) · S. Singh · V. Shrivastava · V. Sharma
Indian Institute of Remote Sensing, Dehradun, India
e-mail: harikeshsngh77@gmail.com

K. Jain et al. (eds.), *Proceedings of UASG 2019*, Lecture Notes in Civil Engineering 51,
https://doi.org/10.1007/978-3-030-37393-1_30

represented by a 3D model and can make entire city easier to understand [8]. 3D building models have proven there applicability for various applications like urban planning, a navigation system, visualization of heritage buildings, augmented reality, property management, and spatial data infrastructures and analysis [2, 4].

With the recent advancement in 3D modeling, the planning of smart cities has taken new trend, like the 2D modeling of water distribution pipes using EPANET [9] is now analyzed in 3D using CityGML. One such application of 3D city modeling is for finding solar potential of the area [7]. For which we require to map the building roof area that can be used for installing solar panels [10]. The amount and pattern in which sunlight falls on them and how the shadow from any obstruction will affect the solar potential can also be analyzed [5].

The main objectives of this study is to generate a 3D city model of the Khanjarpur, Roorkee for estimating the solar potential of the region. Various sub-objectives include extraction of buildings and trees heights, database storage and shadow analysis using various tools and web viewers.

CityEngine (CE) is a rule-based 3D modeling software, a product of ESRI (Environmental Systems Research Institute, Inc.). The modeling principle is described in CE using computer language called Computer Generated Architecture (CGA), so using this, 3D model of the study area is generated. For visualization of 3D model of the study area, CityEngine Web Viewer is used. Shadow analysis of the area is also done using this interface.

Feature Manipulation Engine (FME) is a solution for complex integration between different data formats. It helps in converting various applications and formats using a visual interface. This engine was used to convert the shapefile of the study area into CityGML's XML based format.

CityGML, which is based on Geographic Markup language, is an open data model for the storage, representation and analysis and exchange of semantically enriched 3D city models, it facilitates interoperability of virtual 3D city model. It is an object-oriented data model that helps in generating thorough meaningful 3D vector data out from 2D vector data, non-spatial data and 3D models. CityGML and its thematic models data are finally mapped to the relational database management system (RDBMS), called 3DCity Database. 3DCityDB relational database scheme is used to extend the PostgreSQL/PostGIS to store the CityGML data.

Use of alternative source of energy to cope with the future need of the city will ensure their sustainability [11]. We are on the verge of major energy crises and adaptability with the environment is the only way to ensure our future [3]. Before using any of the alternative energy sources we need a throw analysis of its applicability and potential in the specific area. Using the rapidly evolving trend of UAV (unmanned aerial vehicle) based remote sensing and 3D modeling this objective can be achieved very laudable.

2 Study Area and Data Used

2.1 Study Area

The study area is the Khanjarpur area of Roorkee city (see Fig. 1), India. The area is surveyed by unmanned aerial vehicle (UAV) and lies at 29° 52′0.00″ N and 77° 54′20.00″ E. The area surveyed is 6.17 ha and its 3.9 ha (63.21%) is covered by buildings which have potential to harvest solar energy. Roorkee is one such city of India which has been growing rapidly and the urbanization rate is very high. This study area represents the cities which are under rapid urbanization and will be facing big energy crises soon, if non-conventional sources of energy are not effectively utilized.

2.2 Data Used

RGB point cloud data was obtained from UAV survey conducted by Department of Civil Engineering, IIT Roorkee. The obtained RGB point cloud density was ~1 pt/m^2. RGB DTM at 11 cm resolution and DSM at 11 cm resolution was used in the process of building extraction together with high-resolution optical image at 11 cm resolution.

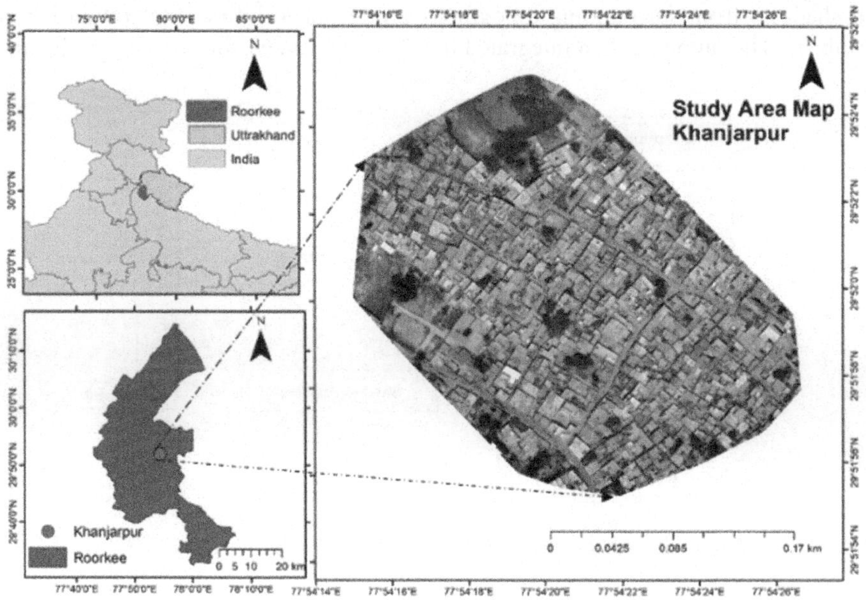

Fig. 1 Study area

3 Methodology

Firstly assigned task comprises of data gathering, data preprocessing, data creation and non-spatial data assembling. Then, organized data of building and road network components is used for data modeling using Rule-Based 3D city Model and CityGML. The methodology is shown in Fig. 2.

Data modeling procedure consists of mapping road network component, to its respective feature classes. Open Geospatial Consortium (OGC) CityGML allows 3D modeling of city objects at 5 discrete levels of details [6]. Then, using the Digital Surface Model (DSM) and Digital Terrain Model (DTM) which are obtained using point cloud data, normalized Differential Surface Model (nDSM) is created. Thereafter, a city information model geodatabase using classification and digitization is created. Using this geodatabase of point cloud data, height information of individual building and tree is generated. Computer Generated Architecture (CGA) rule-based method is used for creating Level of Detail 1 (LOD1) 3D city model of Roorkee region and this model was stored in 3DCity database in CityGML based format. The resultant 3D model with incorporated attributes is then visualized using CityEngine Web Viewer.

The shadow analysis is carried out by converting the LOD1 3D city model into City engine web scene model, which is then visualized in city engine web viewer.

The Solar potential map is then generated from the Solar Radiation Analysis tools, present in the ArcGIS Spatial Analyst extension. The atmospheric effects, slope, site latitude and elevation, aspect, daily and seasonal variation of the sun angle, and effects of shadows from surrounding topography are all taken into consideration for this analysis. The output is then integrated in GIS environment for proper visualization.

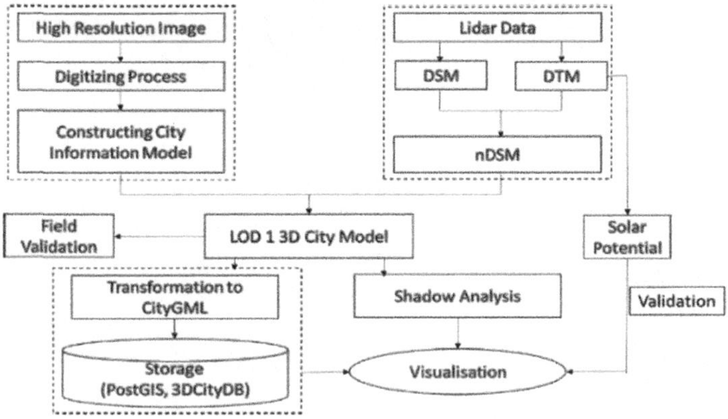

Fig. 2 Flow chart of the working methodology

4 Results and Discussion

4.1 City Information Model Geodatabase

A geodatabase containing information of building geometry, trees roads and road network of the study area is prepared with precise geo-positioning. Around 450 Buildings and 70 trees are identified in Khanjarpur, Roorkee using UAV dataset. In the current study, road network is considered under two categories major roads and minor roads on the basis of the width of roads. The Final result is shown (see Fig. 3).

4.2 Normalized Differential Surface Model (nDSM)

The nDSM (see Fig. 4) is obtained for finding the geometry of all 3 dimensional features of study area. It is prepared by subtracting DTM from DSM hence only those features which are above the ground are left as resulting image.

$$nDSM = DSM - DTM \tag{1}$$

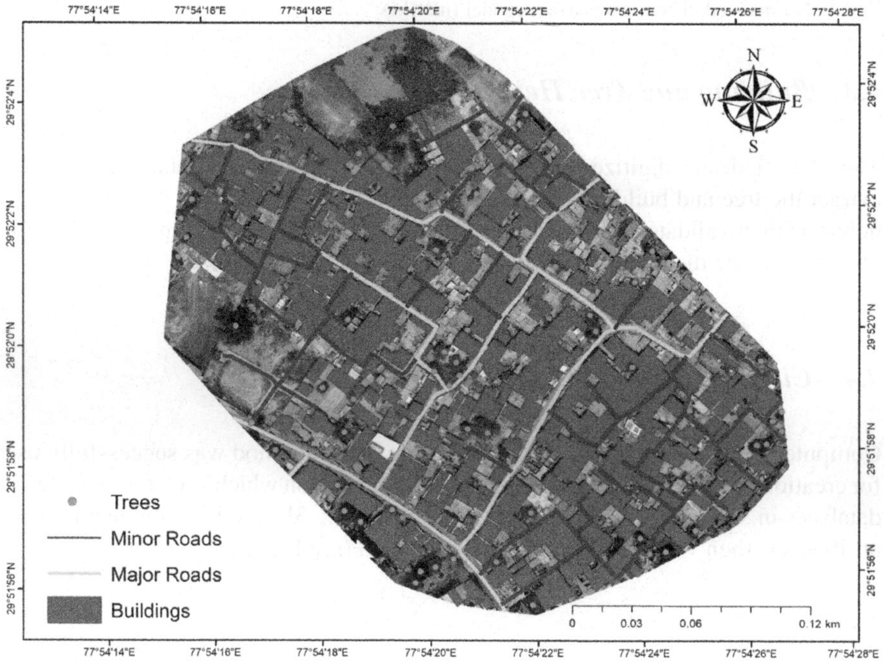

Fig. 3 Digitized buildings and road networks

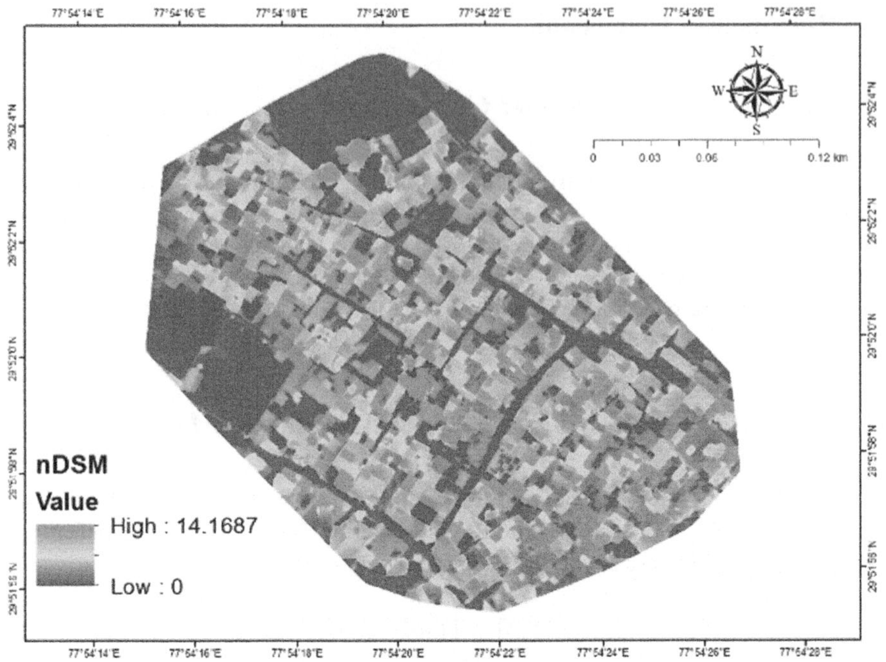

Fig. 4 Normalized differential surface model (nDSM)

4.3 Building and Tree Height Estimation

The classified and digitized city (see Fig. 5) information geodatabase is used to extract the tree and building height information (see Tables 1 and 2). The obtained height is then validated with field observations. The Building height ranged from 3 to 10.5 m while the maximum tree height obtained was 13.3 m.

4.4 CityGML Based 3D Urban City Model

Computer Generated Architecture (CGA) rule based method was successfully used for creation of LOD1 3D city model of Khanjarpur region which was stored in 3DCity database in CityGML based format. The resultant 3D model with incorporated attributes is then visualized using FME as shown (see Fig. 6).

Fig. 5 Building and tree points for height estimation

Table 1 Tree height information

S. no.	Latitude	Longitude	Obtained height (m)
1	29.8660	77.9068	3.045
2	29.8663	77.9071	3.430
3	29.8665	77.9072	3.646
4	29.8676	77.9058	5.428
5	29.8659	77.9053	5.525

Table 2 Building height information

S. no.	Latitude	Longitude	Obtained height (m)
1	29.866	77.9062	4.357
2	29.8658	77.9056	4.399
3	29.8676	77.9047	5.156
4	29.8659	77.9069	5.167
5	29.8668	77.9057	6.116
6	29.8671	77.9057	6.117
7	29.8659	77.9055	7.144
8	29.8675	77.9048	7.215
9	29.8662	77.9062	8.071
10	29.8658	77.9052	8.114

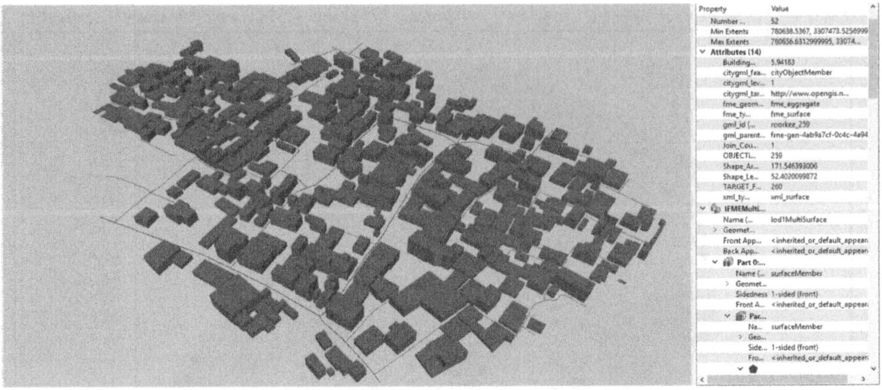

Fig. 6 CityGML based 3D urban city model of Khanjarpur, Roorkee

Fig. 7 Shadow analysis of Khanjarpur, Roorkee

4.5 Shadow Analysis

Shadow analysis of the study region was successfully carried out using CityEngine Web Viewer. The Images (see Fig. 7) analysed from 7 a.m. morning to 6 p.m. for the month of March, depicts that most of the solar radiation occurs from 10.30 a.m. to 3.30 p.m. The result is also validated from field observations.

4.6 Solar Potential Estimation

To calculate the solar potential for the roofs, the ESRI ArcGIS Solar Radiation Analysis tools was implemented, that performs calculations for entire geographical areas in four steps [1] ([CSL STYLE ERROR: reference with no printed form]):

The calculation of an upward-looking hemispherical viewshed based on topography.
Overlay of the viewshed on a direct sunmap to estimate direct radiation.
Overlay of the viewshed on a diffuse skymap to estimate diffuse radiation.
Repeating the process for every location of interest to produce an insolation map.

Global radiation (Globaltot) is calculated as the sum of direct (Dirtot) and diffuse (Diftot) radiation of all sun map and sky map sectors, respectively.

$$\text{Globaltot} = \text{Dirtot} + \text{Diftot} \tag{2}$$

Total direct insolation (Dirtot) for a given location is the sum of the direct insolation ($\text{Dir}_{\theta,\alpha}$) from all sun map sectors:

$$\text{Dirtot} = \Sigma \text{Dir}_{\theta,\alpha} \tag{3}$$

The direct insolation from the sun map sector ($\text{Dir}_{\theta,\alpha}$) with a centroid at zenith angle (θ) and azimuth angle (α) is calculated using the following equation:

$$\text{Dir}_{\theta,\alpha} = \text{SConst} * \beta m(\theta) * \text{SunDur}_{\theta,\alpha} * \text{SunGap}_{\theta,\alpha} * \cos(\text{AngIn}_{\theta,\alpha}) \tag{4}$$

where: SConst—The solar flux outside the atmosphere at the mean earth-sun distance, known as solar constant. The solar constant used in the analysis is 1367 W/m^2. This is consistent with the World Radiation Center (WRC) solar constant, β—The transmissivity of the atmosphere (averaged over all wavelengths) for the shortest path (in the direction of the zenith), m(θ)—The relative optical path length, measured as a proportion relative to the zenith path length (see Eq. 5), $\text{SunDur}_{\theta,\alpha}$—The time duration represented by the sky sector. For most sectors, it is equal to the day interval (for example, a month) multiplied by the hour interval (for example, a half hour). For partial sectors (near the horizon), the duration is calculated using spherical geometry. $\text{SunGap}_{\theta,\alpha}$—The gap fraction for the sun map sector, $\text{AngIn}_{\theta,\alpha}$—The angle of incidence between the centroid of the sky sector and the axis normal to the surface (see Eq. 6), Relative optical length, m(θ), is determined by the solar zenith angle and elevation above sea level. For zenith angles less than 80°, it can be calculated using the following equation:

$$m(\theta) = \text{EXP}(-0.000118 * \text{Elev} - 1.638 * 10^{-9} * \text{Elev2}) / \cos(\theta) \tag{5}$$

where: θ—The solar zenith angle, Elev—The elevation above sea level in meters. The effect of surface orientation is taken into account by multiplying by the cosine of the angle of incidence. Angle of incidence ($\text{AngInSky}_{\theta,\alpha}$) between the intercepting surface and a given sky sector with a centroid at zenith angle and azimuth angle is calculated using the following equation:

$$\text{AngIn}_{\theta,\alpha} = a\cos(\text{Cos}(\theta) * \text{Cos}(\text{Gz}) + \text{Sin}(\theta) * \text{Sin}(\text{Gz}) * \text{Cos}(\alpha - \text{Ga})) \tag{6}$$

where: Gz—The surface zenith angle. Note that for zenith angles greater than 80°, refraction is important, Ga—The surface azimuth angle. For each sky sector, the diffuse radiation at its centroid (Dif) is calculated, integrated over the time interval, and corrected by the gap fraction and angle of incidence using the following equation:

$$\mathrm{Dif}_{\theta,\alpha} = \mathrm{Rglb} * \mathrm{Pdif\,Dur} * \mathrm{SkyGap}_{\theta,\alpha} * \mathrm{Weight}_{\theta,\alpha} * \cos(\mathrm{AngIn}_{\theta,\alpha}) \quad (7)$$

where: Rglb—The global normal radiation (see Eq. 8), Pdif—The proportion of global normal radiation flux that is diffused. Typically it is approximately 0.2 for very clear sky conditions and 0.7 for very cloudy sky conditions, Dur—The time interval for analysis, $\mathrm{SkyGap}_{\theta,\alpha}$—The gap fraction (proportion of visible sky) for the sky sector, $\mathrm{Weight}_{\theta,\alpha}$—The proportion of diffuse radiation originating in a given sky sector relative to all sectors (see Eqs. 9 and 10), $\mathrm{AngIn}_{\theta,\alpha}$—The angle of incidence between the centroid of the sky sector and the intercepting surface. The global normal radiation (Rglb) can be calculated by summing the direct radiation from every sector (including obstructed sectors) without correction for angle of incidence, then correcting for proportion of direct radiation, which equals 1-Pdif:

$$\mathrm{Rglb} = (\mathrm{SConst}\Sigma(\beta m(\theta)))/(1 - \mathrm{Pdif}) \quad (8)$$

For the uniform sky diffuse model, $\mathrm{Weight}_{\theta,\alpha}$ is calculated as follows:

$$\mathrm{Weight}_{\theta,\alpha} = (\cos\theta 2 - \cos\theta 1)/\mathrm{Divazi} \quad (9)$$

where: θ1 and θ2—The bounding zenith angles of the sky sector, Divazi—The number of azimuthal divisions in the sky map, For the standard overcast sky model, $\mathrm{Weight}_{\theta,\alpha}$ is calculated as follows:

$$\mathrm{Weight}_{\theta,\alpha} = (2\cos\theta 2 + \cos 2\theta 2 - 2\cos\theta 1 - \cos 2\theta 1)/4 * \mathrm{Divazi} \quad (10)$$

Total diffuse solar radiation for the location (Diftot) is calculated as the sum of the diffuse solar radiation (Dif) from all the sky map sectors:

$$\mathrm{Diftot} = \Sigma\mathrm{Dif}_{\theta,\alpha} \quad (11)$$

Results from the analysis (see Table 3) shows the effects of the solar radiation for the study area for specified time periods, i.e. from Jan 1 to June 30. The value of solar radiation ranged from 152.82 to 180.35 W-h/m^2 with average value of 171.09 W-h/m^2, which is validated from National Renewable Energy Laboratory, National Solar Radiation Database (NREL-NSRDB) obtained from https://maps.nrel.gov/nsrdb-viewer/ website. This result obtained is then analyzed in Arc Scene (see Fig. 8).

The average value of solar potential obtained from NREL-NSRDB (see Table 4), estimated from 2010 to 2014 was 178.23 W-h/m^2. which was close to the derived value 171.09 W-h/m^2.

Table 3 Solar potential estimation obtained from ESRI ArcGIS

S. no.	Latitude	Longitude	Solar potential (ArcGIS) (W-h/m^2)
1	29.8659	77.9069	152.81
2	29.8664	77.9063	167.43
3	29.8658	77.9052	169.47
4	29.8675	77.9048	169.98
5	29.8676	77.9047	171.12
6	29.8671	77.9057	171.19
7	29.8656	77.9063	173.86
8	29.8662	77.9062	173.96
9	29.8658	77.9056	175.28
10	29.8658	77.9056	175.73

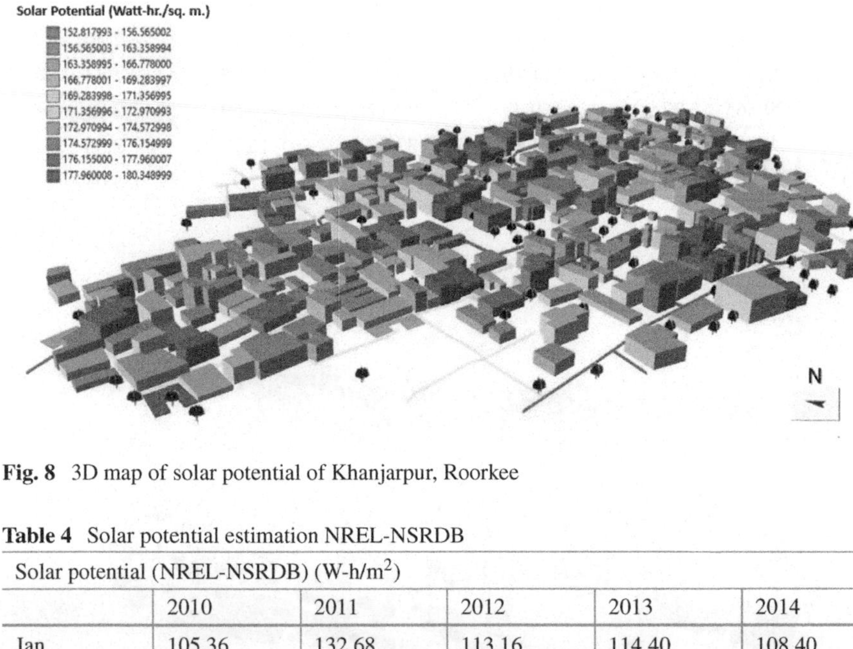

Fig. 8 3D map of solar potential of Khanjarpur, Roorkee

Table 4 Solar potential estimation NREL-NSRDB

Solar potential (NREL-NSRDB) (W-h/m^2)					
	2010	2011	2012	2013	2014
Jan	105.36	132.68	113.16	114.40	108.40
Feb	197.29	163.95	151.61	132.30	137.31
March	229.57	224.17	179.80	230.92	194.96
April	215.60	214.53	216.12	227.24	243.93
May	188.14	206.12	215.02	213.24	221.48
June	172.54	118.78	188.69	116.47	173.00
Average	**184.75**	**176.71**	**177.40**	**172.43**	**179.85**

Building wise solar energy (see Table 5) was calculated by considering effective area of each building to be 50%. The calculated energy ranged from 1.04 kW-h to 9.59 kW-h (see Fig. 9), that depicts that the region has strong solar generation potential.

Table 5 Building wise solar energy estimation

S. no.	Latitude	Longitude	Solar potential (W-h/m^2)	Roof area (m^2)	Effective roof area (m^2)	Solar energy (kW-h)
1	29.865	77.905	176.19	11.77	5.88	1.04
2	29.865	77.905	175.28	17.92	8.96	1.57
3	29.865	77.905	175.73	19.13	9.56	1.68
4	29.866	77.905	176.16	26.67	13.33	2.35
5	29.866	77.906	179.40	27.70	13.85	2.48
6	29.865	77.906	152.81	30.85	15.42	2.50
7	29.867	77.905	171.19	31.89	15.95	2.73
8	29.865	77.905	169.47	33.17	16.59	2.81
9	29.867	77.904	171.12	33.47	16.74	2.86
10	29.867	77.904	169.98	35.94	17.97	3.05

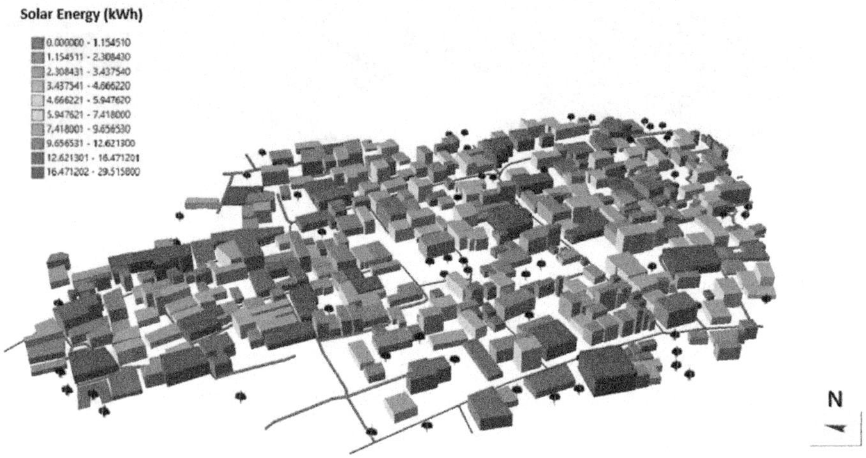

Fig. 9 3D map of building wise solar energy value

5 Conclusion

The present paper aimed at 3D modelling, shadow analysis and solar potential estimation of the urban area of Khanjarpur, Roorkee using UAV based RGB point cloud data. The results obtained from the analysis shows that 3D modelling could be applied for diverse applications along with excellent visualization of complex urban features. For proper estimation of solar potential, high-resolution terrain information is required which could be efficiently obtained from UAV survey.

The results depict that Khanjarpur region has a strong solar potential of average about 170 W-h/m^2 and minimum 1 kW-h and maximum of 9.6 kW-h of energy could be produced by a single unit of the building by harnessing solar energy. If properly harnessed, this would not only reduce the load on the present electricity distribution system but would also lead to a sustainable future.

Acknowledgements We would like to acknowledge Prof. Kamal Jain, Department of Civil engineering, IIT Roorkee to provide us the UAV dataset for this study.

References

1. ArcGIS Help 10.1 [Online]. Available: https://re-sources.arcgis.com/en/help/main/10.1/index.html#/Modeling_solar_radia-tion/009z000000t9000000/. Accessed: 08 Aug 2019
2. Brenner C (2000) Towards fully automated 3D city model generation. Autom Extr Man-Made Objects from Aer Sp Images, 85–92
3. Dubey SR (2015) Energy crisis in India: a commentary on India's electricity sector. Partridge Publishing India
4. Gross H, Thoennessen U, Hansen W (2005) 3D-modeling of urban structures. Evaluation XXXVI:137–142
5. Guo J, Sun B, Qin Z et al (2017) A study of plot ratio/building height restrictions in high density cities using 3D spatial analysis technology: a case in Hong Kong. Habitat Int 65:13–31. https://doi.org/10.1016/j.habitatint.2017.04.012
6. Kolbe TH (2008) Representing and exchanging 3D city models with CityGML. 3D Geo-Inform Sci 15–31. https://doi.org/10.1007/978-3-540-87395-2_2
7. Peronato G, Rey E, Andersen M (2018) 3D model discretization in assessing urban solar potential: the effect of grid spacing on predicted solar irradiation. Sol Energy 176:334–349. https://doi.org/10.1016/j.solener.2018.10.011
8. Salleh S, Ujang U (2018) Topological information extraction from buildings in CityGML topological information extraction from buildings in CityGML
9. Shrivastava V, Jaiswal A, Thakur PK et al (2018) Application of GIS for the design of potable water distribution system in IIRS. ISPRS Ann Photogram Rem Sens Spat Inf Sci 4:87–94. https://doi.org/10.5194/isprs-annals-IV-5-87-2018
10. Waibel C, Evins R, Carmeliet J (2017) Efficient time-resolved 3D solar potential modelling. Sol Energy 158:960–976. https://doi.org/10.1016/j.solener.2017.10.054
11. Yigitcanlar T, Kamruzzaman M, Foth M et al (2019) Can cities become smart without being sustainable? A systematic review of the literature. Sustain Cities Soc 45:348–365. https://doi.org/10.1016/j.scs.2018.11.033

Comparative Computational Analysis on High Stable Hexacopter for Long Range Applications

S. Balaji, P. Prabhagaran, R. Vijayanandh, M. Senthil Kumar and R. Raj Kumar

Abstract The primary lack in the multi-rotor Unmanned Aerial Vehicle (UAV) is unstable platform while flying, which might affect the usage of UAVs. The existing solution methodology for this lack is to increase the propellers has capable to provide more amount of thrust, which supports the stable platform while in the surveillance. Increase the propellers may chance to decrease the operational efficiency of the Hexacopter so in this paper suggested the alternate efficient method. This article recommended the Hexacopter attached with inclined arms, which has high stable platform than the Hexacopter with straight connecting arms. Comparison of these two cases have been completed using numerical simulation also optimization of inclined angle of connecting arms is done with the help of ANSYS Workbench 17.2 in which the physical model is designed by using CATIA V5. As an outcome of this simulation work is efficient Hexacopter, which capable to provide good aerial images for long range applications with low cost.

Keywords Angle · Arms · CFD · Efficiency · Stability

1 Multi-rotor UAV

Nowadays, UAV plays a vital role for obtaining the scientifically study in military and civilian approach depends on the purpose and also for the future condition [1, 2]. Because of its controllability, maneuverability, low-redundancy, multi-rotor UAV plays a predominant role in UAS [3, 4]. In this system, to improve the stability, owing to adding rotor configuration and sensor, making the conceptual model as inclined arm leads to increase the stability and performance [5, 6]. With the help of modeling software and numerical simulation, the proposed conceptual design is analyzed [7, 8].

S. Balaji · P. Prabhagaran · R. Vijayanandh (✉) · M. Senthil Kumar
Department of Aeronautical Engineering, Kumaraguru College of Technology, Coimbatore, Tamil Nadu, India

R. Raj Kumar
Department of Science and Humanities, Kumaraguru College of Technology, Coimbatore, Tamil Nadu, India

K. Jain et al. (eds.), *Proceedings of UASG 2019*, Lecture Notes in Civil Engineering 51,
https://doi.org/10.1007/978-3-030-37393-1_31

1.1 Studies on Inclined Arm Hexacopter

From the survey, the major problem in the Hexacopter have been speculated stability primarily depends on the overall CG and the thrust to the weight by the optimizable rotor configuration, it leads to increase the maintenance issues, decrease the life-time of the system, and might affect the necessary performance under the environmental condition for the long range applications [9]. Instead of concentrating the rotor configuration, creating the conceptual model—Inclined arm might also the stabilize configuration and rate of climb for the layout of the system. By making the arm to certain inclination, Hexacopter can withstand the unstable condition in the long-range application [10, 11]. In this work, to analyzed in aerodynamic perspective to the stabilize system for the long range solicitation with the help of numerical simulation and to predict the optimizable stable condition of the Hexacopter in normal comparing with inclined arm [12].

2 Conceptual Design

2.1 Design Stability

The Stability implies that the forces acting on the Hexacopter (UAVs) are in direction that tends to restore the airframe to it's original equilibrium position after it has been disturbed by a environmental issues. Thus the stability plays a predominant role in static as well as dynamic condition of the Hexacopter. In this approach, the inclined arm UAV is one of the modifications given to the airframe structure. In the view of this Hexacopter the arms are inclined by angle with respect to motors to attain the stable condition [13]. With the help of this modeling, we have to analyze the stability along with the maneuverability, controllability and its performance analysis [14, 15].

2.2 Modeling of Conceptual Design in CATIA

To analyze the constraint parameters via Simulation, we have to model the conceptual design with the help of CATIA. CATIA offers a solution to shape design, visualization to create and analyze the complex innovative shapes. In this designing, to create and visualize with the part design, wireframe and sheet metal, and assembly design for the sophisticated sketching of the inclined Hexacopter with outer surface with their finite dimensions, thus CATIA V5 plays a vital role for visualizing and creating the speculative design [16, 17].

Inclined at 0°

Figure 1 represents the Hexacopter arm where inclined at 0°.

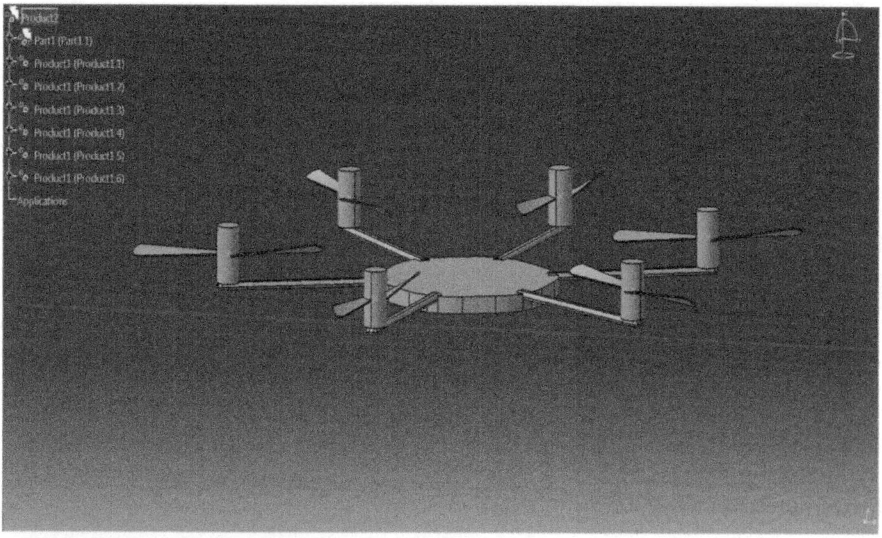

Fig. 1 Hexacopter inclined at 0°

Inclined at 2.5°

Figure 2 represents the Hexacopter arm where inclined at 2.5° from its body.

Inclined at 5°

Figure 3 represents the entire Hexacopter arm where tilted in 5° with respect to the body.

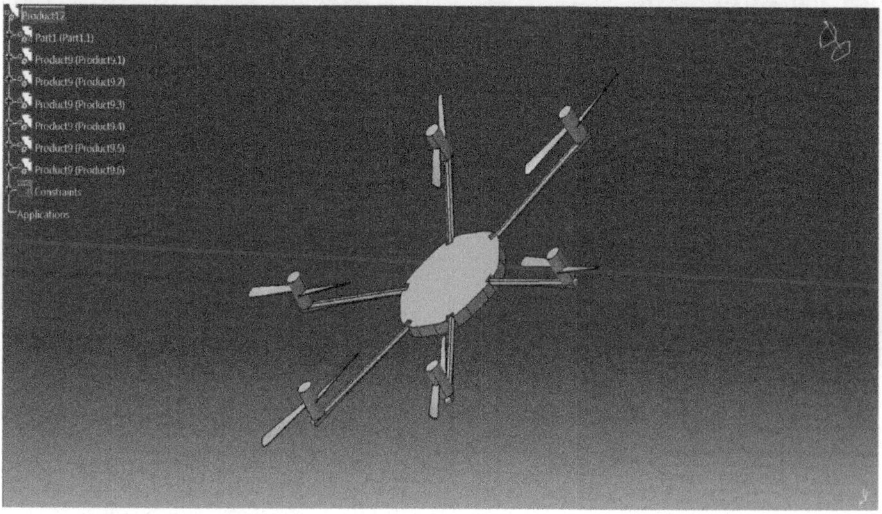

Fig. 2 Hexacopter inclined at 2.5°

Fig. 3 Hexacopter inclined at 5°

Inclined at 7.5°

Figure 4 represents the Hexacopter arm where inclined at 7.5° from its body respectively.

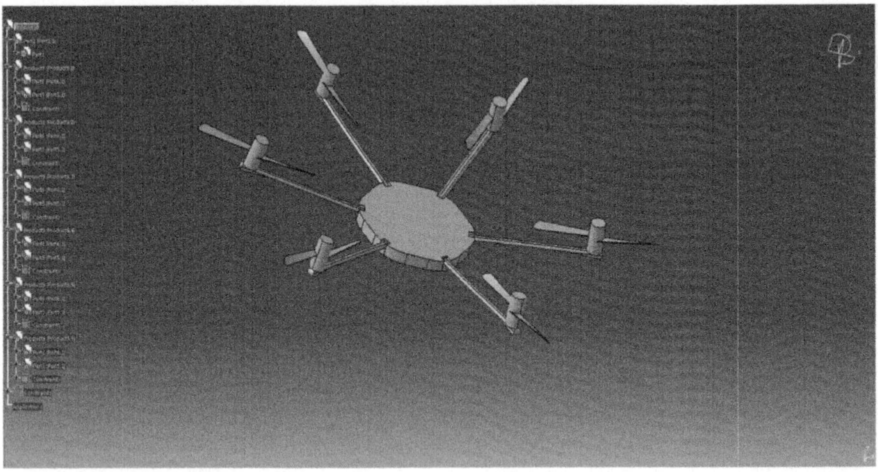

Fig. 4 Hexacopter inclined at 7.5°

3 Numerical Simulation

3.1 Initialization of Numerical Simulation

In numerical simulation, a conceptual model is scrutinizing with the necessary boundary conditions for approaching the theoretical discussion. In this numerical simulation, Fluent and CFX plays a significant role for analyzing the system or model by certain pivotal equation like Navier Stoke equation, Energy equation in the finite volume approach [18, 19]. To analyze this control volume, it examines the elements using the mesh operation. Then the specified equation leads to inspecting and obtaining the necessary solutions with the help of enumerating boundary conditions for the conceptual model or the control volume.

3.2 Boundary Conditions

In numerical simulation, the boundary condition can attribute into artificial and natural condition. The artificial condition contributes on the inlet and outlet [20]. The natural condition contributes on free slip in wall region, no slip in hexacopter region. The remaining contributors in the boundary conditions are solver type, turbulence model, fluid input parameters, solution methods and solution control. In this analysis, pressure based solver is selected for the purpose of representing incompressible flow, which is entered from the inlet velocity [21].

0° arm at 5 m/s

Figure 5 represents the velocity variation and planar view of velocity variation on the entire Hexacopter arm inclined at 0° from its body in the velocity of 5 m/s respectively.

Figure 6 represents the pressure variation for Hexacopter arm at 0° inclination in the velocity of 5 m/s. Similarly, Fig. 7 represents the Hexacopter arm inclined

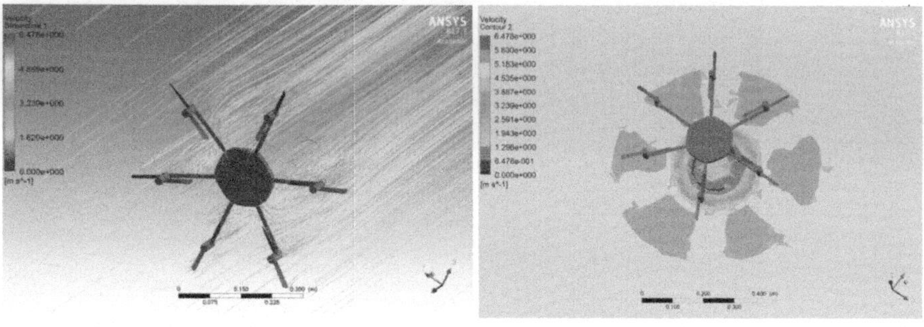

Fig. 5 Velocity variation planar view of velocity variation in condition 1

Fig. 6 Pressure variation in condition 1

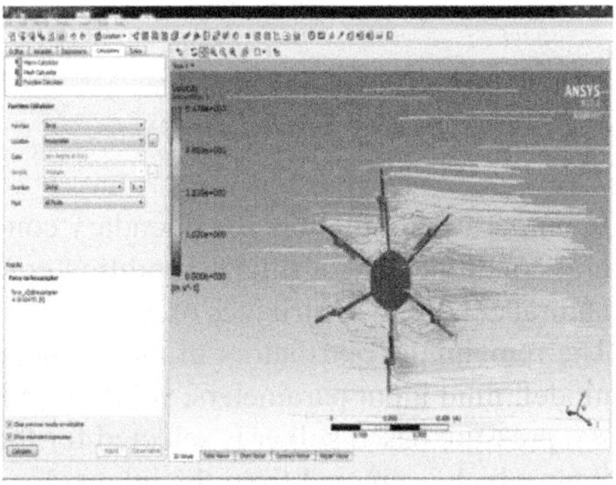

Fig. 7 Force in X-direction (side force) in condition 1

at 0° with the force (side force) acting in the X-direction at the velocity of 5 m/s respectively.

Figure 8 represents the force (drag force) acting in the Z-direction for 0° arm inclination at the velocity of 5 m/s. Similarly, Fig. 9 represents the Hexacopter arm

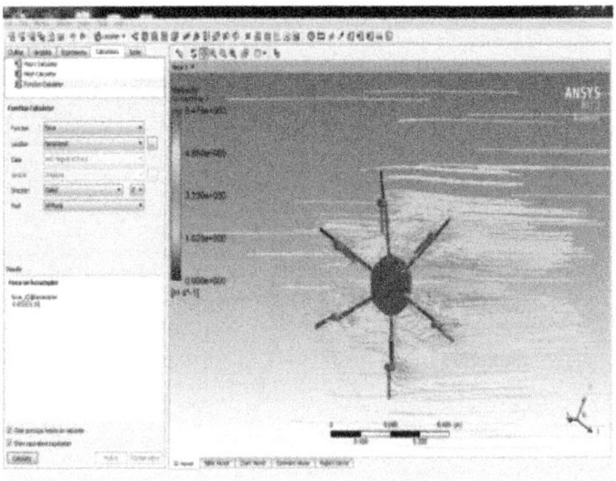

Fig. 8 Force in Z-direction (drag force) in condition 1

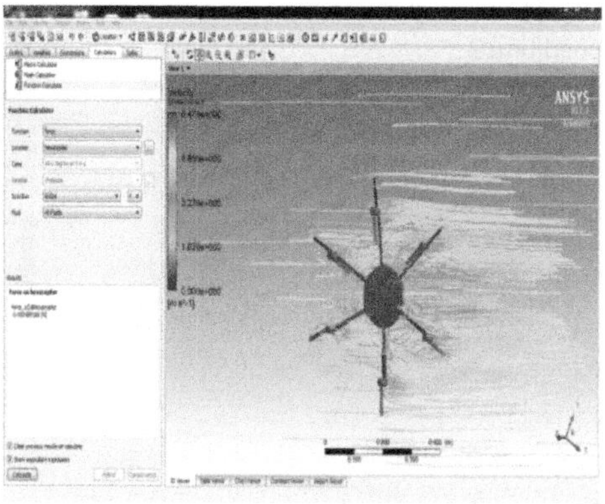

Fig. 9 Force in Y-direction (side force) in condition 1

inclined at 0° with the force acting (side force) in the Y-direction at the velocity of 5 m/s respectively.

2.5° Arm at 5 m/s

Figure 10 represents the velocity variation and its planar view of the Hexacopter arm inclined at 2.5° in the velocity of 5 m/s respectively.

Figure 11 represents the pressure variation for Hexacopter arm at 2.5° inclination. Similarly Fig. 12 represents the pressure variation for the Hexacopter arm which inclined at 2.5° in the velocity of 5 m/s respectively.

Figure 13 represents the Hexacopter arm inclined at 2.5° with the force acting in the Z-direction at the velocity of 5 m/s, whereas Fig. 14 represents the Hexacopter arm inclined at 2.5° with the force acting in the Y-direction at the velocity of 5 m/s respectively.

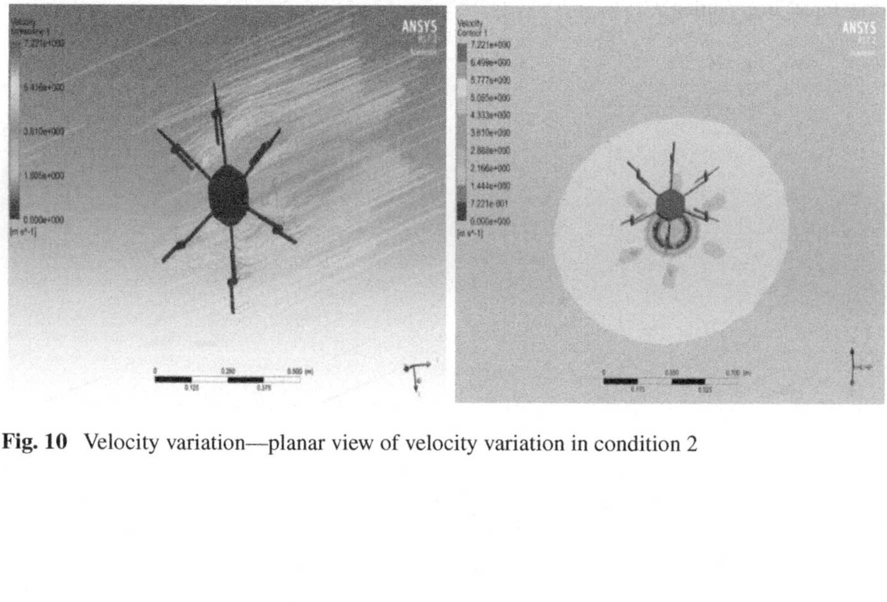

Fig. 10 Velocity variation—planar view of velocity variation in condition 2

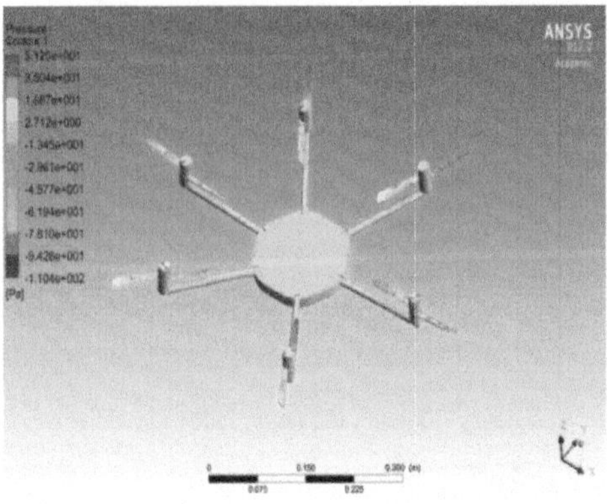

Fig. 11 Pressure variation in condition 2

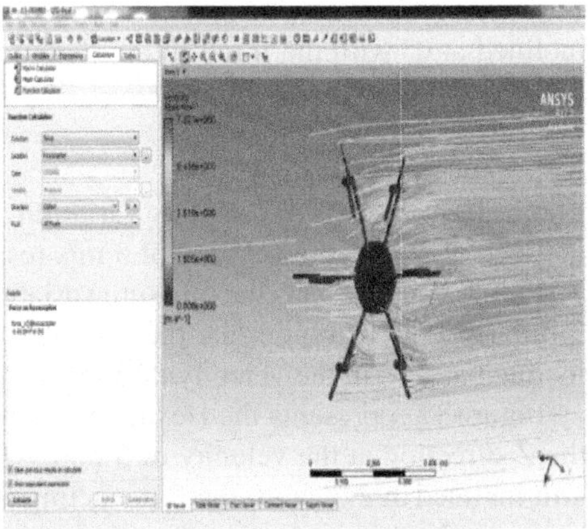

Fig. 12 Force in X-direction (side force) in condition 2

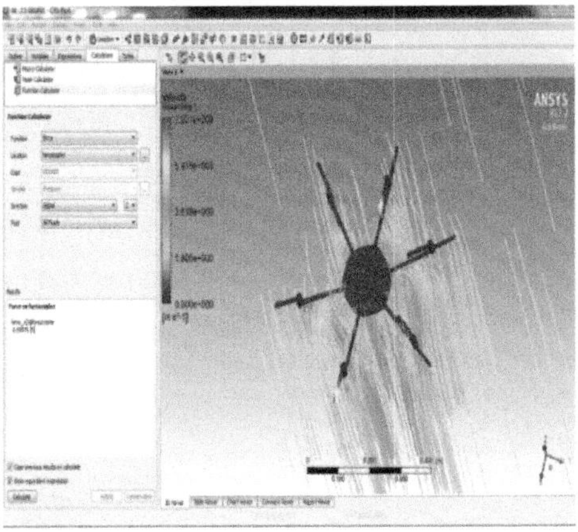

Fig. 13 Force in Z-direction (drag force) in condition 2

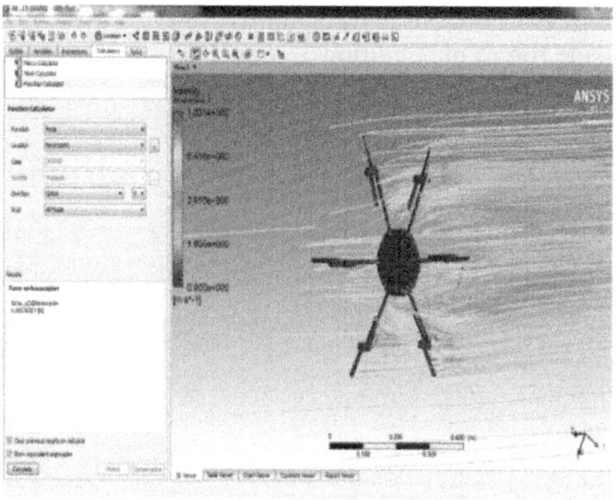

Fig. 14 Force in Y-direction (side force) in condition 2

5° Arm at 5 m/s

Figure 15 represents the velocity variation and the planar view of velocity variation on the entire Hexacopter arm inclined at 5° in the velocity of 5 m/s respectively.

Figure 16 represents the pressure variation for Hexacopter at 5° inclination of arm from its body. Similarly, Fig. 17 represents the Hexacopter arm inclined at 5° with the force (side force) acting in the X-direction at the velocity of 5 m/s respectively.

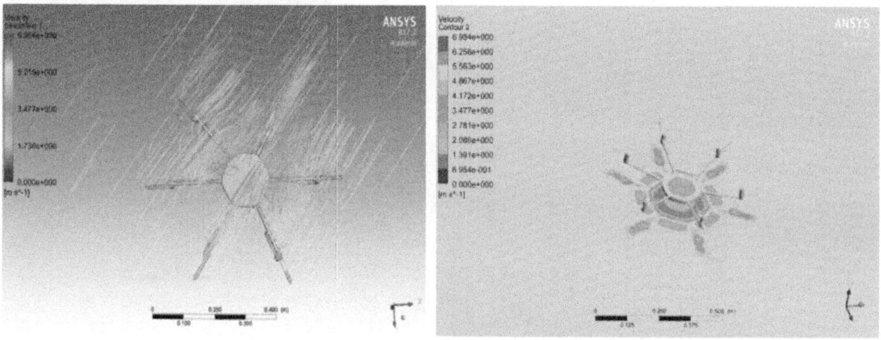

Fig. 15 Velocity—planar view of velocity variation in condition 3

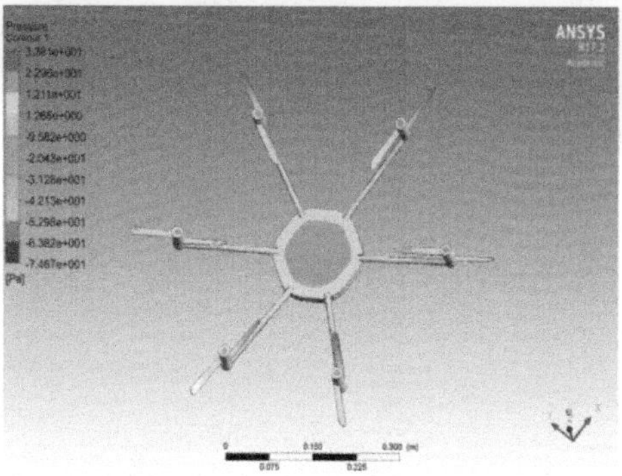

Fig. 16 Pressure variation in condition 3

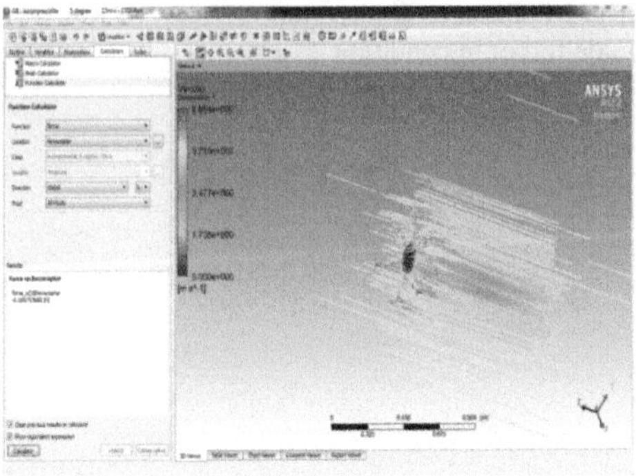

Fig. 17 Force in X-direction (side force) in condition 3

Figure 18 represents the Hexacopter arm inclined at 5° with the force acting in the Z-direction at the velocity of 5 m/s respectively. Figure 19 represents the Hexacopter arm inclined at 5° with the force acting in the Y-direction at the velocity of 5 m/s respectively.

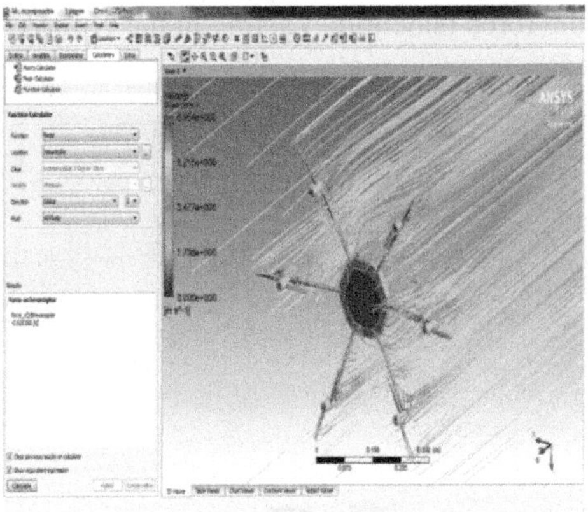

Fig. 18 Force in Z-direction (drag force) in condition 3

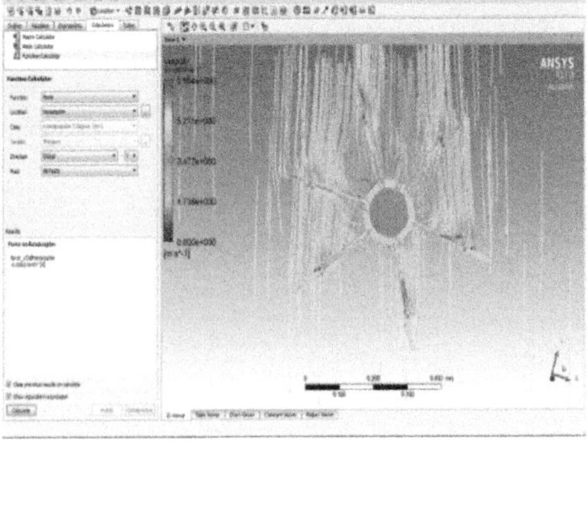

Fig. 19 Force in Y-direction (side force) in condition 3

7.5° Arm at 5 m/s

Figure 20 represents the velocity variation and its planar variation on the entire Hexacopter arm inclined at 7.5° in the velocity of 5 m/s respectively.

Figure 21 represents the pressure variation for Hexacopter arm at 7.5° inclination whereas, Fig. 22 represents the Hexacopter arm inclined at 7.5° with the force acting in the X-direction at the velocity of 5 m/s respectively.

Figure 23 represents the Hexacopter arm inclined at 7.5° with the force acting in the Z-direction at the velocity of 5 m/s, analogously Fig. 24 represents the Hexacopter arm inclined at 7.5° with the force acting in the Y-direction at the velocity of 5 m/s.

Zero at 10 m/s

Figure 25 represents the velocity variation and its planar view of velocity variation on the entire Hexacopter arm inclined at 0° in the velocity of 10 m/s respectively.

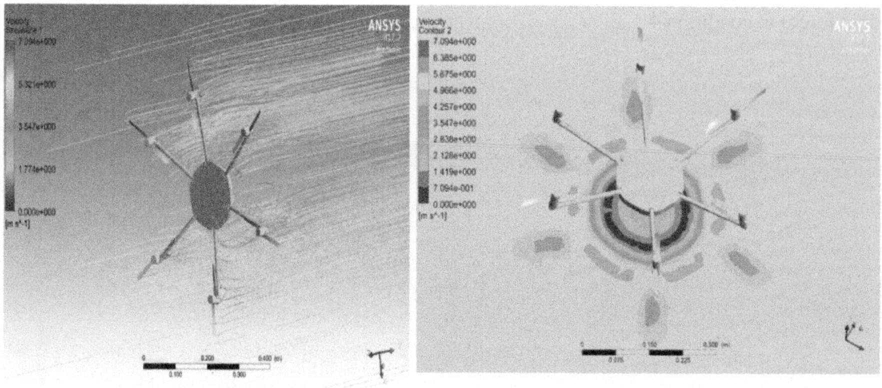

Fig. 20 Velocity variation—planar view of velocity variation in condition 4

Fig. 21 Pressure variation in condition 4

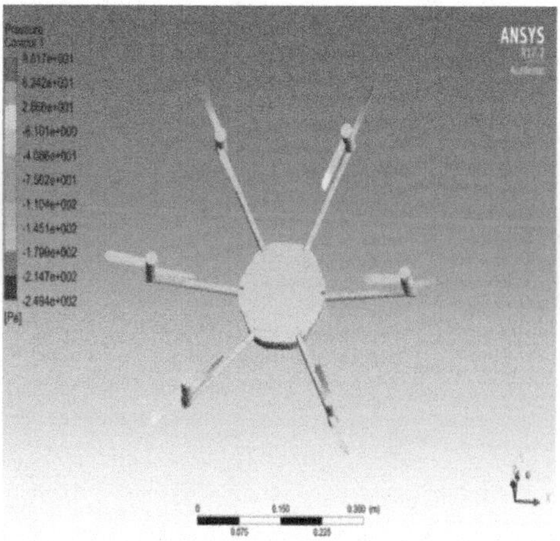

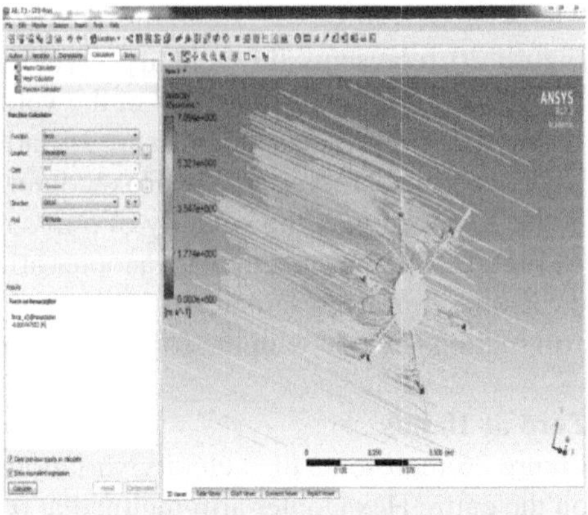

Fig. 22 Force in X-direction (side force) in condition 4

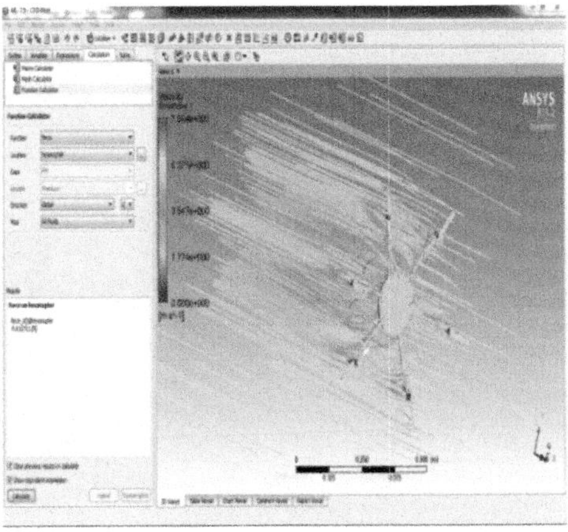

Fig. 23 Force in Z-direction (drag force) in condition 4

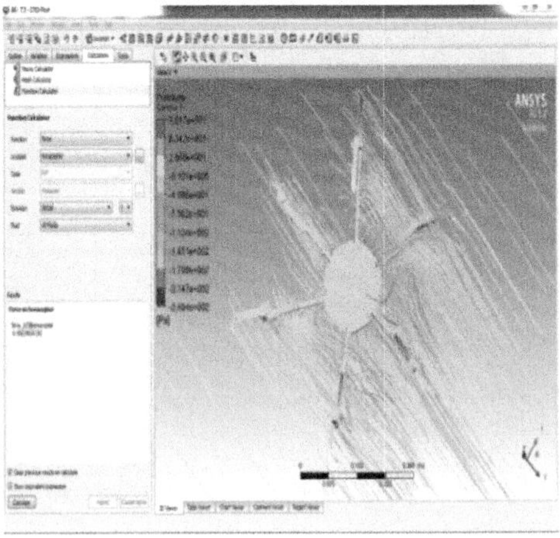

Fig. 24 Force in Y-direction (side force) in condition 4

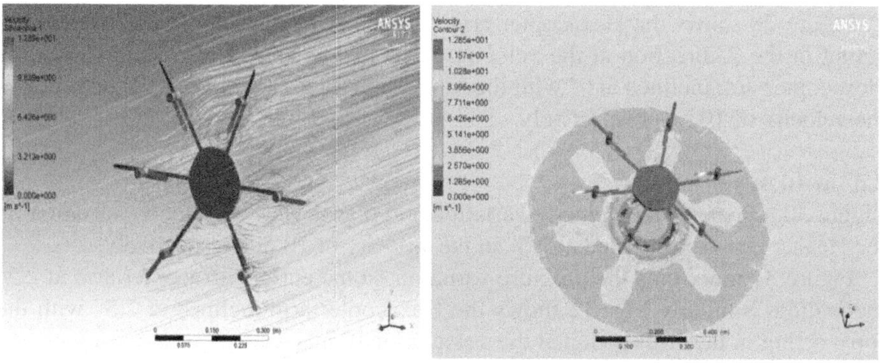

Fig. 25 Velocity variation—planar view of velocity variation in condition 5

Figure 26 represents the pressure variation for Hexacopter arm at 0° inclination. Figure 27 shows the Hexacopter arm inclined at 0° with the force (side force) acting in the X-direction at the velocity of 10 m/s respectively.

Fig. 26 Pressure variation in condition 5

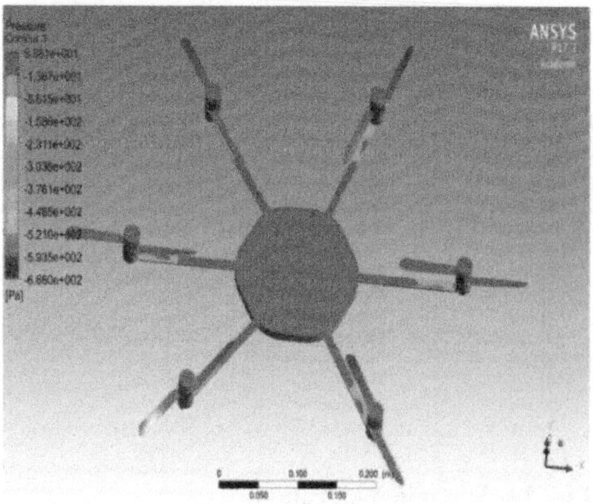

Fig. 27 Force in X-direction (side force) in condition 5

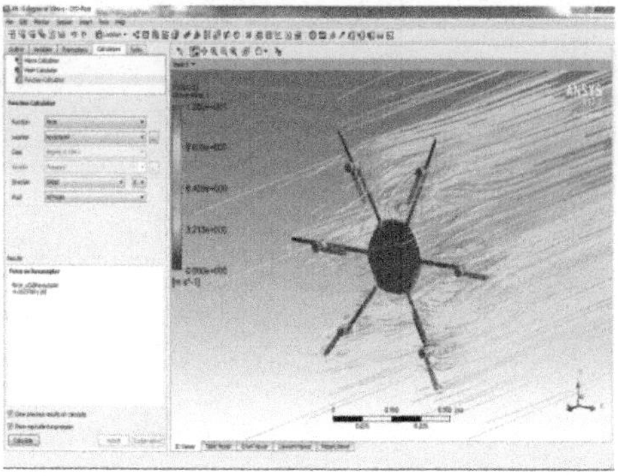

Figure 28 shows the Hexacopter arm inclined at 0° with the force (drag force) acting in the Z-direction at the velocity of 10 m/s whereas Fig. 29 represents the Hexacopter arm inclined at 0° with the force (side force) acting in the Y-direction at the velocity of 10 m/s respectively.

2.5° at 10 m/s

Figure 30 represents the velocity variation and planar view of velocity variation on the Hexacopter arm inclined at 2.5° in the velocity of 10 m/s respectively.

Figure 31 represents the pressure variation on the entire Hexacopter arm at 2.5° inclination. Similarly Fig. 32 shows the Hexacopter arm inclined at 2.5° with the force acting in the X-direction at the velocity of 10 m/s.

Figure 33 represents the Hexacopter arm inclined at 2.5° with the force acting in the Z-direction at the velocity of 10 m/s, where Fig. 34 represents the Hexacopter

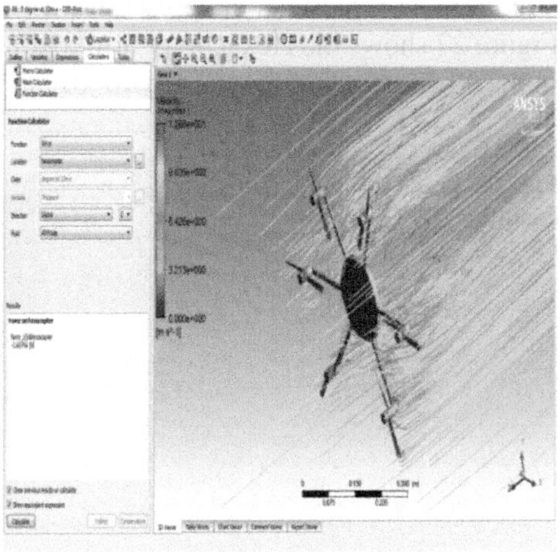

Fig. 28 Force in Z-direction (drag force) in condition 5

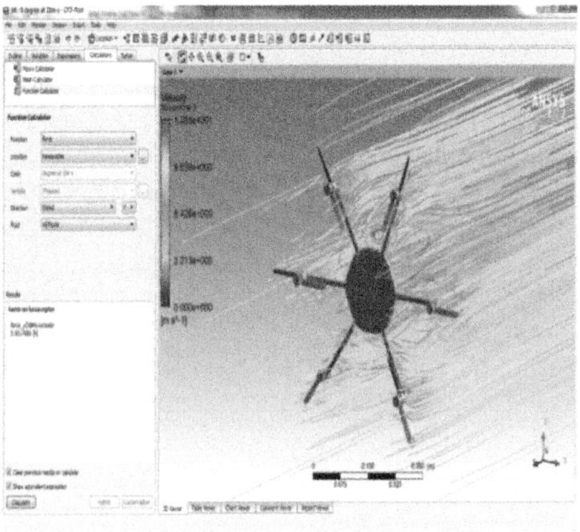

Fig. 29 Force in Y-direction (side force) in condition 5

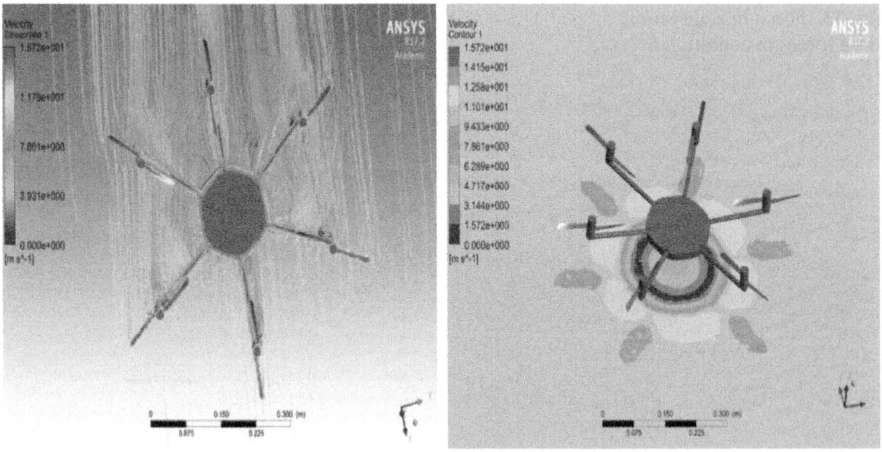

Fig. 30 Velocity variation—planar view of velocity variation in condition 6

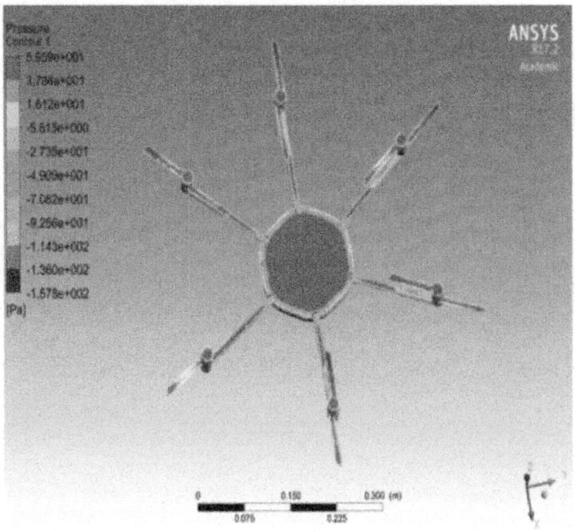

Fig. 31 Pressure variation in condition 6

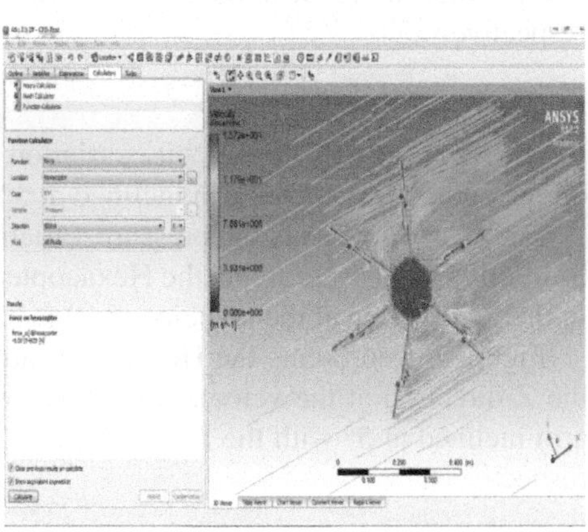

Fig. 32 Force in X-direction (side force) in condition 6

Fig. 33 Force in Z-direction (drag force) in condition 6

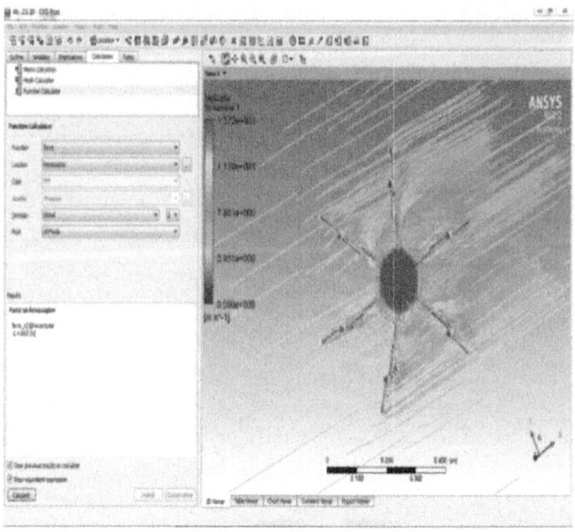

Fig. 34 Force in Y-direction (side force) in condition 6

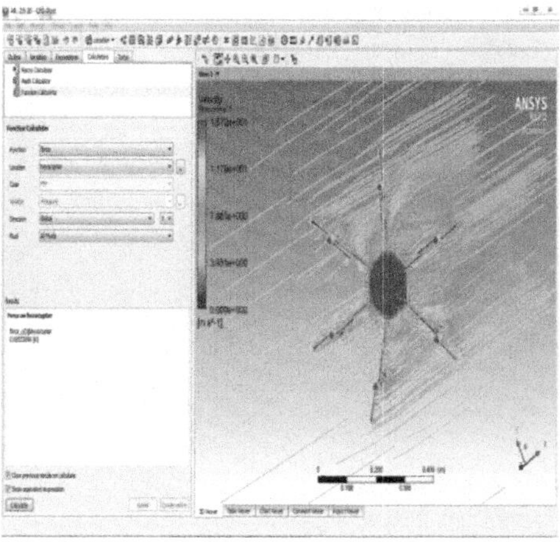

arm inclined at 2.5° with the force acting in the Y-direction at the velocity of 10 m/s respectively.

5° at 10 m/s

Figure 35 represents the velocity variation and its planar view of velocity variation on the entire Hexacopter arm inclined at 5° in the velocity of 10 m/s respectively.

Figure 36 represents the pressure variation for Hexacopter arm at 5° inclination. Similarly Fig. 37 represents the Hexacopter arm inclined at 5° with the force acting in the X-direction at the velocity of 10 m/s.

Figure 38 represents the Hexacopter arm inclined at 5° with the force acting in the Z-direction at the velocity of 10 m/s whereas Fig. 39 exhibits the Hexacopter arm inclined at 5° with the force acting in the Y-direction at the velocity of 10 m/s respectively.

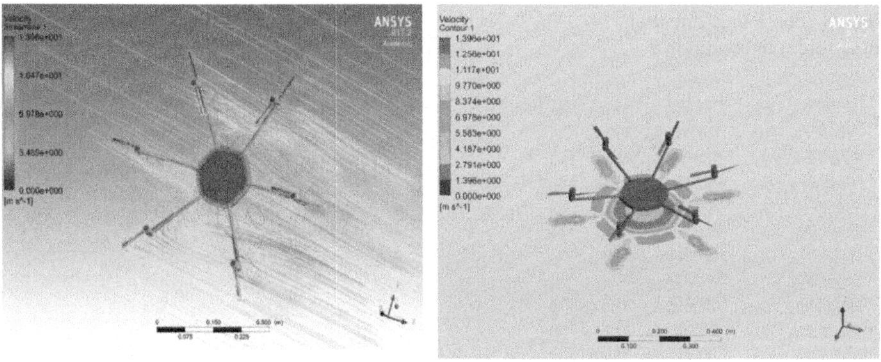

Fig. 35 Velocity variation—planar view of velocity variation in condition 7

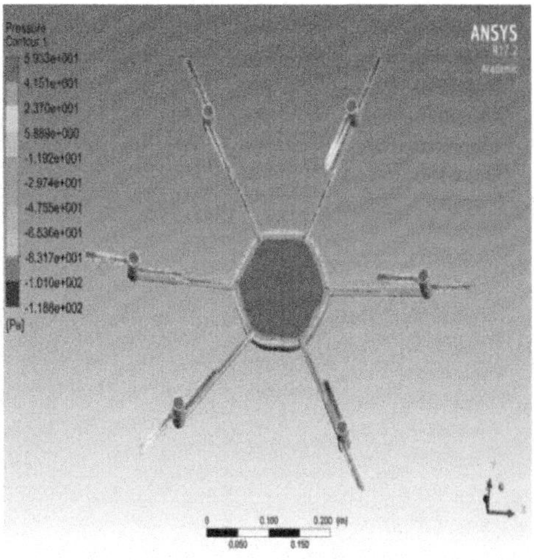

Fig. 36 Pressure variation in condition 7

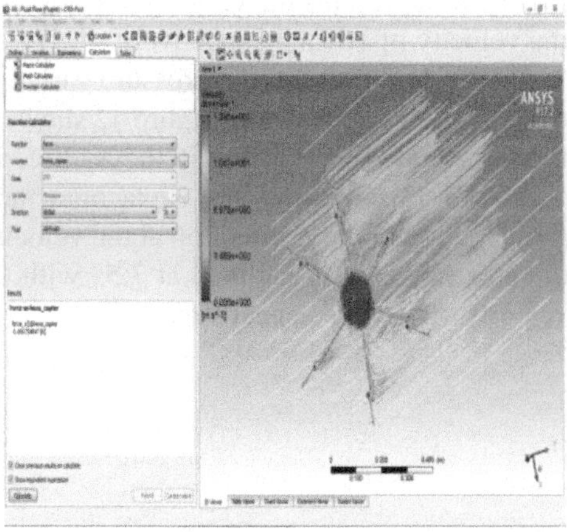

Fig. 37 Force in X-direction (side force) in condition 7

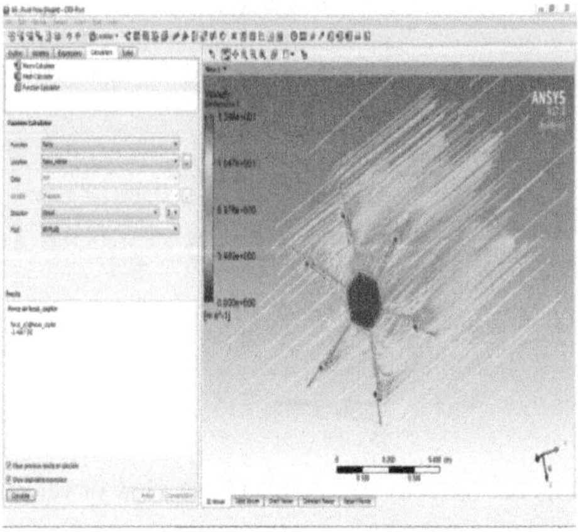

Fig. 38 Force in Z-direction (drag force) in condition 7

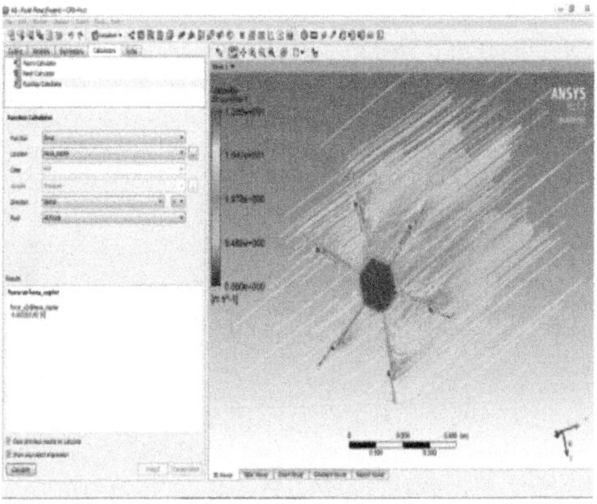

Fig. 39 Force in Y-direction (side force) in condition 7

7.5° at 10 m/s

Figure 40 represents the velocity variation on the Hexacopter arm inclined at 7.5° in the velocity of 10 m/s respectively.

Figure 41 represents the pressure variation for Hexacopter arm at 7.5° inclination from its body respectively. Figure 42 shows the Hexacopter arm inclined at 7.5° with the force (side force) acting in the X-direction at the velocity of 10 m/s.

Figure 43 represents the Hexacopter arm inclined at 7.5° with the force (drag force) acting in the Z-direction at the velocity of 10 m/s. Similarly Fig. 44 represents the Hexacopter arm inclined at 7.5° with the force acting in the Y-direction at the velocity of 10 m/s respectively.

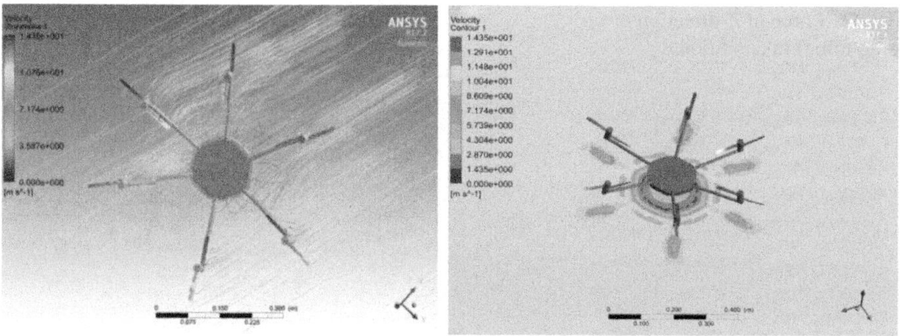

Fig. 40 Velocity variation—planar view of velocity variation in condition 8

Fig. 41 Pressure variation in condition 8

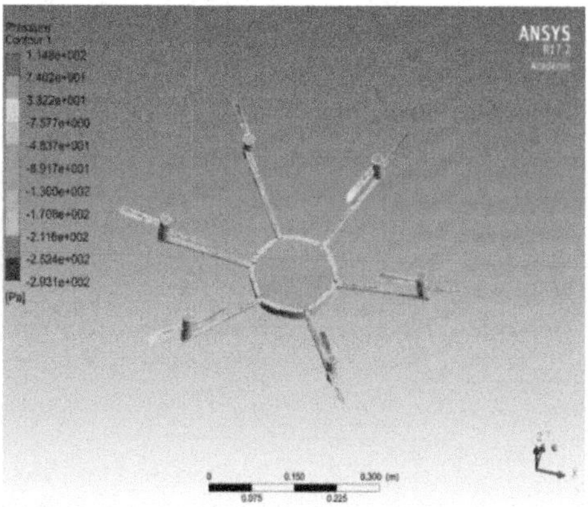

Fig. 42 Force in X-direction (side force) in condition 8

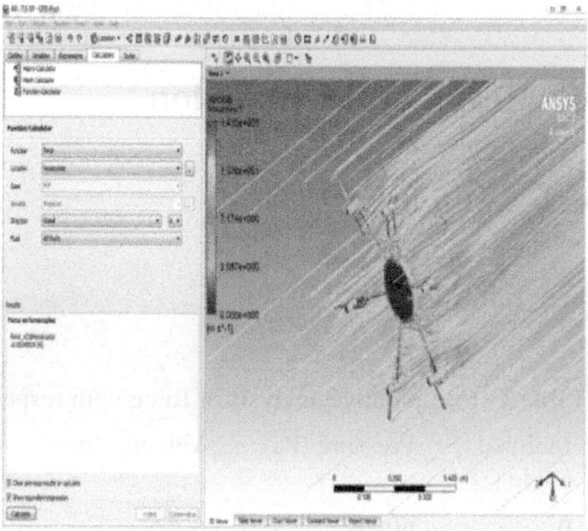

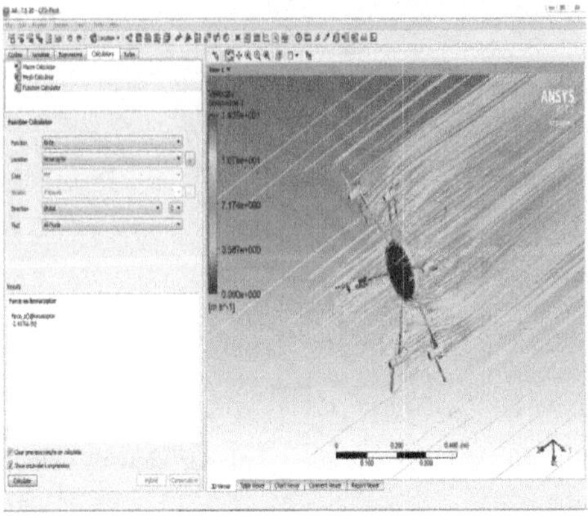

Fig. 43 Force in Z-direction (drag force) in condition 8

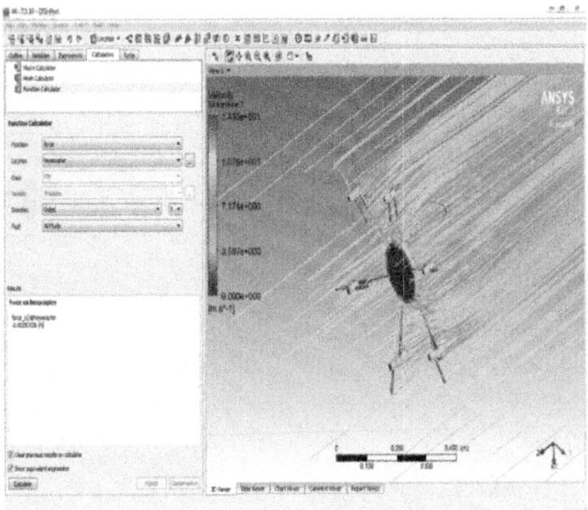

Fig. 44 Force in Y-direction (side force) in condition 8

4 Result and Discussion

4.1 At 5 m/s

See Table 1.

Table 1 Comparative analysis of force with respect to inclined arm at 5 m/s

Inclined angle (°)	Pressure (Pa)	Velocity (m/s)	Force in X (N)	Force in Y (N)	Force in Z (N)
0	60.86	6.478	0.00104751	0.000489188	0.652531
2.5	51.20	7.221	0.00184716	0.00076307	0.60976
5	33.81	6.954	0.000753688	0.00216437	0.629368
7.5	98.17	7.094	0.000447533	0.00024655	0.612711

Table 2 Comparative analysis of force with respect to inclined arm at 10 m/s

Inclined angle (°)	Pressure (Pa)	Velocity (m/s)	Force in X (N)	Force in Y (N)	Force in Z (N)
0	58.81	12.85	0.00237801	−0.0017886	2.60756
2.5	59.59	15.72	0.00194829	−0.00553096	2.41865
5	59.33	13.96	0.000754847	0.000202193	2.4967
7.5	114.8	14.35	0.00245029	0.00250126	2.43766

4.2 At 10 m/s

See Table 2.

Table 1 represents the comparison of force acting on the Hexacopter with inclined arm at different angles in various directions at the velocity of 5 m/s. Table 2 represents the comparison of force acting on the speculative Hexacopter with inclined arm at different angles in various directions at the velocity of 10 m/s. With the help of this comparison, the consequences of Hexacopter with various angles inclination contributes and leads to provide high rate of climb and forward speed compared with the drag force at different velocities. By changing the arm angle with respect to the body of Hexacopter, tends to improve the stability by pushing the air downward inclination compared with the normal (or zero) degree. Due to its stability, the inclined arm Hexacopter remain in its original state of equilibrium, notwithstanding small disturbances by the environment.

5 Conclusion

The conceptual design of the inclined arm Hexacopter is finalized and modeled by CATIA software. With the help of ANSYS Workbench, the numerical simulation is carried out for analyzing the stability by the comparison of force acting on the Hexacopter with respect to various inclinations. In the results, it is inferred that the optimized stable Hexacopter with inclined arm, side force for 7.5° in X and Y are 0.000447533 and 0.00024655 N which is minimal compared to the other angle of inclination at the velocity of 5 m/s, but the velocity at 10 m/s, the side force and pressure are minimal in 5°. This paper also suggested that instead of trial and error method production, designing of inclined arm Hexacopter which is based on the numerical simulation results is optimizable way to provide the high lifetime with high stable for the long time wide applications.

References

1. Vijayanandh R et al (2017) Design, fabrication and simulation of Hexacopter for forest surveillance. ARPN J Eng Appl Sci 12(12):3879–3884 ISSN 1819-6608
2. Sugandi TS, Nathan, Subrata SK, Arifianto O, Moelyadi MA (2018) Prediction of static stability in tandem wing unmanned aerial vehicle. In: 6th international seminar of aerospace science and technology. IOP Publishing IOP conference series. J Phys Conf Ser 1130:012028. https://doi.org/10.1088/1742-6596/1130/1/012028
3. Jacob J, Smith S (2009) Design limitations of deployable wings for small low altitude UAVs. In: 47th AIAA aerospace sciences meeting including the new horizons forum and aerospace exposition, p 745
4. Kim WS, Choi WJ, Nguyen NV, Lee JW, Kim S, Byun YH, Sur JM, Sur JM (2010) Stability analysis of full geometry aircraft through CFD and response surface method. In: 48th AIAA aerospace sciences meeting including the new horizons forum and aerospace exposition, p 1434
5. Mahdi M, Elhassan YA (2012) Stability analysis of a light aircraft configuration using computational fluid dynamics. In: Applied mechanics and materials, vol 225. Trans Tech Publications, pp 391–396
6. Gao L, Li C, Jin H, Zhu Y, Zhao J, Cai H (2017) Aerodynamic characteristics of a novel catapult launched morphing tandem-wing unmanned aerial vehicle. Adv Mech Eng 9(2):1687814017692290
7. Young LA (2015) Conceptual design aspects of three general sub-classes of multi-rotor configurations: distributed, modular, and heterogeneous. In: Presentation at the sixth AHS international specialists meeting on unmanned rotorcraft systems. Scottsdale, AZ, 20–22 Jan 2015
8. Young LA (2006) Future roles for autonomous vertical lift in disaster relief and emergency response. In: Heli-Japan 2006: AHS international meeting on advanced rotorcraft technology and life saving activities. Nagoya, Japan, 15–17 Nov 2006
9. Young LA (2007) Enhanced rescue lift capability. In: 63rd annual forum of the AHS, international. Virginia Beach, VA, 1–3 May 2007
10. Young LA (2010) Rotorcraft and enabling robotic rescue. In: Heli-Japan 2010: AHS international meeting on advanced rotorcraft technology and safety operations. Ohmiya, Japan, 1–3 Nov 2010
11. Rajagopalan RG et al (2012) RotCFD—a tool for aerodynamic interference of rotors: validation and capabilities. In: Future vertical lift aircraft design conference. San Francisco, CA, 18–20 Jan 2012
12. Young LA, Derby MR (2002) Rotor/wing interactions in hover. NASA TM 2002–211392, Apr 2002
13. Driessens S, Pounds PEI (2013) Towards a more efficient quadrotor configuration. In: 2013 IEEE/RSJ international conference on intelligent robots and systems (IROS). Tokyo, Japan, 3–7 Nov 2013
14. Nikaido BE (2013) Ultra portable and rapidly deployable rotorcraft platform for tactical compact communications relay. M.S. project. Mechanical and Aerospace Engineering Department, San Jose State University, San Jose, CA
15. Jiang G, Voyles R (2014) A nonparallel hexrotor UAV with faster response to disturbances for precision position keeping, In: IEEE international symposium on safety, security, and rescue robotics. Hokkaido, Japan, pp 1–5
16. Primicerio J et al (2012) A flexible unmanned aerial vehicle for precision agriculture. Precis Agric 13(4):517–523
17. Zhang C, Kovacs J (2012) The application of small unmanned aerial systems for precision agriculture: a review. Prec Agric 13:693–712
18. de Angelis EL (2018) Stability analysis of a multirotor vehicle hovering condition. Aerosp Sci Technol 72:248–255
19. Gatti M et al (2015) Maximum endurance for battery-powered rotary-wing aircraft. Aerosp Sci Technol, 174–179. https://doi.org/10.1016/j.ast.2015.05.009

20. Bramwell ARS (1960) The longitudinal stability and control of the tandem-rotor helicopter aeronautical research council reports and memoranda, vol 3223, London, Jan 1960, pp 1–2
21. Vijayanandh R et al (2018) Design, fabrication of Tilt-Hexacopter with image processing for critical applications. Int J Pure Appl Math 118(9):935–945 ISSN: 1314-3395

A Summarization of Collision Avoidance Techniques for Autonomous Navigation of UAV

Payal, Akashdeep and C. Raman Singh

Abstract Unmanned Aerial Vehicle (UAV) has received wide interest in recent years because of their promising potentials for several applications that include border surveillance, intrusion, detection, emergent item delivery, wildlife monitoring, rescue operations etc. The growing number of entrances of drones into airspace has emerged a significant interest in the collision and obstacle avoidance in autonomous operation. One of major challenges for safer deployment of UAVs is collision avoidance in order to make UAVs autonomous and self-sustainable. In today's technological era, it is possible for UAVs to avoid obstacles and prevent collisions while navigation by implementing cutting edge computer vision algorithms and visual tracking systems. But it is challenging due to the limited payload in UAV that only permits vision based sensors such as camera, inertial measurement unit such for control and obstacle detection. This paper aims to summarize existing solutions of collision avoidance strategies used in the navigation of UAV. Collision avoidance strategies based on geometry, optimization, sense and avoid, vision-based and force field based methods are investigated in this study.

Keywords Collision avoidance · UAV · Autonomous navigation · Geometry based · Optimization-based methods

1 Introduction

The advance potential and inherent characteristics of cooperativeness of unmanned aerial vehicles (UAV) make them suitable for autonomously carrying out severe

---

Payal (✉) · Akashdeep
UIET, Panjab University, Chandigarh, India
e-mail: payalmittal6792@gmail.com

Akashdeep
e-mail: akashdeep@pu.ac.in

C. Raman Singh
Thapar Institute of Engineering & Technology, Patiala, India
e-mail: raman.singh@thapar.edu

K. Jain et al. (eds.), *Proceedings of UASG 2019*, Lecture Notes in Civil Engineering 51,
https://doi.org/10.1007/978-3-030-37393-1_32

military operations. These small size UAVs have the ability to navigate on low-altitude and make them more robust and less dangerous [1]. Several military tasks, terrain reconnaissance, crime surveillance, target tracking, rescue and atmospheric research such applications of UAVs work for their substantial growth [2, 3]. The 2003 congressional research report of USA conveyed that the current rate of UAVs collision is 100 times that of non-autonomous aircraft [4]. Collision detection is a challenging field since the restricted constraints of UAVs only allow stereoviews from the camera mounted on them and hence, colliding obstacles cannot be identified at an early stage. Since monocular views directly depend on natural light sources such as the sun to light the scene and on texture based features to perceive the environment [5]. To perform military and reconnaissance, vehicles must have a mechanism of avoiding collisions with obstacles while in navigation. One of the major problems faced in navigating missions is the risk of colliding among non-cooperating UAVs. Consequently, considering collision avoidance factors, it become important for the safe navigation of vehicles. In this paper, different approaches of collision avoidance algorithms such as optimization based, potential force-field, geometry-based, vision-based and sense-and-avoid methods have been discussed. The organization of the paper is as follows: Sect. 2 highlighted related work towards collision avoidance in autonomous navigation of UAV and Sect. 3 describes a comprehensive review on collision detection and avoidance approaches. The summarized view of existing avoidance approaches is provided in Sect. 4. Last section concluded with crucial outcomes for further exploration.

2 Related Work

In recent advancement years, there has been a rise in focus given to the problem of collision avoidance techniques for UAVs. Different approaches have been proposed while considering their adaptability, ease of implementation and robustness in dealing with obstacle sensing techniques [6]. The fundamentals of collision avoidance consider the definition of collision and detection term [7]: A collision between a drone and a collision occurring obstacle is defined as follows:

$$||\mathrm{ru} - \mathrm{ro}|| < dC \tag{1}$$

where ru and ro denote the position vectors of the drone and the obstacle, respectively, and dc is the collision radius. A collision occurring obstacle is detected by an aerial vehicle when the following condition is satisfied:

$$||\mathrm{ru} - \mathrm{ro}|| < dRG \tag{2}$$

where dRG denotes a specified detection radius, which can be calculated from the performance of the onboard sensor system. Obstacle detection and collision avoidance are two important aspects of safe flight and in autonomous navigation of UAVs

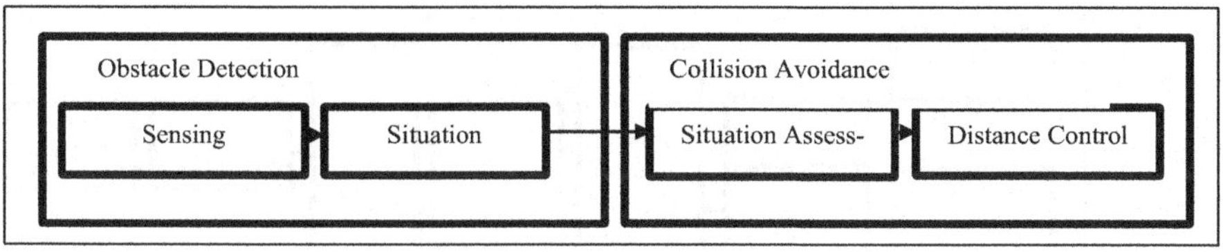

Fig. 1 Overview of obstacle avoidance concept [8]

as depicted in Fig. 1. Obstacle detection is done by sensing the flight environment with the help of sensors such as optical flow (OF) and inertial measurement unit (IMU) and perform situation awareness about the presence of obstacles. Then collision avoidance factors are incorporated by dealing with obstacle situation assessment and respective distance control maneuvers. This collision avoidance factors are embedded in a cooperative flight algorithm through GPS position reporting system and this algorithm can guarantee UAV safety through general rules governing UAV interaction [9]. UAV based civilian applications monitor the performance of real-time navigation environment and relay the output to main processing paradigm to allow timely response [10, 11].

3 Review on Collision Avoidance Methods

Obstacle avoidance is formed by the lack of expertise in surrounding environment knowledge and as a result, an obstacle is formed due to which collision detection and avoidance is a challenging problem. The ability to perceive flight conditions and observe obstacles during navigation is a necessary condition for aircraft [9]. The flight path must be modified to ensure safety while maintaining the optimal path and to ensure collision detection. An important design problem in aerial vehicles is the ability to sense and avoid collisions through cameras mounted on them [13, 14].

The UAV fight based requirements have been acquired such as cooperation between multiple UAVs, long and short range collision avoidance mechanisms [15]. The functional units of collision detection system consist of cooperative and non-cooperative obstacle objects as shown in Fig. 2. UAV maneuver realization has been done by calculating escape trajectory from risk assessment after conflict detection process. The popular mechanisms including the Traffic Collision Avoidance System (TCAS) and Automatic Dependent Surveillance-Broadcast (ADS-B) have been existing to identify the GPS location of the autonomous vehicle [16, 17]. In further sections of the paper, we provide a listing of existing solutions of collision avoidance approaches in detail.

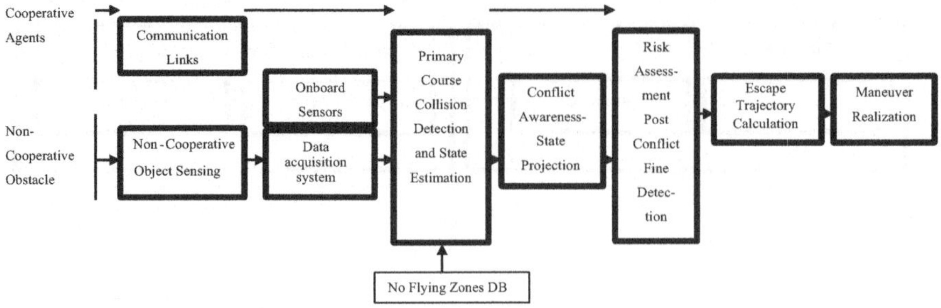

Fig. 2 Collision avoidance system functional units [12]

3.1 Geometry Based Collision Avoidance Methods

UAV flight formation protocol has been designed from Lyapunov approach and algebraic Riccati equation and furthermore, time-varying formation control with switching interaction topologies in the context of UAV is also studied [18, 19]. The close association between multiple UAVs can perform complicated missions effectively through precisely defined geometry when the vortex forces are considered [20]. These aerial vehicles usually perform several maneuvers such as acceleration and deceleration, turning with the help of individual formation members [21]. The UAV flight formation might be useful for quick deployment of troops and vehicles, airborne refueling such tasks [22]. These aerial vehicles complicated the collision avoidance systems problem through dynamic constraints such as minimum speed [12]. Geometric based methods ensure safety in the integrated space that includes avoiding conflicts and collisions amongst existing air traffic [23]. In [24], a novel collision cone based approach proposed to detect and avoid collision between irregularly shaped objects with unknown path trajectories. This collision cone approach effectively determined the collision rate between a robot and an obstacle as well. Radzi et al. [25] presented two-dimensional plane based parametric theorem for collision detection problem. In this approach, collision point, exact time of collision, and distance left to reach the colliding point have been accumulated to perform collision detection. Lin and Saripalli [26] proposed path planning methods based on sampling approach to avoid a collision. The ability to avoid collisions with other aircraft is critical to UAVs. Luo et al. [27] presented formation regrouping control strategy for UAV maneuver leader followed by flight followers. Simulation results demonstrated to conduct obstacle avoidance and formation splitting and regrouping strategies. Belkhouche [28] implemented reactive motion planning methods for drones in a real-world civilian task. Autonomous navigation requires spontaneous actions to detect collision. This dynamic world consisted of virtual space representation that used to solve real-time optimal trajectories. Stochastic based probabilistic methods effectively identify high-dimensional spaces for global planning paths based on the navigation environment accumulated by UAV sensors [29].

3.2 *Sense and Avoid Collision Avoidance Methods*

To implement see and avoid collision avoidance methods, a schematic overview of the available airspace safety layers is shown in Fig. 3. UAV is an intruder for airspace so it has to follow some procedures for autonomous navigation. For controlled airspace, there is also air traffic management available which organizes all users to navigate in a controlled manner. There exist transponder based technologies to make UAV visible to others. And the last safety layer is always the UAV itself who has to look outside and search for other colliding aerial objects. This principle is formally known as See and Avoid [30].

Sense and Avoid Methods [14] have low computational requirements and short response time in regards of information collected from sensors. The techniques based on stereo views to estimate the closeness of obstacle from UAV [31], novel vision-based obstacle detection algorithm for ground vehicles [32] and sensor fusion techniques with weighted filter were proposed in literature studies. Kim et al. [33] introduced a collision-free reinforced barrier that allows a navigating vehicle to provide optimal flight coverage with no conflict of interest. This approach generates potential adjustable positions within UAVs with the help of a novel algorithm that acquire a zone-based approach but also adjustable potential field positions. Ma et al. [34] presented a generic framework in which reinforcement learning module to develop reactive obstacle avoidance algorithm based on actor-critic approach for a UAV. A saliency detection algorithm designed in this framework using deep convolutional neural networks to extract visual cues. The reinforcement learning module used an actor-critic structure through radial basis function to approximate control policy in continuous action spaces.

3.3 *Optimization-Based Collision Avoidance Methods*

Optimization-based and geometric methods are contrary to approaches and have presented diverse solutions for collision avoidance. It attempts to find an input parameter that minimizes the obstacle related performance index. Some examples include predictive control (nonlinear model) [35], swarm optimization based methods to

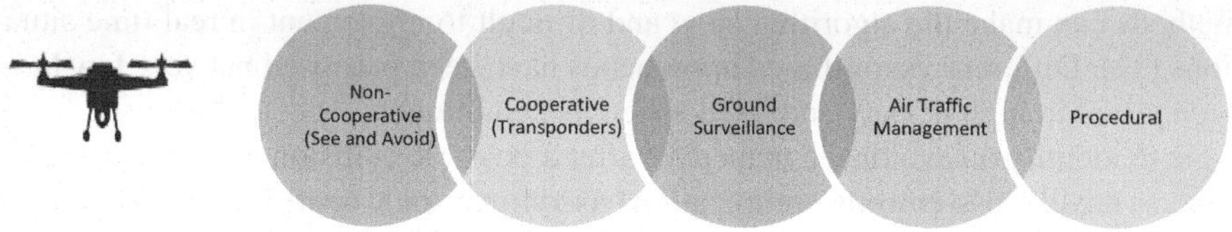

Fig. 3 Schematic overview of airspace safety layers for see and avoid methods [30]

avoid detected non-stationary obstacles [36], controlling the high velocity of obstacles based methods [27] and optimal route planning based methods [28]. Another geometric approach known as a rapid random tree (RRT) utilizes a collision cone algorithm to yield a closed-form solution for closed aim problems [3]. state time based probabilistic roadmap [31], ant colony algorithm [37], genetic algorithm [38], geometric optimization [39] and Markov decision process [40] also falls in this category but existing optimization based methods demand a high rate of computation and require a definite choice of a collision termination criteria.

3.4 Potential Field Collision Avoidance Approaches

The artificial potential field based collision avoidance approaches aim to solve an obstacle avoidance problem for autonomous navigation and robotics applications. In [41], the collision avoidance algorithm with space-based trajectory controller proposed for aerial vehicle tracking. Such a null approach based controller was designed using a behavioral-based approach with hierarchical objectives. Then a positive potential function also designed for the movement of obstacles in the air environment. Other approaches include an alternate route through expanding elliptical search [42], cross entropy based trajectory space findings [41] and artificial potential field [43]. Some limitations such as local minima, the narrowness of the channel, trajectory jitter and target unreachability exist in case of potential field approaches of collision avoidance.

3.5 Vision Based Collision Avoidance Methods

Vision-based methods include image processing for data captured by sensors and neural networks for processing. Some techniques such as redundant sensor techniques [44], colored petrinets [45], using convolution neural networks [46], saliency-based methods [34], Optical flow based approach [47], estimate depth map using multiple images [47] and using motion parallax [48]. A vision-based obstacle avoidance algorithm for a UAV is presented in [31] for the perception of the environment. Stereo vision is a computationally demanding process and hence use of complex analysis methods can make the algorithm slow and difficult to implement in real time situations [32]. Different vision-based approaches have been proposed but very few have been tested in high density and fixed speed test conditions. Fu et al. [49] used fuzzy logic to identify an avoidance maneuver after a possible collision is detected. In this paper, a novel cross-entropy optimization based fuzzy logic controller for safe drones to expand its collision avoidance capabilities in the GPS-denied surroundings was developed.

4 Summary

In this paper, the main objective is to discuss various collision avoidance methods in the context of autonomous navigation of UAV. Advancement in artificial intelligence approaches such as human-based learning, neural networks, fuzzy logic etc. that makes drone robust at the time of preventing it by hitting an obstacle while navigating. Optimization-based and geometric methods are contrary to approaches and based on optimal path planning based methodology. Sense and avoid methods are a subset of vision methods approaches and make use of advanced algorithms in computer vision. Artificial potential force field based methods develop a map where a trajectory waypoint generates an attraction towards obstacles whereas simultaneously generate repulsive forces. Based on the result of the force strategy, a potential field based algorithm generates a collision-free navigating path. Moreover, all existing solutions provide a way for collision situation assessment problem and take necessary precautionary measures to avoid the collision.

5 Conclusion

This paper gives a preliminary but fairly in-depth study of existing collision avoidance approaches in a dynamic real-world environment. The collision detection and avoidance strategies have been implemented into a realistic situation where UAV tries to move around several moving obstacles while attempting to reach a specified goal position. The survey of collision avoidance approaches is based on the main collision avoidance system design parameters formed by scientific communities in the context of autonomous UAV collision detection approaches.

References

1. De Croon G, De Clercq KME, Ruijsink R, Remes B, De Wagter C (2009) Design, aerodynamics, and vision-based control of the DelFly. Int J. Micro Air Veh 1(2):71–97
2. Murray RM (2007) Recent research in cooperative control of multi-vehicle systems. J Dyn Syst Meas Control 129(5):571–583
3. Zhu L, Cheng X, Yuan F-G (2016) A 3D collision avoidance strategy for UAV with physical constraints. Measurement 77:40–49
4. Osborne RW, Bar-Shalom Y, Willett P, Baker G (2011). Design of an adaptive passive collision warning system for UAVs. IEEE Trans Aerosp Electron Syst 47(3):2169–2189
5. Mori T, Scherer S (2013) First results in detecting and avoiding frontal obstacles from a monocular camera for micro unmanned aerial vehicles. In: 2013 IEEE international conference on robotics and automation (ICRA), pp 1750–1757
6. Hromatka M (2013) A fuzzy logic approach to collision avoidance in smart UAVs (2013). Honors Theses, pp 1963–2015, 13. https://digitalcommons.csbsju.edu/honors_theses/13
7. Seo J, Kim Y, Kim S, Tsourdos A (2017) Collision avoidance strategies for unmanned aerial vehicles in formation flight. IEEE Trans Aerosp Electron Syst 53(6):2718–2734

8. Gageik N, Benz P, Montenegro S (2015) Obstacle detection and collision avoidance for a UAV with complementary low-cost sensors. IEEE Access 3:599–609
9. Ryan A, Zennaro M, Howell A, Sengupta R, Hedrick JK (2004) An overview of emerging results in cooperative UAV control. In: 43rd IEEE conference on decision and control, 2004. CDC, vol 1, pp 602–607
10. Rathinam S, et al (2007) Autonomous searching and tracking of a river using an UAV. In: 2007 American control conference, pp 359–364
11. Li Z, Ding J (2007) Ground moving target tracking control system design for UAV surveillance. In: 2007 IEEE international conference on automation and logistics, pp 1458–1463
12. Albaker BM, Rahim NA (2009) A survey of collision avoidance approaches for unmanned aerial vehicles. In: 2009 international conference for technical postgraduates (TECHPOS), pp 1–7
13. Zsedrovits T, Bauer P, Zarandy A, Vanek B, Bokor J, Roska T (2014) Error analysis of algorithms for camera rotation calculation in GPS/IMU/camera fusion for UAV sense and avoid systems. In: 2014 international conference on unmanned aircraft systems (ICUAS), pp 864–875
14. Ma Z, Hu T, Shen L, Kong W, Zhao B (2015) A detection and relative direction estimation method for UAV in sense-and-avoid. In: 2015 IEEE international conference on information and automation, pp 2677–2682
15. Accardo D, Fasano G, Forlenza L, Moccia A, Rispoli A (2013) Flight test of a radar-based tracking system for UAS sense and avoid. IEEE Trans Aerosp Electron Syst 49(2):1139–1160
16. Holdsworth R, Lambert J, Harle N (2001) Inflight path planning replacing pure collision avoidance, using ADS-B. IEEE Aerosp Electron Syst Mag 16(2):27–32
17. Stark B, Stevenson B, Chen Y (2013) ADS-B for small unmanned aerial systems: case study and regulatory practices. In: 2013 international conference on unmanned aircraft systems (ICUAS), pp 152–159
18. Dong X, Zhou Y, Ren Z, Zhong Y (2016) Time-varying formation control for unmanned aerial vehicles with switching interaction topologies. Control Eng Pract 46:26–36
19. Dong X, Zhou Y, Ren Z, Zhong Y (2017) Time-varying formation tracking for second-order multi-agent systems subjected to switching topologies with application to quadrotor formation flying. IEEE Trans Ind Electron 64(6):5014–5024
20. Lavretsky E (2002) F/A-18 autonomous formation flight control system design. In: AIAA guidance, navigation, and control conference and exhibit, p 4757
21. Kim Y, Choi J (2008) Fuel-efficient formation flight-control design based on energy maneuverability. J Guid Control Dyn 31(4):1145–1150
22. Zou Y, Pagilla PR, Ratliff RT (2009) Distributed formation flight control using constraint forces. J Guid Control Dyn 32(1):112–120
23. Jenie YI, van Kampen E-J, Ellerbroek J, Hoekstra JM (2017) Taxonomy of conflict detection and resolution approaches for unmanned aerial vehicle in an integrated airspace. IEEE Trans Intell Transp Syst 18(3):558–567
24. Chakravarthy A, Ghose D (1998) Obstacle avoidance in a dynamic environment: a collision cone approach. IEEE Trans Syst Man Cybern A Syst Hum 28(5):562–574
25. Radzi NFM, Mubin M, Rahim NA (2011) Collision detection algorithm for UAVs based on parametric theorem approach. In: 2011 IEEE 3rd international conference on communication software and networks (ICCSN), pp 493–496
26. Lin Y, Saripalli S (2017) Sampling-based path planning for UAV collision avoidance. IEEE Trans Intell Transp Syst 18(11):3179–3192
27. Luo D, Zhou T, Wu S (2013) Obstacle avoidance and formation regrouping strategy and control for UAV formation flight. In: 2013 10th IEEE international conference on control and automation (ICCA), pp 1921–1926
28. Belkhouche F (2017) Reactive optimal UAV motion planning in a dynamic world. Rob Auton Syst 96:114–123
29. Larson J, Bruch M, Ebken J (2006) Autonomous navigation and obstacle avoidance for unmanned surface vehicles. In: Unmanned systems technology VIII, vol 6230, p 623007

30. Nussberger A, Grabner H, Van Gool L (2014) Aerial object tracking from an airborne platform. In: 2014 international conference on unmanned aircraft systems (ICUAS), pp 1284–1293
31. Cieśluk J, Gosiewski Z (2013) A stereovision system for real time obstacle avoidance by unmanned aerial vehicle. Solid State Phenom 198:159–164
32. Chakraborty A, Srinivasan B (2015) A novel stereo based obstacle avoidance system for unmanned aerial vehicles. Int J Comput Appl 131(13):39–42
33. Kim H, Ben-Othman J, Bellavista P (2017) Collision-free reinforced barriers in UAV networks. J Comput Sci 22:289–300
34. Ma Z, Wang C, Niu Y, Wang X, Shen L (2018) A saliency-based reinforcement learning approach for a UAV to avoid flying obstacles. Rob Auton Syst 100:108–118
35. Zhang Y, Fairley G (2013) Multiple-trajectory-prediction (MTP) algorithm for UAS's sense and avoid (SAA) operation. In: 2013 integrated communications, navigation and surveillance conference (ICNS), pp 1–11
36. Senanayake M, Senthooran I, Barca JC, Chung H, Kamruzzaman J, Murshed M (2016) Search and tracking algorithms for swarms of robots: a survey. Rob Auton Syst 75:422–434
37. Cekmez U, Ozsiginan M, Sahingoz OK (2016) Multi colony ant optimization for UAV path planning with obstacle avoidance. In: 2016 international conference on unmanned aircraft systems (ICUAS), pp 47–52
38. Tsai C-C, Huang H-C, Chan C-K (2011) Parallel elite genetic algorithm and its application to global path planning for autonomous robot navigation. IEEE Trans Ind Electron 58(10):4813–4821
39. White BA, Shin H-S, Tsourdos A (2011) UAV obstacle avoidance using differential geometry concepts. In: IFAC world congress, pp 6325–6330
40. Ragi S, Chong EKP (2013) UAV path planning in a dynamic environment via partially observable Markov decision process. IEEE Trans Aerosp Electron Syst 49(4):2397–2412
41. Santos MCP, Rosales CD, Sarcinelli-Filho M, Carelli R (2017) A novel null-space-based UAV trajectory tracking controller with collision avoidance. IEEE/ASME Trans Mechatron 22(6):2543–2553
42. Tang J, Fan L, Lao S (2014) Collision avoidance for multi-UAV based on geometric optimization model in 3D airspace. Arab J Sci Eng 39(11):8409–8416
43. Sun J, Tang J, Lao S (2017) Collision avoidance for cooperative UAVs with optimized artificial potential field algorithm. IEEE Access 5:18382–18390
44. Alexopoulos A, Kandil A, Orzechowski P, Badreddin E (2013) A comparative study of collision avoidance techniques for unmanned aerial vehicles. In: 2013 IEEE international conference on systems, man, and cybernetics, pp 1969–1974
45. Olsson P-M, Kvarnström J, Doherty P, Burdakov O, Holmberg K (2010) Generating UAV communication networks for monitoring and surveillance. In: 2010 11th international conference on control automation robotics & vision, pp 1070–1077
46. Chaudhary K, Zhao M, Shi F, Chen X, Okada K, Inaba M (2017) Robust real-time visual tracking using dual-frame deep comparison network integrated with correlation filters. In: 2017 IEEE/RSJ international conference on intelligent robots and systems (IROS), pp 6837–6842
47. Goerzen C, Kong Z, Mettler B (2010) A survey of motion planning algorithms from the perspective of autonomous UAV guidance. J Intell Robot Syst 57(1–4):65
48. Mohanan MG, Salgoankar A (2018) A survey of robotic motion planning in dynamic environments. Rob Auton Syst 100:171–185
49. Fu C, Olivares-Mendez MA, Suarez-Fernandez R, Campoy P (2014) Monocular visual-inertial slam-based collision avoidance strategy for fail-safe UAV using fuzzy logic controllers. J Intell Robot Syst 73(1–4):513–533

Developing Intelligent Fire Alarm System and Need of UAV

Girish Joshi, Bikash Pal, Iltaf Zafar, Shruti Bharadwaj and Susham Biswas

Abstract Accidental fire is one of the leading cause of unnatural deaths in India. While there are a wide variety of fire alarm systems available in the market, most Indian homes are still unequipped with them. In commercial space, fire alarm systems are installed due to regulatory requirement but they are expensive, inadequate or often fail due to the problem of the false alarm, no battery, or switch off conditions. An intelligent Smoke, CO Alarm system with CCTV input is developed keeping the above limitations in mind. The device works in the IoT platform and can connect wirelessly to a mobile phone to notify in case of fire, abnormal CO concentrations, smoke or bad battery health at any time and from anywhere. With an early heads-up warning with the location of fire, it is now possible to cross-check where and what the trouble is before the alarms start hooting and informing the outside people. And in-case it was a false alarm, it can silence it using a mobile phone. The detailed positional information of fire in the building is planned to be utilized for safeguarding. UAV will be navigated to the outdoor building location to capture images of fire and smoke and communicate it to mobile phone to enable firefighting and initiate immediate rescuing by local people. Potential access of real-time images on the background of google map with the demarcation of building's entry and exit points in relation to the location of fire is demonstrated on RGIPT's campus as a successful intelligent fire alarm system.

Keywords UAV · Image · Smoke · CO detector · CCTV · IoT · Mobile

G. Joshi · B. Pal · I. Zafar · S. Bharadwaj · S. Biswas (✉)
Rajiv Gandhi Institute of Petroleum Technology, Jais, India
e-mail: susham@rgipt.ac.in

G. Joshi
e-mail: epe15014@rgipt.ac.in

B. Pal
e-mail: epe15010@rgipt.ac.in

I. Zafar
e-mail: mzafar@rgipt.ac.in

S. Bharadwaj
e-mail: pgi17001@rgipt.ac.in

K. Jain et al. (eds.), *Proceedings of UASG 2019*, Lecture Notes in Civil Engineering 51,
https://doi.org/10.1007/978-3-030-37393-1_33

1 Introduction

Detection of fire in the building, issuing warning, monitoring, initiating firefighting and rescuing are important components of a fire alarm system. Fire accidents are common in India and the world and very often causing significant damage to property and life. Public and private offices, residential complexes need the fire alarm system to safeguard themselves from fire hazards. Existing systems use various sensors, such as temperature, smoke, flame etc. to initiate the alarm with wired or wireless connections. On the top of indoor sensors, camera-based outdoor sensors have come up for surveillance and monitoring in real time mode. UAV with a camera provides an excellent opportunity to capture photographs at the time of fire or smoke to instigate fire alarm system.

It is now becoming mandatory to incorporate fire monitoring system at various buildings. Traditionally buildings used to get equipped with a wired system to sense the possible onset of fire. A wired system is not getting replaced with the wireless system. These systems primarily focused on detecting early sign of fire, smoke, temperature rise etc. These systems are kept at indoor fail to indicate changes outdoor. Further, use of CCTV based system, which if covered at a restricted place, limits its effectivity at indoor. The predominant use of UAVs for natural disaster management is focused on rescuing individuals entrapped in disaster. These systems are not very useful for building fire monitoring [6]. Further, they also do not provide the option to transfer the monitoring data to a centralized server to initiate preventive measures directly.

2 Background

UAV based system can provide an opportunity to access areas difficult to approach at the time of emergency. Security agencies, regulatory authorities etc. frequently use it for surveillance, monitoring and rescuing at the time of disaster. Chen et al. [3] proposed a dynamic routing technique, where UAV recorded the data from the disaster site to analyze it for decision making. Choi and Lee [2] worked on a composite UAV based aerial and ground-based components to record the data. Sensor data from the environment was transferred to ground component for processing control commands in real time. Alp's landslide data from Rotolon catchment area in Italy was recorded by Frigerio et al. [4] with UAV for continuous web-based monitoring. Ueyama et al. [7] used a UAV based system to set up a wireless sensor network for river monitoring.

People have worked on building evacuation system. Aedo et al. [1] worked on personalized building notification with evacuation routes in a smartphone. However, the system cannot automatically identify the user's location, thus, in case of building getting collapsed, it may not work to its desired efficacy. Wu et al. [8], also worked on designing emergency evacuation system for a large building in case of fire. The paper accounted for use of various sensors such as fire detector, video camera, RFID tag, mobile terminal etc. The system keeps on monitoring during regular time. In case of emergency, the system keep on analyzing the number of people present in different rooms, evacuation potential of entrances, capacity of roads etc. RFID based active system also provided the potential to guide the building occupants their personalized evacuation path. Liu et al. [5] proposed a similar system for indoor evacuation, which can transmit a simple navigational message. It uses location beacons inside the building and environmental condition inside to determine personalized one step evacuation route.

3 Developing Intelligent Fire Alarm System

RGIPT has a conventional wired fire alarm system which faces several challenges of the false alarm, inappropriate message, incomplete information and limitation of a wired system. All the above weaknesses are tried to be overcome with the development of IoT based intelligent fire alarm system working wirelessly and needing UAV assisted real-time monitoring. The UAV-Assisted Intelligent Fire Alarm System (UAV-IFAS) (Fig. 1) consists of an intelligent fire detector (IFD), unmanned aerial vehicle (UAV), central server database (CSD), security manager (SM), closed-circuit television (CCTV), and user (U). Fire detector, CCTV is planned to provide necessary initial information of fire. These are planned to transmit messages to server wirelessly. Security manager and users will interact smartly with the server for initiating necessary response at the emergency situation.

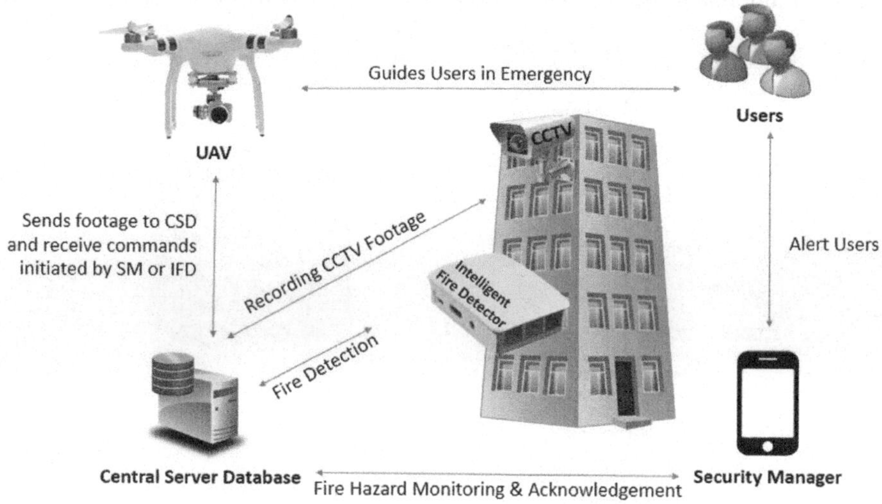

Fig. 1 Overview of UAV assisted fire alarm system

Each IFD is planned to check the surrounding temperature, humidity, smoke level, and CO level. In case of abnormal situations, the sensor values are noted in the CSD. In normal conditions the UAV rests on its stand. The stand also acts as the charger for the UAV. CCTV is used to monitor the space for potential fire hazards as well. The CCTV footage is stored on the server and can be monitored by SM. The SM can acknowledge the fire at any instant using the interface application. To acknowledge the fire SM can rely on: **a** CCTV footage **b** UAV footage and **c** Manual Inspection.

When abnormal conditions are detected by IFD, the CSD sends a signal to the UAV to fly to the hazard location. The UAV flies to the location, captures the footages, and uploads it on the CSD. The SM can acknowledge fire using this footage. Also, there is a fire detection classifier that can do the job of SM. In case the alarm is false the UAV is headed back to its stand. No action is taken and everything restores to normalcy. However, in case, the SM acknowledges the fire using either the UAV footage, CCTV footage, or manual inspection. The sprinkler system and alarms are activated. The UAV now attempts to guide the users with the way out. The footage recorded by the UAV is uploaded on the map of the building. This footage along with the footage of CCTV helps the fire fighting team accessing the scale of the fire, number of users in the building, and to device rescue strategy.

The IFD, UAV, SM, and CCTV are all connected to the Central Server Database via the wireless network. The following (Fig. 2) figure explains the main components of our UAV-IFAS:

The core of the UAV-IFAS is the Intelligent Fire Detector, from now mentioned as IFD. The IFD is a smart fire detector which has temperature, humidity, smoke level, and CO level sensors. It is connected to the wireless network. The IFD communicates with the CSD by transmitting its sensor values. The CSD knows the geo-location of IFD. IFD is designed to overcome the problems faced by conventional fire detectors. The conventional fire detectors lack because the extensive wiring is required to install

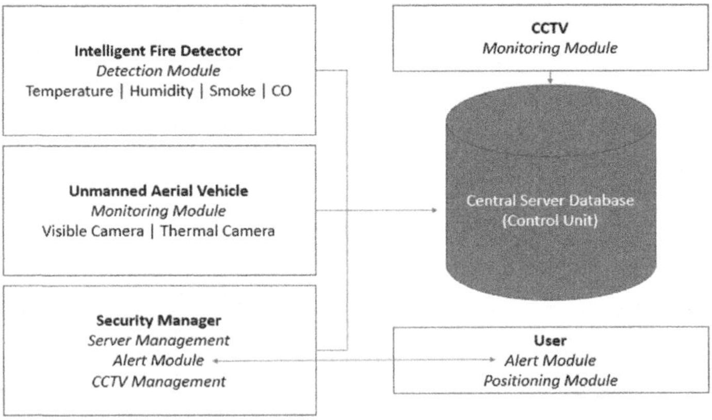

Fig. 2 Components of UAV-IFAS

such systems which significantly increases the project cost. The solution for this would be to *make wireless fire alarms.*

Further, the following limitations of the conventional fire alarm system are tried to be overcome in the proposed intelligent fire detection system.

There is *no control to the immediate user* which instils the fear of unknown. In case the fire alarm is dead, users can not know. If the fire alarm starts buzzing in the middle of the night, it is required to pull out the ladder. The solution for this would be to make fire alarms that can be controlled via a mobile application.

The conventional fire alarm systems are not smart enough. They don't inform the user:

- About the source of the fire.
- About what do when there is fire, how to exit? Shortest routes for the exit.
- About the life of the battery, health of the sensors and actuators.

The number of sensors in the IFD (Fig. 3) helps in eliminating false alarm cases. For example when both humidity and temperature sensor gives high values then that would indicate steam not fire. So IFD is smart enough to check for steam. Moreover, IFD can detect smoke, CO, perform health checks, and give location based alarms.

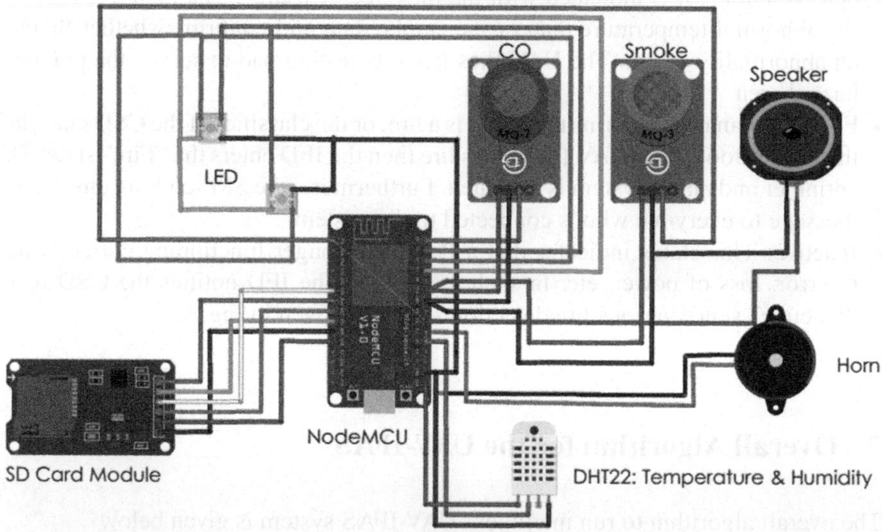

Fig. 3 Circuit diagram for intelligent fire detector, connecting various sensors on IoT platform circuit schematic of intelligent fire detector

4 Technical Specification of the Prototype of IFD Is as Follows

Microcontroller + IoT Module

- NodeMCU ESP8266 Wi-Fi Development Board Sensors
- Ionization Smoke Sensor: MQ2
- Ionization CO Sensor: MQ7
- Temperature & Humidity Sensor: DHT22 Speaker, Horn, & RGB LEDs for Audio & Visual Aids
- SD Card Module to Store Custom Sounds.

The IFD can indicate four states: (1) "Active" indicates that the IFD is functioning correctly; (2) "Unsafe": indicates when the local temperature, humidity, smoke level, or/and CO level is beyond an established threshold; (3) "Fire": indicates when a fire has been detected; (4) "Inactive": indicates that the IFD is no longer functioning.

- Active: When IFD is functioning correctly or if it detects that a fire has ended, it indicates this by sending an "Active" message to the CDS.
- Unsafe: If IFD detects an abnormal temperature or the smoke detection system detects smoke, it is indicated with the message "Unsafe". The manager receives the abnormal temperature data or the smoke data and confirms whether there is an abnormality or not. The UAV flies from its resting pad to survey the potential hazard area.
- Fire: If a manager confirms that there is a fire, or the classifier of the CSD classifies the CCTV footage or UAV footage as fire then the IFD enters the "Fire" state. The sprinkler and alarm system is activated. Furthermore, the SM sends an emergency message to everyone who is connected to the system.
- Inactive: This status indicates that a IFD is no longer functioning correctly due to error, loss of power, etc. In such a situation, the IFD notifies the CSD of its "Inactive" state. Various fire detection states are given in Fig. 4.

5 Overall Algorithm for the UAV-IFAS

The overall algorithm to run intelligent UAV-IFAS system is given below

State: Active, Inactive, Unsafe, Fire
Tt: Temperature Value in time t
Tm: Normal threshold of temperature
Ht: Humidity Value in time t
Hm: Normal threshold of humidity
St: Smoke Value in time t
Sm: Normal threshold of smoke
Ct: CO Value in time t

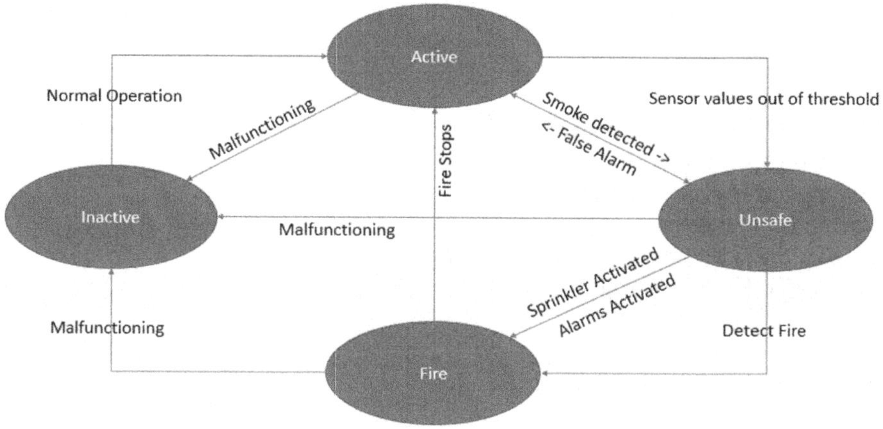

Fig. 4 State diagram of intelligent fire detector

Cm: Normal threshold of CO
UAVimage: Image taken by UAV at hazard location
CCTVimage: Image taken by CCTV.

Connect to IFD

if IFD == "Inactive" **then**
Transmit "Inactive" message to SM
Schedule IFD maintenance/repair program
end

while IFD == "Active" **do**
Receive Tt, Ht, St, Ct from IFD

if Any Variable Change == True **then**
Save Tt, Ht, St, Ct to CSD

Transmit Warning Message to SM

if Tt < Tm && Ht < Hm && St < Sm && Ct < Cm **then**
Receive "Safe" message from SM
Run Normally
end

if Tt >= Tm || St >= Sm || Ct >= Cm **then**
if Tt >= Tm && Ht >= Hm **then**
Receive "Safe" message from SM
Steam Check
Run Normally
else
State = "Unsafe"

Broadcast “Unsafe” message to SM

while State == “Unsafe” **do**

Launch UAV to the Hazard Site
Turn CCTVs towards Hazard Site
Live stream Hazard Site to SM
Update UAVimage and CCTVimage to CSD
Class = fireDetectionClassifier (UAVimage, CCTVimage)
Receive Tt, Ht, St, Ct from IFD
Update Tt, Ht, St, Ct to CSD
Receive State from SM
if SM fails to acknowledge fire **then**
State = Class
end
if State == “Fire” **then**
Broadcast fire message to Users
Run Sprinkler System
Activate Alarm System
end
end
end
end
end
end

6 Results and Discussion

The IFD will keep on monitoring the various buildings of RGIPT from inside and CCTV will also be giving video footage of rooms round the clock. Comprehensive coverage of CCTV network is made inside the campus to monitor possible fire situations in almost every indoor location. In the case of initiation of fire or fire like conditions IFD (Fig. 5) will send an immediate message to control server and control room through a wireless connection (Fig. 6). Once the emergency message has been received it is transmitted to control room. Security managers keep on monitoring it in real time and check the messages coming from temperature or smoke or CO detectors crossing the respective threshold values. Temperature (rate of rising) or other parameters of nearby rooms are compared to eliminate possible false alarm cases. CCTV footages are also corroborated to negate the possible occurrence of false alarm (Fig. 7). In case of any false alarm, it will be deactivated and no firefighting message will be communicated by the server. On the other hand, genuine fire like messages to be first verified with messages (temperature or smoke or CO etc.) coming from fire sensors of nearby locations and with CCTV images.

Fig. 5 Developed IFD prototype

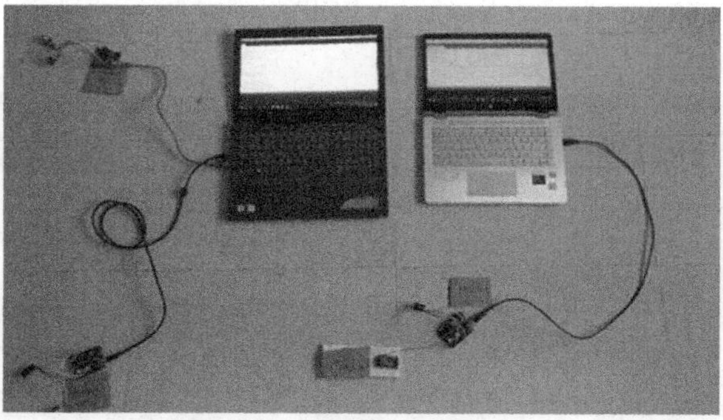

Fig. 6 Intelligent fire alarm system transmitting data to CSD in case of fire

Once fire signals are confirmed, server pass the message to all the concerned users connected with fire location and its immediate neighbour. Further, coordinate information of fire location is programmed to UAV kept at control room, for its instantaneous fly and navigate to fire location. Camera mounted on UAV then enable capturing the video from emergency site and pass on the real time photographs to server wirelessly.

UAV mounted camera here enables streaming of real time videos to all the concerned users. Use of UAV here is found to be advantageous to capture the firing site from outdoor at close range. Thus, it will give the detailed photograph or video of fire at the site in real time, unlike the CCTV footages at indoor. In an example case of fire at Room No AB1 of RGIPT (501 550,815.3 mE 2,905,185.1 mN 372 ft), upon confirmation of fire, server will initiate flying of UAV-quadcopter from control room (551,164.1 mE 2,905,217.8 mN 365 ft) in autopilot navigation mode (Figs. 8 and 9).

Fig. 7 CCTV monitoring of indoor locations at RGIPT to find fire or smoke

Fig. 8 Autopilot navigation of UAV from RGIPT's control room to fire site at room no AB1-501

Registered users will get intimation of fire with address or ID of house, along with its video to know about the status of fire in real time (Fig. 10).

7 Conclusion

A prototype of an intelligent fire detection alarm is tried to be made. The system needs fire detection sensors (temperature, CO, smoke, RH) to pass on messages of the possible onset of fire to the server in wireless mode. The server is also equipped

Fig. 9 Testing of UAV: navigation of quadcopter to prospective fire location

Fig. 10 Intimation of fire alarm in mobile with specific address of building and its video of fire site

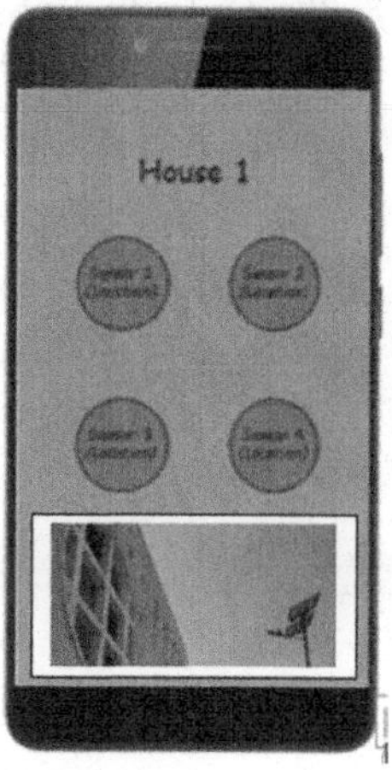

to gather CCTV footage (through several CCTV cameras placed at various locations) inside the buildings of the campus. Upon confirmation of initiation of fire with building address, messages are communicated to all the users inside the building and to the people outside who can approach the building on fire for rescuing operations. IoT based wireless system reduce the limitation of the wired system. Further, messages from server help navigating the UAV to fire-affected building site automatically to enable close range photography or videography to indicate the situation of fire in real time to control it effectively.

Acknowledgements Authors acknowledge the contributions of S&T club students of RGIPT (especially of first year) and other (e.g., A Pachauri) who have worked tirelessly towards developing and building the intelligent fire alarm system at RGIPT.

References

1. Aedo I, Yu S, Diaz P, Acuna P, Onorati T (2012) Personalized alert notifications and evacuation routes in indoor environments. Sensors 12:7804–7827
2. Choi K, Lee I (2011) A UAV based close-range rapid aerial monitoring system for emergency responses. Int Arch Photogram Rem Sens Spat Inf Sci 38:247–252
3. Chen D, Liu Z, Wang L, Dou M, Chen J, Li H (2013) Natural disaster monitoring with wireless sensor networks: a case study of data intensive applications upon low-cost scalable systems. Mob Netw Appl 18:651–663
4. Frigerio S, Schenato L, Bossi G, Cavalli M, Mantovani M, Marcato G, Pasuto A (2014) A web-based platform for automatic and continuous landslide monitoring: the Rotolon (Eastern Italian Alps) case study. Comput Geosci 63:96–105
5. Liu JWS, Lin FT, Chu ETH, Zhong JL (2016) Intelligent indoor emergency evacuation systems. In: Proceedings of the future technologies conference (FTC), San Francisco, CA, USA, 6–7 Dec 2016, pp 600–609
6. Seo SH, Choi J, Song J (2017) Secure utilization of beacons and UAVs in emergency response systems for building fire hazard. Sensors 17:2200
7. Ueyama J, Freitas H, Faiçal BS, Geraldo Filho PR, Fini P, Pessin G, Gomes PH, Villas LA (2014) Exploiting the use of unmanned aerial vehicles to provide resilience in wireless sensor networks. IEEE Commun Mag 52:81–87
8. Wu ZY, Lv W, Yu K (2016) A framework of intelligent evacuation guidance system for large building. In: Proceedings of the 5th international conference on civil, architectural and hydraulic engineering (ICCAHE 2016), Zhuhai, China, 30–31 July 2016, pp 695–701

Smart Agriculture: The Age of Drones in Agriculture

Vaishnavi Gautam and Sagar Sarkar

Abstract The purpose of this project is to provide a helping hand to our farmers in their difficult work of farming. The idea is to use swarm technology (one main drone rest 3–4 worker drones) in the agricultural field to solve problems related to farming. Today AI sensors have made the work for humans very easy. Similarly, in this project, drones will be equipped with different sensors like hyperspectral, thermal and LiDAR. These sensors will be used to solve problems like weed detection, drought condition detection, water stress, nitrogen content etc. Hyperspectral sensors are more precise and accurate than multispectral sensors in detecting minerals and vegetation, which will help in maintaining the plant nutrient status, identifying plant disease, water quality assessment and surface chemical composition. Thermal sensors measure the surface temperature of land and objects and create their thermal images for further analysis [1]. Then theses created thermal images are analyzed for identifying any heat stress, water stress, and plant metabolism from their canopy temperature. LiDAR sensors are laser equipped and use a laser beam to create 3D models of crops to identify drought stress and to optimize water use. LiDAR sensors are used to measure the vegetation level, the topography of the ground underneath. As an initiative, authors tried to apply this project in a small scale but if this initiative become successful, this can be applied in a large scale to provide more and more help to our farmers.

Keywords Swarm · Hyperspectral · Drone · LiDAR · Mother drone · Worker drone · Thermal

V. Gautam (✉) · S. Sarkar
University of Petroleum and Energy Studies, Dehradun, Uttarakhand, India
e-mail: vaishnavii1312@gmail.com

S. Sarkar
e-mail: sagarsarkar2702@gmail.com

K. Jain et al. (eds.), *Proceedings of UASG 2019*, Lecture Notes in Civil Engineering 51,
https://doi.org/10.1007/978-3-030-37393-1_34

1 Introduction

1.1 LiDAR

Light Detection and Ranging (LiDAR), a remote sensing method used for measuring distance. It uses laser pulses combined with other important aspect collected by the airborne system to generate three-dimensional projection to the distant surfaces, objects etc. LiDAR generally consist of three components: Scanner, GPS and Inertial Measurement Unit (IMU). When laser light from an airborne system is targeted over an area or an object, the reflected laser beam is sensed to measure the range. The GPS and IMU tell the current location and orientation of the airborne system respectively. When all these three data are combined together, a set of elevation points called "point clouds" is generated. These point clouds are used to produce digital elevation models, canopy models, building models and structure.

Why use LiDAR in agriculture? The use of LIDAR in the agriculture sector is limitless. LIDAR is generally used to map the water flow, determination of soil type, land mapping, prevention of soil erosion, crop analysis.

1.2 Multi-spectral and Hyper-spectral

There are two types of spectral imaging: Multi-spectral and Hyper-spectral. Both of the sensors can detect infrared and ultraviolet lights which is not visible to the human eye [2]. The basic difference between multi-spectral and hyperspectral is the number of bands and bandwidth [2]. Hyper-spectral generally composed of about 100–200 contiguous spectral brands of narrow bandwidths (5–10 nm), while multi-spectral are usually composed of about 5–10 bands of large bandwidth (70–400 nm). Which one should be preferred? Well, the answer depends on the application. With the use of hyper-spectral imaging, it is easy to differentiate minerals, plants etc. of much smaller spectral difference [3]. Moreover, with the use of hyperspectral, you can tell if a plant is suffering any stress and what is the cause of the stress. On the other hand with multi-spectral imaging, you can tell if an area has vegetation or not.

1.3 Thermal

Every object emits infrared heat energy as a function of their temperature. This energy emitted by the objects is called "Heat Signature". A thermal sensor is basically a heat sensor which detects tiny differences in temperature. It collects all the infrared radiation and creates an electronic image based on the information about the temperature difference [4].

In agriculture, thermal imaging can help to identify: water stress, heat stress, plant metabolism by measuring canopy temperature, soil salinity, seedling viability etc.

Drone swarm technology—the ability of drones to communicate and make decisions based on shared information—is one of the biggest inventions of this century. The swarm drone concept can change the image of the modern industry. Mainly in the defense sector. The proper implication has not yet done. Lots of project and research has been already done and is still going on this technology. One such project is the "Swarm Robotics for Agriculture Application (SAGA) project" [5]. The aim of this project was to demonstrate the application of swarm robotics for precision farming. Swarm of drones can help farmers to identify the percentage of weed and can help to improve crop yields.

2 Advantages

Technology has always shown its advantages in every field. And so does it play its role even in the field of agriculture. The use of drones in fields is apt for large scale farming. With the use of drones precision agriculture is possible, it can help in achieving more yields by using resources effectively. Today, everyone wants to complete his/her work in minimum time as possible, with the help of drones, the task which takes hours to be completed can be done in just a button push. Operating a drone is not too difficult, it just needs some basic training. The health of the crops is monitored with all the data collected. Farmers can even know the environmental conditions and take precautions.

3 Prior Research

Agriculture is one of the most advancing science of this era. Agriculture is basically the science and art of cultivating plants and livestock. With the advancement of technology, even the agriculture field is also benefitted. It has a positive trend in agriculture as it solves a number of problems faced by our farmer class. Drones in agriculture have been a big boom in improving the efficiency of agriculture. They surely have proven to compensate lack of skilled human resources and also sometimes use of heavy machines and tools. In the last past decades drones have found great use in the agriculture sector. From irrigating and spraying pesticides to complete thermal images of the farms, in order to get the exact details of the canopy, water deficient areas, weeds popping up in the various regions of the farms. With the use of various sensors, drones have got the ability to tell the temperatures and detect their own path.

Below are some examples of currently used agricultural drones.

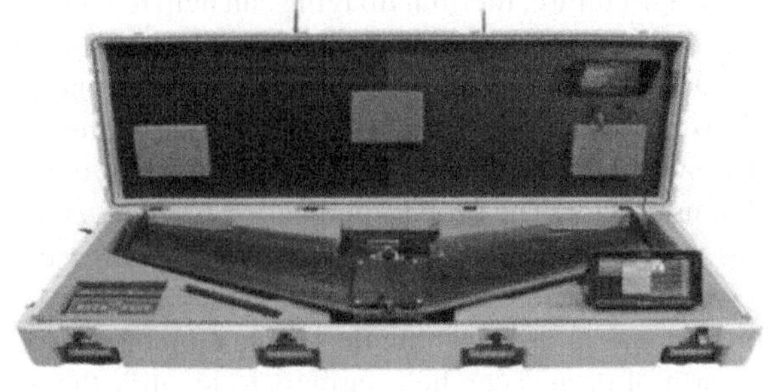

Fig. 1 Honeycomb AgDrone

Table 1 Specifications of Honeycomb AgDrone

Parameters	Values
Drone type	Fixed wing type
Build material	Kevlar composite
Wingspan and battery	490 in; 800 mAh LiPo
Area coverage	858 acres
Special feature	Programmed dual camera signal system
Flight specification	Cruise flying speed: 46.7 km/h Maximum flying Speed: 82 km/h

3.1 Honeycomb AgDrone System

It is a 49 in fixed wing drone, build of Kevlar composites which provide it high durability. It covers a total area of 858 acres/per h [6] (Fig. 1; Table 1).

3.2 EBEE SQ-SenseFly

It is designed to monitor crops from their initial phase of planting till their harvesting period [7]. It captures crop data across four multispectral bands, plus RGB imagery [8]. It covers a large area as compared to other quadcopters drones in a single flight. It is enabled with AG mapping software to create NVDI maps for crop fields and detect problems [7, 9] (Fig. 2; Table 2).

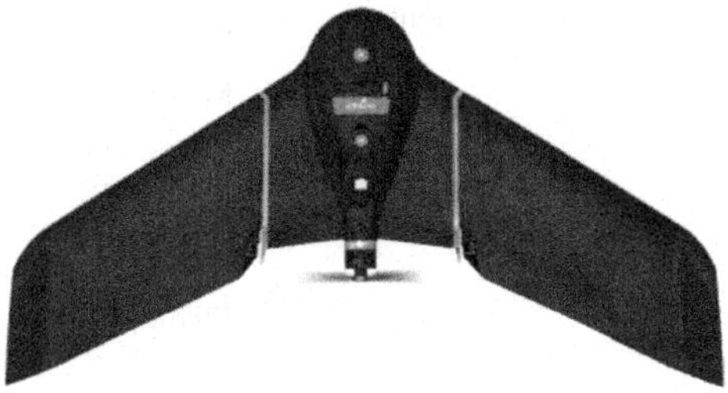

Fig. 2 EBEE SQ-SenseFly

Table 2 Specifications of EBEE SQ-SenseFly

Parameters	Values
Drone type	Separable wings with low-noise electric motors
Flight operations	Flight time: 55 min Linear landing ~5 m Software: eMotion Ag
Sensors	Spectral sensors, GPS, IMU, Magnetometer, SD Card
Camera	4–1.2 MP spectral camera 1 fps 13 MP RGB camera

3.3 DJI Agras MG-1

It is an octocopter specially designed to assist farmers to spray fertilizers, pesticides, and herbicides. Its powerful propulsion system allows it to carry heavy liquid payloads up to 10 kg. It can cover an area of 4000–6000 m^2 in just 10–20 min. It has an incorporated divergent cooling framework to keep the air streaming to each piece of the installed electronic system. It has a Y-shaped folding design. It has foldable motor system. Its frame is made of high strength carbon fibre [10] (Fig. 3; Table 3).

Fig. 3 DJI Agras M-1

Table 3 Specifications of DJI Agras M-1

Parameters	Values
Material	High performance engineered plastics
Liquid tank	10 kg (payload), 10 L (volume)
Nozzle	4
Battery	MG-12,000
Flight parameters	Allowable weight: 24.5 kg Working speed: 8 m/s Max flying speed: 22 m/s

4 Concept and Methodology

Author tried to bring the swarm drone concept in agriculture as a tool to help farmers in their daily difficult work of farming. In this smart agriculture concept, there will a group of 5 drones: one mother (main) drone and four worker drones. The four worker drones will be such that, each one will be assigned one specific work, for example, the work of drone number 1 will be to sprinkle water wherever there will be any water stress.

The mother drone will have all the sensors: LIDAR, HYPERSPECTRAL and THERMAL. The mother drone will scan the field for any water stress, mineral deficiency, weed percentage, any pest related problems and ground/plant canopy temperature. The mother drone will scan the field in a pre-planned manner as shown in Fig. 4. Accounting the total payload weight and field size the mother drone will scan the field in an approx. about 50 to 60 min duration. During the scanning, the mother drone will keep on updating its database about the location of the region where any defect is identified. After scanning it will return to the take-off spot. The

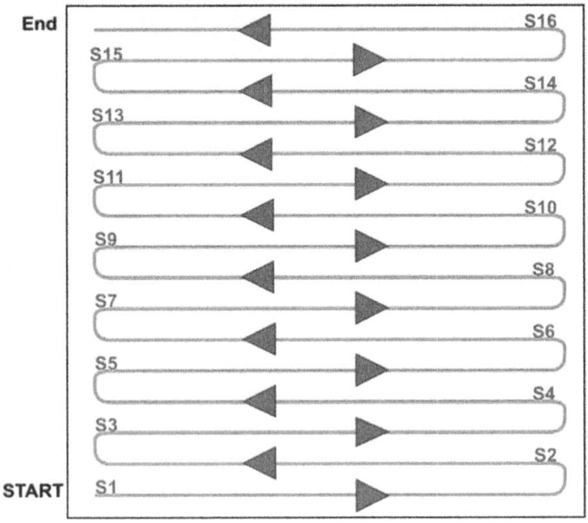

Fig. 4 Proposed scanning map for mother drone (the field is divided into several lanes, S1, S2, S3, etc.)

working of mother drone is explained as a block diagram in Fig. 5. After scanning the captured images from all the sensors will be collected and will be analyzed in software in ground station. Here the ground station can be any mode of computation: laptop, mobile, tablet. Prior to the data analysis, pre-processing of raw data is required to correct for any distortion due to the imaging system and imaging conditions.

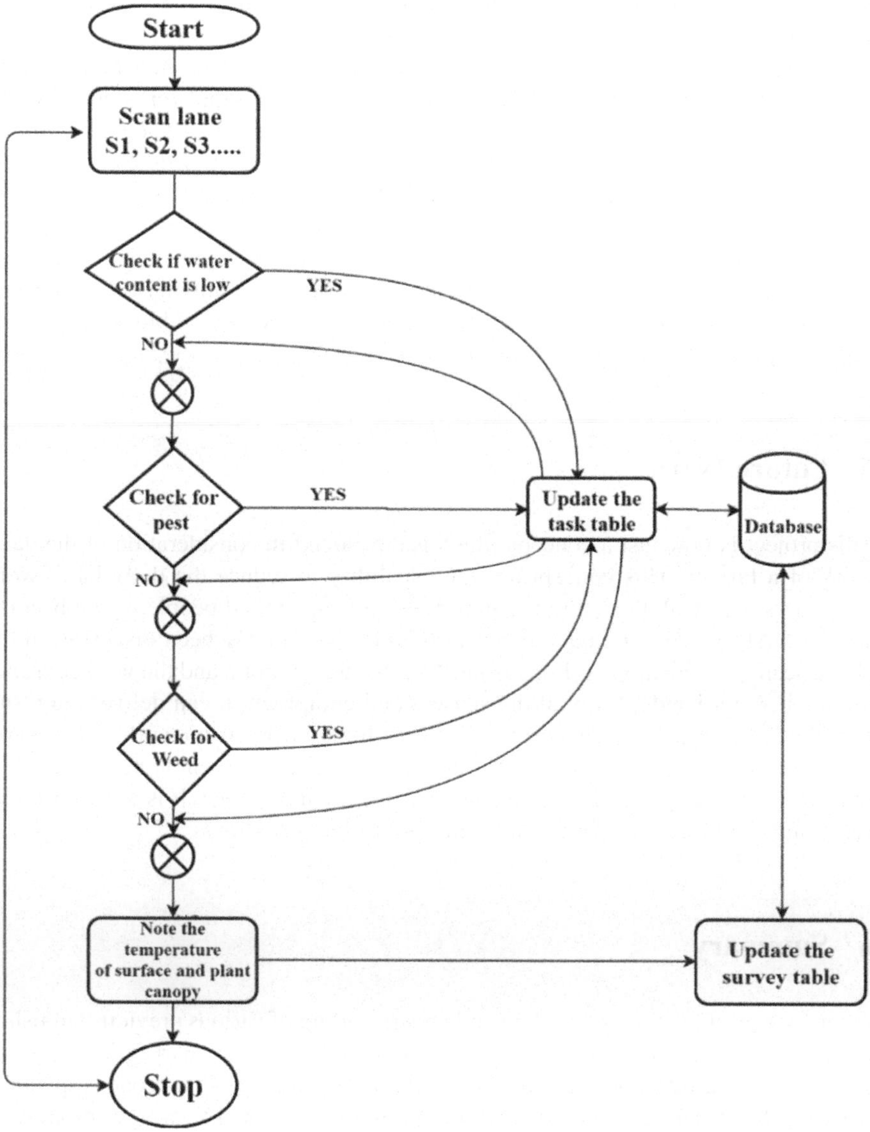

Fig. 5 Working of mother drone (it will scan for water stress, pest, weed and record the surface & canopy temperature. During the scan it will keep on updating its task table)

Like, radiometric correction for uneven sensor response and geometric correction for geometric distortion due to earth's rotation [11]. After all the correction/pre-processing is performed the corrected image will then analyzed for generating final result i.e. the location of regions affected with water stress, mineral deficiency, pest, and weeds.

After the analysis is done respective works will be assigned to the worker drones i.e. location to regions with any kind of defect will be sent to their corresponding worker drone. For example: let S4 and S8 lane were identified by the mother drone as regions having pest related issues. So the location of these lanes will be sent to the worker drone which is assigned for pest issues. Now as all the works get assigned the worker drones are now ready to get to their job. The initial command will be given from the base station only, after that the drones will be work on their own, communicating and co-operating with each other. Initially, the four worker drones will be sent to different lanes as per their works. This is important because during their work there can be situations when two working drones can come on one spot and situation of collision can occur. To avoid this situation initially the drones are sent to different lanes and the after procedure is explained as a block diagram in Fig. 6.

5 Future Work

This project is now just a concept which is developed in consideration of the daily work of a farmer. This concept has the capability to reduce the daily labor work of farmers. But with the help of required sensors and skilled people, it can become a reality. All the sensors, equipment, techniques are already been discussed in the above sections. The mother drone is just like the human brain and the worker drones are the helping hands. With all the sensors and equipment it can deliver the result for which it is designed. This project is aimed for very precision work. This swarm concept has the capability to take the agriculture sector to a whole new level. Any improvement/changes as per the user requirement can be done. It is a user-friendly setup and can be easily operated after the 2–3 training session.

6 Summary

After a series of continuous reading and understanding of various previous published papers and analysis of currently used agriculture drones, a new smart agriculture concept of swarm drone based farming is proposed. The main/mother drone is specially designed for scanning the fields and identifying problems. The worker drones, on the other hand, are designed to solve these problems. All together they will help to achieve the goal of smart and precision agriculture.

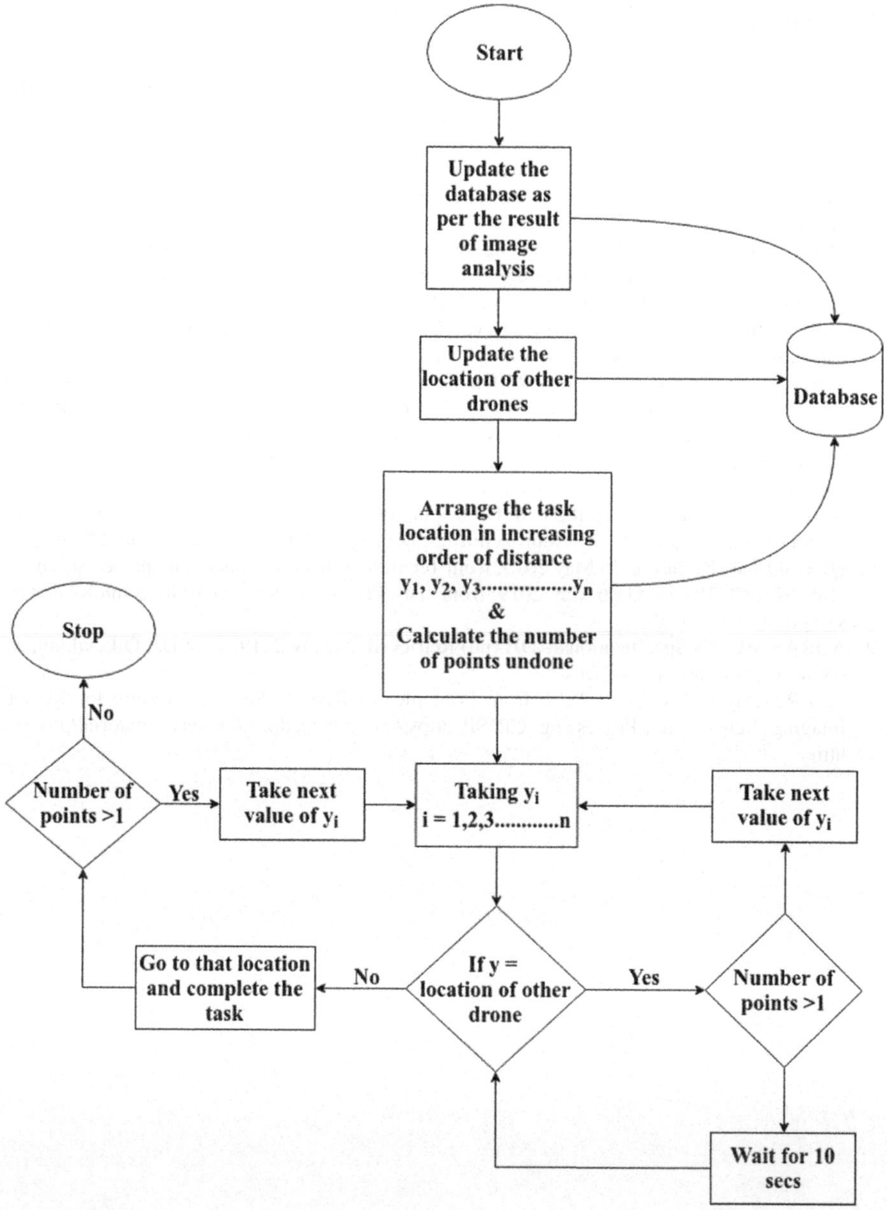

Fig. 6 Working of each worker drone for a single cycle (after the scanning and analysis, the data is transferred to all the worker drones. Every single drone will be sent to a different lane. Each drones will update their database according to the data shared and location of other drones. After that it will arrange its task locations in increasing order as per its current location. Then it will choose a task location and check if that location is occupied by another drone or not. If yes, it will check whether it have some other location to complete or not. If yes, it will choose another location and will go there. Or if not, it will wait 10 s and let that drone complete its work. And if no other drone is present at the selected location it will just simply go to that location and execute its work)

References

1. Sensors 101: the basics of LiDAR, thermal, hyperspectral, and multispectral technology (2019) Retrieved from PrecisionHawk. https://www.precisionhawk.com/blog/media/topic/sensors-101-basics-lidar-thermal-hyperspectral-multispectral-technology, 26 May 2019
2. Multispectral vs. Hyperspectral Imagery Explained (2019) Retrieved May 2019, 26, from GIS Geography. https://gisgeography.com/multispectral-vs-hyperspectral-imagery-explained/, 25 Apr 2019
3. The top 10 questions about airborne hyperspectral imaging (2018) Retrieved 26 May 2019, from Specim, Spectral Imaging Ltd. http://www.specim.fi/the-top-10-questions-about-airborne-hyperspectral-imaging/, 10 Dec 2018
4. Thermal Imaging Scan (IR) (nd) Retrieved 26 May 2019, from inspection advantage. http://inspectionadvantage.com/thermal-inspection/
5. Albani D, Ijsselmuiden J, Haken R, Trianni V (2017) Monitoring and mapping with robot swarms for agricultural applications. In: 2017 14th IEEE international conference on advanced video and signal based surveillance (AVSS). https://doi.org/10.1109/avss.2017.8078478
6. AgDrone System™ Specifications (nd) Retrieved 26 May 2019, from HoneyComb. http://www.honeycombcorp.com/specifications
7. Agricultural Drones | Using the Best Drones in Agriculture (2019) Retrieved May 26, 2019, from Dronethusiast. https://www.dronethusiast.com/agricultural-drones/, 29 Mar 2019
8. eBee SQ (nd) Retrieved 26 May 2019, from Tecnitop. https://tecnitop.com/en/ebee-sq-en/
9. eBee SQ (nd) Retrieved 26 May 2019, from senseFly. https://www.sensefly.com/drone/ebee-sq-agriculture-drone/
10. AGRAS MG-1's Specifications—DJI (nd) Retrieved 26 May 2019, from DJI Official. https://www.dji.com/mg-1/info#specs
11. (nd) Retrieved May 2019, 2019, from Principles of Remote Sensing—Centre for Remote Imaging, Sensing and Processing, CRISP. https://crisp.nus.edu.sg/~research/tutorial/process.htm

The manufacturer's authorised representative in the EU is Springer Nature Customer Service Centre GmbH, Europaplatz 3, 69115 Heidelberg, Germany. If you have any concerns regarding our products, please contact ProductSafety@springernature.com

Printed and bound by CPI Group (UK) Ltd, Croydon, CR0 4YY
30/09/2026
02225453-0019